Eco² Cities

Eco² Cities

Ecological Cities as Economic Cities

Hiroaki Suzuki
Arish Dastur
Sebastian Moffatt
Nanae Yabuki
Hinako Maruyama

THE WORLD BANK

Washington, DC

Eco² Cities : Ecological Cities as Economic Cities is available as an interactive textbook at http://www.worldbank.org/pdt.
The electronic version allows communities of practice, and colleagues working in sectors and regions, as well as students
and teachers, to share notes and related materials for an enhanced, multimedia learning and knowledge-exchange
experience.

ISBN 978-0-8213-8046-8
eISBN 978-0-8213-8144-1
DOI 10.1596/978-0-8213-8046-8

Cataloging-in-Publication data for this title is available from the Library of Congress.

Cover photo: Ricardo Almeida/SMCS
Back cover photo: Arish Dastur
Cover design: Naylor Design, Inc.

TABLE OF CONTENTS

BOXES

FIGURES

MAPS

TABLES

Foreword

Urbanization in developing countries is a defining feature of the 21st century. About 90 percent of global urban growth now takes place in developing countries, and, between 2000 and 2030, the entire built-up urban area in developing countries is projected to triple. Urbanization has enabled economic growth and innovation across all regions, currently accounting for three-quarters of global economic production. At the same time, urbanization has also contributed to environmental and socioeconomic challenges, including climate change, pollution, congestion, and the rapid growth of slums.

Global urban expansion poses a fundamental challenge and opportunity for cities, nations, and the international development community. It sets forth before us a once-in-a-lifetime opportunity to plan, develop, build, and manage cities that are simultaneously more ecologically and economically sustainable. We have a short time horizon within which to affect the trajectory of urbanization in a lasting and powerful way. The decisions we make together today can lock in systemic benefits for current and future generations.

The Eco² Cities Initiative appears at a critical historic juncture in relation to this challenge and opportunity. This book, which marks the launch of the Eco² Cities Initiative, sends a positive message. The knowledge and expertise to resolve these challenges exist, and forward-thinking cities in developed and developing countries have already applied this knowledge to make the most of the opportunities. Many cities have shown that cost is not a major barrier to accomplishing urban sustainability.

The Eco² Cities Initiative is an integral part of the new World Bank Urban Strategy that was launched in Singapore in November 2009. The Eco² Cities Initiative is also complementary to the ongoing efforts the World Bank and its development partners have undertaken in sustainable development and climate change.

Cities are now on the front line of the management of change and are playing a leading role in the global development agenda. It is only through cities that the challenges of poverty reduction, economic growth, environmental sustainability, and climate change may be addressed together. Sustainable city planning, development, and management can unite these objectives and link them to activities at the local, regional, national, and global levels. We believe the Eco² Cities Initiative will en-able cities to make the most of their opportunities in effective, creative, and holistic ways, thereby ensuring a more meaningful and sustainable future.

James W. Adams
Vice President
East Asia and Pacific Region
The World Bank

Katherine Sierra
Vice President
Sustainable Development
The World Bank

Preface

This book provides an overview of the World Bank's Eco² Cities: Ecological Cities as Economic Cities Initiative. The objective of the Eco² Cities Initiative is to help cities in developing countries achieve a greater degree of ecological and economic sustainability.

What Do We Mean by Ecological Cities?

Ecological cities enhance the well-being of citizens and society through integrated urban planning and management that harness the benefits of ecological systems and protect and nurture these assets for future generations.

Ecological cities strive to function harmoniously with natural systems and value their own ecological assets, as well as the regional and global ecosystems on which we all depend. Through their leadership, planning, policies, regulations, institutional measures, strategic collaborations, urban design, and holistic long-term investment strategies, they drastically reduce the net damage to the local and global environment, while improving the overall well-being of their citizens and the local economy. Ecological cities also learn from and incorporate management and design solutions that arise from the efficient and self-organizing strategies used by ecosystems.

What Do We Mean by Economic Cities?

Economic cities create value and opportunities for citizens, businesses, and society by efficiently using the tangible and intangible assets of cities and enabling productive, inclusive, and sustainable economic activity.

Often, when people talk about economic cities, they are referring to a narrower definition of productive cities that is driven by a singular emphasis placed on the indicator of GDP. While productivity is certainly an attribute of economic cities, it is not the only attribute, and the short-term and excessive pursuit of productivity often displaces fundamental social and cultural considerations and may undermine longer-term economic resilience. In some cases, an overemphasis on productivity overshadows our basic value systems and exposes us to substantial and systemic risk, as evidenced in the causes and consequences of the current global economic crisis. We propose a more balanced notion of economic cities whereby the emphasis falls on sustainable, innovative, inclusive, and resilient economic activity within the context of a larger cultural and value system.

So What Do We Mean by an Eco² City?

As the name implies, an Eco² city builds on the synergy and interdependence of ecological sustainability and economic sustainability and the fundamental ability of these to reinforce and strengthen each other in the urban context.

Innovative cities have demonstrated that, supported by the appropriate strategic approach, they are able greatly to enhance their resource efficiency by realizing the same value from a much smaller and renewable resource base, while decreasing harmful pollution and unnecessary waste. By achieving this, they improve the quality of the lives of their citizens, enhance their economic competitiveness and resilience, strengthen their fiscal capacity, provide significant benefits to the poor, and create an enduring culture of sustainability. Urban sustainability of this kind is a powerful and enduring investment that will pay compounding dividends. In a rapidly paced and uncertain global economy, such cities are most likely to survive shocks, attract businesses, manage costs, and prosper.

It is for the purpose of enabling cities in developing countries to realize this value and take on a more rewarding and sustainable growth trajectory that the Eco² Cities Initiative has been developed.

How Does the Eco² Cities Initiative Work?

The World Bank's Eco² Cities Initiative is a broad platform that provides practical, scalable, analytical, and operational support to cities in developing countries so they may harness the benefits of ecological and economic sustainability.

The publication of this book marks the completion of the first phase of the initiative: the development of the analytical and operational framework. This framework may be applied by cities in developing countries to work systematically toward the positive results we have outlined earlier and throughout the book. As a framework, it provides a point of departure and needs to be customized to the particular context of each city.

Following careful assessments of cities that have benefited tremendously by undertaking this sort of approach and also following detailed examinations of the major challenges that have prevented most other cities from accomplishing similar achievements, the framework has been structured around four key principles that have been found to be integral to lasting success. These principles are the foundation upon which the initiative is built. They are (1) a city-based approach enabling local governments to lead a development process that takes into account their specific circumstances, including their local ecology; (2) an expanded platform for collaborative design and decision making that accomplishes sustained synergy by coordinating and aligning the actions of key stakeholders; (3) a one-system approach that enables cities to realize the benefits of integration by planning, designing, and managing the whole urban system; and (4) an investment framework that values sustainability and resiliency by incorporating and accounting for life-cycle analysis, the value of all capital assets (manufactured, natural, human, and social), and a broader scope for risk assessment in decision making.

A set of core elements has been derived through these principles. Each city may transform the core elements into a series of concrete action items or stepping stones that should take into account local conditions and follow a logical sequence. Together, these stepping stones enable a city to develop its own unique action plan and sustainability pathway.

In this context, the ideal situation arises when a city adopts the four key principles, applies the analytical and operational framework to its particular context, and, by doing so, develops and begins to implement its own sustainability pathway.

Cities may begin incrementally by engaging in capacity building and data management and by initially targeting their most critical priorities through the development and implementation of a catalyst project. Unlike stand-alone projects in resource efficiency, a catalyst project is distinguished by an explicit objective and ability—beyond the immediate project scope and objectives—to drive the city forward on its sustainability pathway by catalyzing the process of change.

Challenges Need to Be Overcome

It is important to understand the many challenges that cities will face in trying to adopt a more well integrated, long-term approach. They include technical, administrative, and financial capacity constraints, coupled with the chronic problems faced by city management; institutional barriers ranging from the fragmentation of responsibilities and incentives across a wide a range of stakeholders to short-term and narrow accounting frameworks for decision making; the challenges of political economy, governance, and individual political agendas; locked-in relations among networks of public and private institutions and suboptimal technological systems and operating systems; misconceptions and misinformation about the true, complete, and long-term costs and benefits; and a general human inertia that is resistant to change.

The list is daunting, but these challenges are precisely why a more systematic approach such as Eco^2 is needed. Clearly, taking on all the challenges at the same time will not be possible for most cities, and they will need to adopt an incremental, phased approach. Often, a multisectoral approach will need to be crafted in stages by building upon a sectoral intervention. Cities will need to be creative in how they navigate their transformations.

It is reassuring to note that many cities, including those in developing countries, have grappled with similar issues and managed to overcome them over time through strategic, incremental, and purposefully chosen interventions. It is only by considering these challenges, together with the valuable ground-level lessons from best practice cities, that we have framed our strategic response.

Building on a Rich Legacy

The Eco^2 Cities Initiative builds on a rich and diverse legacy and seeks to reinforce successful concepts of city building and urban management found in every region of the world. In many instances, ancient cities and settlements in Africa, Asia, Europe, the Middle East, and South America were characterized by a strong understanding of and respect for nature. The industrial revolution of the 19th century made possible a great expansion in urban areas and a fabulous increase in material wealth, but there were many negative consequences for the environment and the quality of life. It triggered the birth of modern urban planning. The ideas of Ebenezer Howard and Patrick Geddes in the 1890s represented attempts to define how rapidly growing cities might achieve greater harmony with surrounding regional ecologies and improve social conditions at the same time. Howard's Garden City solution is perhaps the most enduring planning concept of the 20th century. Since then, there have been many pioneers across the world responsible for various movements pertaining to this theme: regional planning; new towns; greenbelt cities; design with nature; ecological planning; the new urbanism; green infrastructure; and, most recently, the shift toward Local Agenda 21, triple bottom-line full-cost accounting, and low carbon cities. ICLEI–Local Governments for Sustainability, founded in 1990, has emerged as a major international force in this area.

The specific term *eco city,* in use since the early 1970s, has been loosely used to refer to cities that adopt any combination of environmentally progressive measures, such as achieving a greater percentage of green space for residents, constructing a pedestrian- and transit-friendly transportation system, or requiring buildings to become more energy efficient. In an attempt to define more clearly what eco city means, some countries, including China, have now codified standards for green buildings and for ecological cities.

The many waves of interest in ecological cities have helped the concept mature and evolve. From this perspective, Eco^2 city is a useful term for recognizing a new generation of eco cities that move beyond individual, stand-alone green measures to a systems perspective supported by long-term, full-cost accounting. It requires that a city be understood as a whole and that design solutions incorporate some of the complex, multipurpose features of natural ecologies. Eco^2 refers not only to a fusion of economic and ecological strategies, but to a step forward in the long path toward a complete and lasting approach to sustainable development. It is an evolving concept, and we hope to collaborate with cities around the world for ideas and perspectives on ways to improve and deepen this concept.

How Will the Eco^2 Cities Initiative Evolve?

The implementation phase of the Eco^2 Cities Initiative has begun with the release of this book. The Initiative will focus on the application of the framework in specific pilot cities and the creation of a community of practice enabling practitioners at the city, country, regional, and global levels to learn from one another. It will also include scaling up and mainstreaming the approach through national programs and capacity building.

Application in the real world will initially require effort and commitment. It will require political will, leadership, capacity building, collaboration, institutional reform, and even a new process for creative design and decision making. Ideally, reform-minded city leaders will strive to undertake a comprehensive approach. Other cities may seek to start a change process through strategic and catalytic actions. The deeper and more comprehensive the approach, the more profound the changes. The successful application of such an initiative may prove to be transformative in a city. Such transformations have already occurred in the inspiring cases the reader encounters through this book. The Eco^2 initiative is intended to provide the support that cities need to make their own transitions.

As we begin to apply the framework, diversity among the conditions and contexts of the first set of pilot cities (city size, national context, geographical conditions, socioeconomic conditions, institutional frameworks, fiscal capacity, and so on) will provide a broad and rich platform for assessing the value of the framework in different circumstances, and we will continue improving our approach based on the feedback and the experience.

It is evident that a city-by-city approach is important as we test Eco^2 and learn from the ground-level experiences of each case. However, given the magnitude and rate of urbanization, we will not be able to achieve the desired global impact within the window of opportunity currently open to us if we limit ourselves to a city-by-city approach. Accordingly, we will aim at mainstreaming and scaling up the Eco^2 Cities Initiative through programmatic national approaches.

Projects are not the only opportunity for mainstreaming. A critical step forward in mainstreaming will involve deepening and customizing the ownership of the agenda in each country through supportive national policy and sustained capacity building. This may include arrangements with a range of stakeholders, including local planning institutes around the world, such as the Institute for Research and Urban Planning of Curitiba, Brazil.

As we continue to work toward our common objectives, the Eco^2 Cities Initiative will evolve and grow as new knowledge, methods, tools, and resources become available. As we forge new partnerships and work with more cities, new possibilities and innovative ideas will emerge. The Eco^2 Cities Initiative will constantly work to incorporate these in an inclusive, iterative, and purposeful way.

Acknowledgments

The World Bank's Eco² Cities Initiative has been conceived and developed and is managed by Hiroaki Suzuki (Team Leader) and Arish Dastur (Co–Team Leader), who authored, compiled, and edited this book, together with Sebastian Moffatt, Nanae Yabuki, and Hinako Maruyama.

The other contributing authors are Feng Liu, Jas Singh, Georges Darido, Khairy Al Jamal, Charles W. Peterson, Alain Bertaud, Nobue Amanuma, Malin Olsson, Karolina Brick, Maria Lennartsson, Claire Mortimer, and Bernd Kalkum. The peer reviewers are Stephen Karam, Robert Taylor, Neeraj Prasad, Josef Leitmann, and Sam Zimmerman. Important comments and suggestions have been received from Alan Coulthart, Tim Suljada, Brian Dawson, Carly Price of AusAID (formerly the Australian Agency for International Development), Thomas Melin of the Swedish International Development Cooperation Agency (SIDA), and Sumter Lee Travers and Fang Chen of the International Finance Corporation, as well as Jas Singh, Victor Vergara, Shomik Mehndiratta, William Kingdom, Jan Bojo, Paul Kriss, Rohit Khanna, Peter Ellis, Habiba Gitay, Mir Altaf, Rama Chandra Reddy, Monali Ranade, Axel Baeumler, Nat Pinnoi, Masato Sawaki, and Johannes Heister of the World Bank. Geoffrey Payne, Örjan Svane, and Richard Stren provided valuable suggestions at the early stages of Eco², and Yuko Otsuki supplied strong support to the team.

The book has benefited from the guidance of Keshav Varma, Sector Director of Urban Development for the East Asia and Pacific Region of the World Bank. Strong support has also been provided by John Roome, Christian Delvoie, Abha Joshi-Ghani, Eleoterio Codato, Ede Jorge Ijjasz-Vaquez, and Amarquaye Armar.

Elisabeth Mealey advised us on our communications strategy. Claudia Gabarain contributed to the Web design and online strategy. Inneke Herawati, Iris David, Bobbie Brown, Vellet Fernandes, and Sandra Walston supplied important logistical support. Dean Thompson provided editorial support on an earlier version of this book. Patricia Katayama and Mark Ingebretsen of the Office of the Publisher gave guidance to the team on publishing and handled the final editing process. Naylor Design, Inc., provided the cover design and layout. Many of the graphics in the book were produced by Sebastian Moffatt and the Sheltair Group.

The team acknowledges the valuable contributions of the Sheltair Group, the Energy Sector Management Assistance Program (ESMAP, jointly sponsored by the World Bank and the United Nations Development Programme), and the World Bank's Finance, Economics, and Urban Development Department. The team also acknowledges the guidance received from key decision makers in the following cities: (1) the City of Curitiba, Brazil, especially Carlos Alberto Richa, Mayor, Curitiba; Eduardo Pereira Guimarães, Secretary of International Relations, Curitiba; Cléver Ubiratan Teixeira de Almeida, President, Institute for Research and Urban Planning of Curitiba, and Priscila Tiboni, Foreign Affairs Advisor, Institute for Research and Urban Planning of Curitiba; (2) the City of Stockholm, especially Malin Olsson, Head of Section, City Planning; and Klas Groth, City Planning; (3) the City of Vancouver, especially Brent Toderian, Director of Planning; (4) the City of Yokohama, especially Toru Hashimoto, Senior Project Manager, Co-Governance and Creation Taskforce, and Yoshihiro Kodama, Co-Governance and Creation Taskforce; and (5) the City of Brisbane, especially David Jackson, Manager, Economic Development; John Cowie, Senior Project Officer, Economic Development; and Lex Drennan, CitySmart Project Director.

This publication has been made possible through scaled-up funding by the World Bank's East Asia and Pacific Region and generous co-funding from AusAID.

The Structure of This Book

The book is divided into three parts.

Part 1 describes the Eco² Cities Initiative framework. It describes the approach, beginning with the background and rationale. Key challenges are described, and lessons are drawn from cities that have managed to turn these challenges into opportunities. A set of four key principles is introduced. The description of the program is then developed around those four principles. Each of the principles is addressed in separate chapters that present the core elements of the program and the stepping stones each city may follow as it develops its own unique Eco² pathway. Part 1 concludes with an overview of some of the ways in which cities may draw on the resources of various development partners as they embark on their unique pathways.

Part 2 presents a city-based decision support system that introduces core methods and tools to help cities as they work toward applying some of the core elements and stepping stones outlined in part 1. Part 2 looks into methods for collaborative design and

decision making and methods to create an effective long-term framework able to help align policies and the actions of stakeholders. Part 2 also examines material flow analysis and the use of layered maps to facilitate an integrated approach to urban infrastructure and spatial planning. Techniques for life-cycle costing are described, and specific tools are referenced. Finally, part 2 introduces methods that may be useful in conducting forecasting workshops and resiliency planning. It is expected that, as the Eco² initiative grows, a greater depth of information will be generated to enrich the city-based decision support system.

Part 3 consists of the Field Reference Guide. The guide contains background literature designed to support cities in developing more in-depth insight and fluency with the issues at two levels. It provides a city-by-city and sector-by-sector lens on urban infrastructure. It begins with a section on a series of case studies from best practice cities around the world. Each city offers a separate example of how various elements of the Eco² approach may be applied. The next section comprises a series of sector notes, each of which explores sector-specific issues in urban development. The sectors include energy, water, transportation, and solid waste. This section includes a note on the management of the spatial structure of cities. Together, these sector notes provide insights on the functioning of each sector and on the current interrelationships among the sectors. As we view these issues through a city-by-city and sector-by-sector lens, we start to see a bigger picture. Part 3 also includes a final section on relevant specific financial instruments of the World Bank.

While part 1 and part 2 address the Eco^2 Cities Initiative directly, the Field Reference Guide provides the background on current best practices and a full scope of policies, specific measures, and institutional measures that need to be considered. Together, these three parts provide cities with an up-to-date survey of the terrain and guidance on how to move forward on their own pathways. This book lays out the scope of Eco^2 and should be viewed as an evolving document, particularly parts 2 and 3. The Eco^2 Cities Initiative Web site, at http://www.worldbank.org/eco2, provides detailed, updated information.

Abbreviations

ASF	Auckland Sustainability Framework (New Zealand)
BRT	bus rapid transit
CBD	central business district
CDM	clean development mechanism
CO_2	carbon dioxide
CPF	Carbon Partnership Facility
CTF	Clean Technology Fund
CY	current year
DAC	Development Assistance Committee (OECD)
DPL	development policy lending
DSM	demand-side management
DSS	decision support system
ELP	environmental load profile
ER	emission reduction
FAR	floor area ratio
FY	fiscal year
GDP	gross domestic product
GEF	Global Environment Facility
GHG	greenhouse gas
GIS	geographic information system
IBRD	International Bank for Reconstruction and Development (World Bank)
IDA	International Development Association (World Bank)
IFC	International Finance Corporation (World Bank)
IPPUC	Institute for Research and Urban Planning of Curitiba (Brazil)
LCC	life-cycle costing
LFG	landfill gas

LIBOR	London interbank offered rate
MDB	Multilateral Development Bank
MIGA	Multilateral Investment Guarantee Agency (World Bank)
O_2	oxygen
OECD	Organisation for Economic Co-operation and Development
PUB	Public Utilities Board (Singapore)
RGS	regional growth strategy
SCF	Strategic Climate Fund
SIP	Small Investment Program
SO_2	sulfur dioxide
UNDP	United Nations Development Programme
UNFCCC	United Nations Framework Convention on Climate Change

Note: All dollar amounts are U.S. dollars (US$) unless otherwise indicated.

Executive Summary

Challenges and Opportunities

Urbanization in developing countries may be the most significant demographic transformation in our century, restructuring national economies and reshaping the lives of billions of people. It is projected that the entire built-up urban area in developing countries will triple between 2000 and 2030, from 200,000 square kilometers to 600,000 square kilometers. These 400,000 square kilometers of new urban built-up area that are being constructed within a mere 30 years equal the entire world's total built-up urban area as of 2000. One might say we are building a whole new urban world at about 10 times the normal speed in countries with serious resource constraints (natural, fiscal, administrative, and technical). We are doing this in an increasingly globalized context characterized by many new, constantly fluctuating, interlinked, and uncontrollable variables.

What is driving this massive rate of urbanization? Historically and across most regions, urbanization has propelled the growth of national economies. On average, about 75 percent of global economic production takes place in cities, and in developing countries, the corresponding share is now rapidly increasing. In many developing countries, the urban share of national GDP already surpasses 60 percent. In most regions of the world, the opportunities provided by urbanization have enabled large segments of populations to lift themselves out of poverty.

However, urbanization at this rate and scale is certain to be accompanied by unprecedented consumption and loss of natural resources. Calculations already show that, if developing countries urbanize and consume resources as developed countries have done, an ecological resource base as large as four planet Earths would be needed to support their growth. But, of course, we have only one Earth. Because the underlying resource base required to sustain such a transition does not exist, cities in developing countries and in developed countries must find more efficient ways to meet the needs of their populations.

It is clear that, if we are to absorb and sustain this powerful wave of urbanization, while continuing to manage the existing built stock, we will need a paradigm shift. The following fundamental questions must be addressed: How can cities continue to effectively harness the opportunities for economic growth and poverty reduction offered by urbanization, while also mitigating the negative impacts? How can cities accomplish this given the speed and the scale at which this urbanization is progressing and given their own capacity constraints? How can ecological and economic considerations be dovetailed so that they result in cumulative and lasting advantages for cities? How do we transition from Eco versus Eco to Eco^2 cities?

Innovative cities have demonstrated that, supported by the appropriate strategic approach, they may greatly enhance resource efficiency by realizing the same value from a much smaller and renewable resource base, while decreasing harmful pollution and unnecessary waste. By achieving this, they have improved the quality of the lives of their citizens, enhanced their economic competitiveness and resilience, strengthened their fiscal capacity, provided significant capacity to the poor, and created an enduring culture of sustainability. Urban sustainability of this kind is a powerful and enduring investment that will pay compounding dividends. In a rapidly paced and uncertain global economy, such cities are the most likely to survive shocks, attract businesses, manage costs, and prosper.

Most encouraging about the efforts made by these innovative cities is the fact that many of the solutions are affordable even if budgets are limited, and they generate returns, including direct and indirect benefits for the poor. At the same time, much of the success may be achieved by using existing, well-tested methods and technologies and by focusing on local, home-grown solutions.

The challenge that lies ahead is to take full advantage of the many opportunities created by the rapid rates of change and by successful innovations. Inappropriate institutional structures and mind-sets are commonly cited as the single greatest challenge whenever cities try to

take advantage of such opportunities. Best practices exist for long-term planning and regional growth management, and the emergence of new tools for systems analysis and mapping offers potential for more well integrated, practical, and rigorous analysis and planning. Methods for collaborative design and decision making among key stakeholders have also proven effective. Realizing that successful cities are often fundamental to successful nations, the higher levels of government are becoming key partners in helping cities take the initiative.

There is also growing commitment at the international level to support cities and help finance longer-term investments within cities. New funding opportunities have emerged for cities in developing countries that are willing to implement actions to achieve sustainable urban development, particularly measures promoting energy and resource efficiency that lead to reductions in greenhouse gas emissions. New accounting methods for estimating the full costs and benefits of various policy, planning, and investment options are also being used (for example, life-cycle costing). At the same time, accounting for all capital assets (manufactured, natural, social, and human) and the services they provide offers a more holistic and complete incentive framework to cities. Channeling these opportunities toward the massive scale and accelerating the pace of urban development are creating the potential for a tremendously positive impact.

The Eco^2 Cities Initiative has been developed for the purpose of enabling cities in developing countries to benefit from the promise of a more rewarding and sustainable growth trajectory while the window of opportunity is still open to them.

The Analytical and Operational Framework

The Eco^2 analytical and operational framework is rooted in four key principles. Cities face challenges in trying to adopt new approaches. These challenges have been carefully anticipated in the framework, and together with the valuable ground-level lessons from best practice cities, they have helped frame our strategic response: the key principles that define the Eco^2 Cities Initiative. Each has been elevated to the status of a principle because it is widely applicable, critical to success, and frequently ignored or underappreciated.

These four principles are (1) a city-based approach enabling local governments to lead a development process that takes into account specific circumstances, including the local ecology; (2) an expanded platform for collaborative design and decision making that accomplishes sustained synergy by coordinating and aligning the actions of key stakeholders; (3) a one-system approach enabling cities to realize the benefits of integration by planning, designing, and managing the whole urban system; and (4) an investment framework that values sustainability and resiliency by incorporating and accounting for life-cycle analysis, the value of all capital assets (manufactured, natural, human, and social), and a broader scope for risk assessments in decision making.

The four principles are interrelated and mutually supportive. For example, without a strong city-based approach, it is difficult to fully to engage key stakeholders through an expanded platform for collaborative design and decision making. And without this expanded platform, it is difficult to explore creative new approaches to the design and management of integrated systems and to coordinate policies for implementing the one-system approach. Prioritization, sequencing, and the effectiveness of investments in encouraging sustainability and resiliency will be greatly enhanced if the city is appreciated as a single system and the platform for collaboration is expanded.

A set of core elements that define the Eco^2 framework has been derived through these four key principles. Cities are encouraged to realize the core elements through a series of concrete action items or stepping stones that take into account local conditions and follow a logical sequence. Together, these stepping stones enable a city to develop its own unique action plan and sustainability pathway. The Eco^2 Cities Initiative also introduces cities to methods and tools that will lead to more effective decision making through powerful diagnostics and scenario planning. These methods and tools may also be used to realize the core elements and implement the stepping stones.

In this context, the ideal situation arises when a city adopts the four key principles; applies the analytical and operational framework to its particular context; and, by doing so, develops and begins to implement its own sustainability pathway.

PRINCIPLE 1: A city-based approach

A city-based approach is the first principle, and it carries two complementary messages. First, it recognizes that cities are now on the front lines in managing change and leading an integrated approach. Cities not only embody the engines of economies and the homes of citizens, but also are responsible for a majority of the resource and energy consumption and the harmful emissions. Only at the city level is it possible to integrate the many layers of site-specific information and to work closely and rapidly with the many stakeholders whose input may influence the effectiveness of a sustainability pathway and who have a stake in its successful implementation. In addition, fiscal and administrative decentralization has brought important decision making and management responsibility to local governments. Cities may exercise proactive leadership and thereby trigger a process of change.

Second, a city-based approach serves to emphasize the importance of the incorporation of the unique aspects of place, especially ecological systems. In this sense, a city-based approach responds to the opportunities and constraints of local ecologies. How might development fit into the topography of the area so that water is provided by gravity and drainage is provided by natural systems (reducing the need for expensive infrastructure investments and the related operating costs)? How might a city protect its water recharge areas and wetlands so that water capacity and quality are sustained? How do we distribute populations and design cities so that local or regional renewable energy—windy sites, forests, solar access—is sufficient to meet basic needs? These types of questions may ultimately provide urban professionals with their most exciting design challenge: how to fit cities into the landscape in ways that respect and complement the natural capital and that ensure the availability of ecological services for present and future generations.

A city-based approach is thus place specific and focuses on enabling local leadership, local ecologies, and the broader local context. In fact, one of the first stepping stones of a city will involve reviewing and adapting the Eco^2 framework to the local context.

PRINCIPLE 2: An expanded platform for collaborative design and decision making

Cities are increasingly experiencing a splintering of infrastructure responsibilities, the overlapping and intersection of jurisdictions, and an increase in the private sector ownership of key assets. If cities are to lead in the process of urban development, especially in the context of rapid urbanization, they must get ahead of this curve.

A city may lead a collaborative process on at least three tiers of an expanded platform. At the first tier, projects may be completely within the realm of control of the city administration, meaning that the city must get its own house in order (for example, by supporting an energy efficiency upgrade for all municipally owned buildings, or a ride-share program for employ-

ees, or energy and transportation peak load management through adjustments in working hours). At the second tier, projects will involve the city in its capacity as a provider of services and include its formal planning, regulatory, and decision-making powers. This may include water provision, land use planning, or transit development. At this level, greater collaboration is warranted with other stakeholders (including the private sector and consumers) who may influence and may be affected by the outcomes. The third tier of the expanded platform will entail collaboration at the scale of the entire urban area or region. This may pertain to issues such as the development of new land or metropolitan management and may necessarily involve senior government officials, key private sector partners, and civil society.

A core element of the three-tier platform for collaboration is a shared long-term planning framework for aligning and strengthening the policies of city administrations and key stakeholders and for guiding work on future projects. In this way, three-tier collaboration may encourage everyone to row in the same direction.

PRINCIPLE 3: A one-system approach

The one-system approach aims to take full advantage of all opportunities for integration by promoting a view of the city and the urban environment as a complete system. Once we see the city and the urban environment as a system, it is easier for us to design the elements to work well together. This may mean enhancing the efficiency of resource flows in an urban area through integrated infrastructure system design and management. For example, the looping and cascading of energy or water through a hierarchy of uses may satisfy many demands with the same unit of supply.

The one-system approach also includes integrating urban form with urban flows by coordinating spatial development (land use, urban design, and density) and the planning of infrastructure systems. For instance, new development may be directed to those locations that possess a surplus of water, energy, and transit. Urban form and spatial development also establish the location, concentration, distribution, and nature of the demand nodes that affect the design of infrastructure system networks. With this effect, they establish the physical and economic constraints and parameters for infrastructure system design, capacity thresholds, technology choices, and the economic viability of the various options. This has tremendous implications for resource use efficiency.

Integrating the planning of flows and forms and rendering initiatives operational are both a challenge and a huge opportunity for any city. The one-system approach also focuses on how to implement projects using a more well integrated procedure. This means sequencing investments so that the city sets the correct foundation by addressing the long-lasting, cross-cutting issues first. This also means creating a policy environment that enables an integrated approach, coordinating a full range of policy tools, collaborating with stakeholders to align key policies, and targeting new policies to reflect the different circumstances involved in urbanization in new areas and to improve existing urban areas.

Integration may apply to the elements within a sector or across sectors. It may apply to implementation policies, the collaboration of stakeholders and their plans, the sequencing of financing mechanisms, and all of these in combination. In every case, the integration of elements tends to reveal opportunities for greater efficiency, synergy, and increased utility from a given investment, with corresponding improvements in ecological and economic performance.

By applying the one-system approach, cities and their surrounding natural and rural areas can strive to coalesce into a functional system that works well as a new whole.

PRINCIPLE 4: An investment framework that values sustainability and resiliency

The simple concept of investing in sustainability and resiliency in cities has become extremely difficult to put into action. Policies, plans, and projects tend to be assessed on their short-term financial returns or on an economic valuation based on narrowly structured cost-benefit analyses from the perspective of a single stakeholder or project objective. Investments are valued in monetary terms, and what cannot be monetized is either ignored or addressed as externalities. Decisions are dominated by immediate capital costs, despite the fact that over 90 percent of the life-cycle costs of typical infrastructure are often expended in operations, maintenance, and rehabilitation.

Few cities worldwide have a real knowledge of the impact of new development on their long-term fiscal condition. Life-cycle costs are often backloaded, which means that future generations will have a massive infrastructure deficit because they must face the costs of the repair and replacement of infrastructure without any prior capitalization.

At the same time, ecological assets, the services they provide, and the economic and social consequences of their depletion and destruction are not accounted for in most government budgets. Because these assets are not measured, their value is treated as zero, and the valuable services they provide go unaccounted for.

Principle 4 requires that cities adopt a new framework for making policy and investment decisions. The framework has multiple elements. A new range of indicators and benchmarks must be adopted to assess and reward the performance of all stakeholders. The family of indicators must address the needs of all categories of decision making (for example, strategy evaluation versus operations). Longer time horizons are needed, and life-cycle cost-benefit analysis must be applied to understand the full implications of policies and investment options. All four categories of capital assets (manufactured, natural, human, and social) and the services they provide must be appropriately valued or priced and monitored through indicators. The combination of indicators should be viewed as a whole so that the qualitative dimensions of city life (cultural, historic, and aesthetic) are not ignored in assessing costs and benefits.

At the same time, investing in sustainability and resiliency will entail broadening our scope of risk assessment and management to include managing the many indirect, difficult-to-measure risks that nonetheless threaten the viability of an investment or even the city as a whole.

The principles described above underlie the Eco² approach. Using the analytical and operational framework, a city may apply the principles through a set of core elements and use these elements to create a phased, incremental Eco² pathway (see the diagram). The sustainability pathway of each city will be designed in consideration of the city's own needs, priorities, and capacities. While the analytical and operational framework enables a city to chart out its sustainability pathway, the city-based decision support system introduces the methods and tools that provide cities with the capacity to undertake more well integrated development and navigate this pathway more effectively.

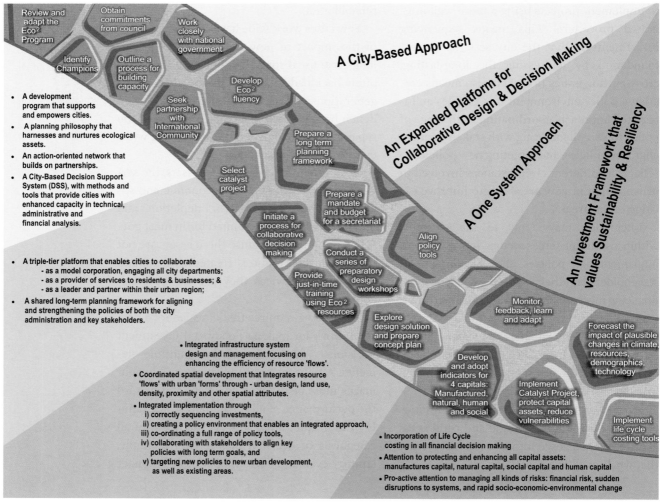

Source: Author compilation (Sebastian Moffatt).

A City-Based Decision Support System

The city-based decision support system introduced in part 2 introduces methods and tools that enable cities to more effectively develop their capacity to realize some of the core elements of the Eco² initiative. It comprises a few core methods that, together, provide cities with a greater ability to implement the core elements of the four principles listed above.

The fundamental purpose of these methods is to simplify the process of analysis, assessment, and decision making. They provide practical ways for cities to take leadership, collaborate, and analyze and assess various ideas for projects. All the methods are well-tested approaches to accomplishing the work. They are expected to remain relevant for many years.

The methods support the typical planning process at different times and in different ways. Some methods may be used repeatedly. For example, meta diagrams that summarize resource flows may be used, first, as a way to establish a baseline for assessing how a location is currently performing and then, later, to help with diagnosing, target setting, scenario development, and cost assessment.

As an illustration, the Methods for Analyzing Flows and Forms reveal the important relationships between spatial attributes of cities (forms) and their physical resource consumption and emissions (flows). The combination of these analytical methods helps cities develop a transdisciplinary platform for analyzing current situations and forecasting scenarios (see the diagram).

One of the first stepping stones for a city may be to plan a process for capacity building. Reviewing the decision support system is a good place to begin. While this book introduces the core methods, the capacity-building plans of a city may include obtaining more information, acquiring specific tools, obtaining outside technical support, and applying the methods to a catalyst project.

A Field Reference Guide

The Eco² Field Reference Guide provided as part 3 is a technical resource especially tailored to building ground-level and technical knowledge. It contains background literature designed to support cities in developing more in-depth insights and fluency with the issues at two levels. It provides a city-by-city and sector-by-sector lens on urban infrastructure. It begins by exploring a series of case studies from best practice cities around the world. Each city offers a separate example of how various elements of the Eco² approach may be applied.

The Field Reference Guide also provides a series of sector notes, each of which explores sector-specific issues that pertain to urban development. As cities develop their sustainability pathways, surveying issues through the lens of each urban infrastructure sector will help. Ideally, this will lead to a kaleidoscopic view of the city, in which each perspective may be compared with the next so that the relationships among energy, water, transportation, and solid waste may be understood in the context of the city.

As we study these sectors, it becomes clear that many of the operational and jurisdictional boundaries impede innovation and creativity in the effort to achieve better outcomes. It is also clear that investments made in one sector may result in savings in another sector (for example, investments in water efficiency usually result in large energy cost savings) and that pooling scarce resources to invest in multifunctional and multipurpose common elements may benefit everyone (for instance, through single-purpose underground infrastructure corridors).

The sector notes shed light on critical sector-specific issues that have an impact on city sustainability, but are not under the direct control of city authorities. These issues may need to be addressed on a sector-by-sector basis in collaboration with key stakeholders, particularly the higher levels of government. Identifying the critical pressure points beyond

Combining Flows and Forms to Create a Transdisciplinary Platform

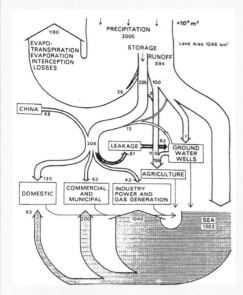

This flow diagram summarizes all the water flow through Hong Kong (China), and is one of the first illustrations of an urban metabolism.

Source: Boyden, Millar, and Newcombe (1981).

Source: Copyright © ESRI, used by permission, http://www.esri.com/.

Flows: Material Flow Analysis and Sankey Diagrams

Material flow analysis and Sankey diagrams are a method for calculating and illustrating the flow of resources through an urban area of any size. Inputs and outputs are determined as resources are extracted from nature; processed by infrastructure; consumed by homes and businesses; treated by infrastructure; and, finally, returned for reuse or delivered back to nature as waste. Colorful, but simple diagrams are used to educate everyone on the resource flows and the effectiveness of their use, all on a single page.

Forms: Layering Information on Maps

Maps are especially useful in collaboration because they speak so well to so many. (A picture is worth a thousand words.) The layers of information make it possible immediately to interrelate the various features and qualities of the landscape and also easily to quantify important spatial relationships. Layering is an old technique that has become more powerful as a result of computer technology and satellite imagery.

Integrating Forms and Flows: A Transdisciplinary Platform

Because diagrams and maps may be easily understood and shared by a broad range of professionals and decision makers, they help to bring stakeholders and experts together, facilitating a common understanding of integrated approaches to design and decision making. Forms and flows should be analyzed and understood for current and future scenarios. In combination, the methods represent a transdisciplinary platform for understanding the spatial dynamics of a city and its physical resource flows, elements that are interdependent, but difficult to integrate because they involve such different skills and stakeholders.

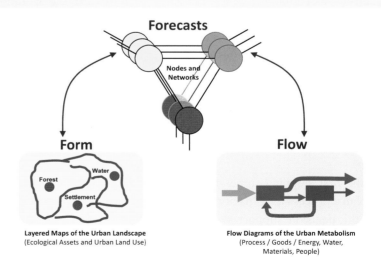

A platform is needed to integrate the design concepts for urban form with the corresponding resource flows.

Source: Redrawn and adapted from Baccini and Oswald (1998).

the direct control of city authorities is also important in devising an expanded platform for collaboration.

The guide also provides a strategy for managing the spatial structure of cities and important lessons on how spatial planning and land use regulations may powerfully affect mobility and affordability.

Moving Forward Together

As forward-looking cities in developing countries identify and implement their sustainability pathways, support may be available from best practice cities worldwide, and from the international community, including development agencies, and academia. Cities are encouraged to tap into the unique resources of each of these partners. In this context, the World Bank Group, together with other development partners, may be in a position to provide technical assistance, as well as capacity-building and financial support, to cities that demonstrate strong political will and commitment to implementing the Eco² initiative.

References

Baccini, Peter, and Franz Oswald. 1998. *Netzstadt: Transdisziplinäre Methoden zum Umbau urbaner Systeme*. Zurich: vdf Hochschulverlag.

Boyden, Stephen, Sheelagh Millar, and Ken Newcombe. 1981. *The Ecology of a City and Its People: The Case of Hong Kong*. Canberra: Australian National University Press.

PART 1

The Framework

Opportunities and Challenges, Principles and Pathways

CHAPTER 1

Ecological Cities
as Economic Cities

This chapter outlines key issues driving the urgent need for a new approach to urban planning, development, and management. While all the transformations that are now occurring may be seen as threats, they may also be perceived as opportunities for the rapid and widespread adoption of a new approach to design, decision making, and investment. In its review of selected case studies, the chapter demonstrates the tangible benefits of cost-effective approaches that have led to greater ecological and economic sustainability in cities. It also clarifies commonly held misconceptions about urban sustainability and concludes that cities should invest in and capitalize on opportunities. If acted on correctly, the changes now under way offer fresh opportunities to achieve sustainability and resiliency in urban areas for generations to come.

Challenges and Opportunities

The scale and rate of urbanization are unprecedented

Urbanization in developing countries may be the most significant demographic transformation in our century as it restructures national economies and reshapes the lives of billions of people. It is projected that the entire built-up urban area in developing countries will triple between 2000 and 2030 from 200,000 square kilometers (km^2) to 600,000 km^2 (Angel, Sheppard, and Civco 2005). These 400,000 km^2 of new urban built-up area that

will be constructed within only 30 years equal the total built-up urban area throughout the world as of 2000 (Angel, Sheppard, and Civco 2005). One might say that we are building a whole new urban world at about 10 times the speed in countries with serious resource constraints (natural, fiscal, administrative, and technical). We are doing so in an increasingly globalized context characterized by many new, constantly fluctuating, interlinked, and uncontrollable variables.

For the first time in history, more than half the world's population, or 3.3 billion people, resides in urban areas. This portion of the world's population living in cities is expected to grow to almost 5 billion by 2030 (UN-Habitat 2008). Over 90 percent of the urban growth is taking place in developing countries. By the middle of the century, Asia alone will host 63 percent of the global urban population, or 3.3 billion people (UN-Habitat 2008). Cities in East Asia housed about 739 million people in 2005 (Gill and Kharas 2007). They will need to accommodate another 500 million by 2030 (Gill and Kharas 2007).

The growth of the world's urban population is being accompanied by an increase in the number and size of cities. There were about 120 cities with populations over 1 million each in 2000. The number is projected to rise to more than 160 by 2015 (World Urbanization Prospects Database). The world will have 26 megacities—cities with populations of more than 10 million—by 2025. Developing countries in Asia will host 12 of these cities (UN-Habitat 2008). An important element in this growth is the fact that 50 percent of the overall urban increase is occurring in medium and smaller cities of less than 500,000 people. Over the next decade, half of the expansion in the urban population in East Asia is projected to be absorbed by such cities (Gill and Kharas 2007).

These population statistics imply a massive investment in manufactured capital, including the building stock and urban infrastructure. The urban strategies that frame decision making and shape policies and investments within the next few years will no doubt have consequences for generations to come.

Cities are engines of economic growth

What is driving the massive rates of urbanization described above? Historically and across most regions, urbanization has propelled the growth of national economies. On average, about 75 percent of global economic production takes place in cities, and in developing countries, this share is now rapidly rising (World Bank 2009). In many developing countries, the urban share of GDP already surpasses 60 percent (World Bank 2009).

The competitiveness of cities is determined by a variety of factors, including geography, national policies, local leadership, market forces, and capital inflows. Historically, nature and geography (altitude, topography, and climate, as well as proximity to coasts, rivers, borders, and natural resources) have often been the trigger in the development of cities. National policies play an important role in inducing and facilitating the growth of cities by determining the location, quality, and connectivity of key infrastructure investments, which, in turn, influence the investment and location decisions of private capital. Together, these generate and enable economic diversification and a range of economic activities that lead to population increases through rural-urban migration and productivity gains. In the context of expanding globalization, the role of trade and foreign investment is being recognized as an additional factor driving the growth of cities. It has also been observed that, as they transition swiftly to a high-value, knowledge-based economy, cities benefit from their critical competitive advantage in their ability to attract, retain, and invest in human capital (Florida 2002). Under these circumstances, a critical determinant of the growth of cities has been a city's capacity to provide a business-enabling environment (such as good infrastructure, policies favorable to reductions in the cost of doing business, and connectivity to external markets); a good quality of life; and an environment that attracts and retains human capital by providing strong social infrastructure and a clean, affordable, and livable environment.

As organizational systems, urban agglomerations provide unique opportunities because of economic and spatial scale. The supply of critical infrastructural services (physical and social), as well as the institutional and administrative organization on which much economic development and social welfare are predicated, becomes financially viable and reaches economies of scale in cities. At the same time, geographical proximity reduces transaction costs and generates economic efficiencies by creating concentrated markets for labor, capital, and goods. This encourages growth, diversification, and innovation in the provision of a wide range of goods and services and the spillover of the knowledge and skills critical to the creation of new ideas. Cities also serve as concentrated markets for the agricultural output of their rural hinterlands.

It is not simply the concentration of activities that makes cities attractive; it is also the diversification and intensification of activities that eventually make cities resilient, competitive, and dynamic. Moreover, in addition to the spatial dimension, cities have a temporal dimension. To remain relevant and competitive, successful cities must continue to evolve. The waterfront manufacturing land area of many older industrial cities has now been converted into high-end residential and financial sector real estate. Through the vastly improved global telecommunications and Internet infrastructure, it is now possible for large segments of the service sector (the products and services of which may be transferred instantly across the globe) to access consumer and labor markets at the push of a button. This will create new and interesting possibilities for human settlement and employment. Although global economies based on cities have emerged at various times throughout history, never before has a global economy achieved the current reach: no city today operates outside the global economic system, and every city has found a place in the network of cities.

It is because of the transformative forces of agglomeration economies that countries in East Asia are experiencing a major shift in economic activities and employment patterns from agriculture to industry and services that is being accompanied by economic diversification within sectors. The concentration of economic production in urban areas is particularly significant in East Asia. The more dynamic coastal regions of China produce more than half the country's GDP on less than one-fifth of the country's land area (World Bank 2008). The concentration of GDP is not proportional to the concentration of population. For example, in Thailand, Bangkok accounts for 40 percent of national GDP, but only 12 percent of national population. Such imbalances are commonly observed in other major Asian cities, for instance, Ho Chi Minh City (29 percent of national GDP, but 6 percent of national population); Manila (31 percent and 13 percent, respectively); and Shanghai, China (11 percent and 1 percent, respectively) (World Bank 2003).

Poverty within and around cities is a challenge

In most regions, the opportunities provided by urbanization have enabled large segments of the population to lift themselves out of poverty. The United Nations Population Fund has examined this relationship between greater opportunity and declining poverty in 25 countries and concluded that urbanization has contributed significantly to poverty reduction. For instance, 28.3 percent of Bolivia's poverty reduction between 1999 and 2005 was attributable to urbanization (UNFPA 2007). It is no wonder the poor are continuing to migrate to urban areas in search of better lives. However, while urbanization has led to economic growth and helped reduce poverty, urbanization alone has not been able to eradicate poverty. Urban poverty and inequality exist despite the concentration of income in cities.

Slums represent the worst form of urban poverty. Individuals and communities living in slums face severe inadequacies in the most basic human requirements, such as shelter, land, water, safe cooking fuel and electricity, heating, sanitation, garbage collection, drainage, paved roads, footpaths, and streetlighting. Largely because of insufficient supplies of serviced land at affordable prices, often caused by unrealistic regulations imposed on land and chronic administrative deficiencies, poor households are unable to gain access to land and housing through legal channels. The poor are thereby forced to live in ramshackle and flimsy settlements on environmentally sensitive areas (slopes and low-lying areas), along roads and railway lines, close to hazardous industrial facilities, and often near the ecological resources of cities. Moreover, because basic urban services are not provided in slums, slum dwellers often live in the worst conditions and have no choice but to pollute surrounding land and water resources. Industries frequently pollute freely and unchecked in slum areas because the disenfranchised residents have little legal, financial, or political recourse. In many cases, the conditions in slums are life threatening; slums are significantly more susceptible to floods, landslides, diseases, exposure to toxic industrial waste, indoor air pollution, fires, and so on.

Slums expanded substantially in the 1990s, when the urban populations in developing countries were growing more rapidly than the capacity of cities to support them. More than 810 million people, or more than one-third of the urban population in developing countries, were living in slums in these countries in 2005 (UN-Habitat 2008). About 64 percent of these slum dwellers, or 516 million people, live in Asia (UN-Habitat 2008). The United Nations Human Settlements Programme (UN-Habitat), has projected that if no firm and concrete actions are taken, the number of slum dwellers will increase to about 2 billion people in the next 25 years (UN-Habitat 2003).

Slum areas are a visible symbol of social exclusion, and they also threaten the well-being of the city by compromising collective ecological assets and increasing the risks of waterborne and infectious diseases.

Migration to urban areas is increasing, driven by the promise of a better future. While cities have had a significant impact on economic productivity, they need to do more to address the crucial issue of urban poverty, particularly the problem of slums.

The flip side of urban migration is the loss in populations in many rural areas and hinterland communities. While people are being pulled from the countryside by the promise of wealth, they are being pushed from their traditional communities because of uncontained urban growth and an almost complete absence of effective, complementary rural planning. Indeed, the problems of slums and the excessive pace of urban growth are also symptoms of poor rural planning and inadequate investment in rural development. The solution is to adopt a more well integrated approach spatially, engaging rural areas in a long-term collaborative exercise to create rural-urban links and promote the management of urban growth.

Continued urbanization is impossible if it is based on standard practices

Urbanization at the rate and scale described above is certain to be accompanied by an unprecedented consumption and loss of natural resources. Calculations already show that, if developing countries urbanize and consume resources at the same degree and scope as developed countries, a resource base equivalent to four planet earths will be needed to support the growth (Rees 2001). Even more surface area will be necessary if farmers are required

Source: National Aeronautics and Space Administration.

to fallow their fields and regenerate soils and if biodiversity is to be sustained. But, of course, we have only one Earth. The resource base essential for sustaining the rural-urban transition will not be available unless the cities in developing countries and in developed countries find more efficient ways to meet the needs of their populations.

In addition to resource inefficiencies, business as usual in urbanization and economic growth generates enormous waste and pollution that impose heavy environmental, social, and economic costs on the local scale and the global scale. Many of these costs are paid by the cities themselves through significantly diminished human health and well-being related to the pollution of air, water, and land; the destruction of ecological assets; the growing fiscal burden; and the reduced long-term economic competitiveness. It is often the poor who suffer most from localized pollution and unhealthy living conditions because they do not have access to safe housing and safe neighborhoods. Such issues are of immediate concern to city leaders who wish to improve the well-being of all citizens, provide a stable and attractive environment for businesses, protect and capitalize on urban ecological assets, and enhance the fiscal strength of cities.

The inadequate management of wastewater and solid waste has led to major environmental and health hazards in cities in many developing countries. In addition, the World Health Organization estimates that more than 1 billion people in Asia are exposed to outdoor air pollutant levels that exceed the organization's guidelines. A recent joint study by the Chinese government and the World Bank has estimated that the cost of ambient air pollution in China's urban areas amounted to about US$63 billion in 2003, equivalent to 3.8 percent of China's GDP during that year (World Bank 2007).

The globalized cost of business-as-usual urbanization is also substantial. It is estimated that cities consume about 67 percent of all global

A few forward-looking cities are now taking the issue of climate change seriously. For instance, city authorities in Brisbane, Australia, are comprehensively addressing this issue through their CitySmart Program. Brisbane officials hope their experiences will pave the way for other cities. (See part 3 for more on Brisbane's initiatives.)

energy and account for well over 70 percent of greenhouse gas emissions, the main contributor to climate change.[1] Heating and lighting in residential and commercial buildings generate nearly 25 percent of greenhouse gas emissions globally. This is equivalent to the amount generated through all agricultural and industrial activities combined. Transport contributes 13.5 percent of the global greenhouse gas emissions, while road transport contributes 10 percent (UN-Habitat 2008). These emissions cause irreversible climate change, which seriously affects global ecosystems, the global economy, and, especially, poorer nations.

According to the Stern Review on the Economics of Climate Change, business-as-usual scenarios could lead to a 5 to 10 percent loss in global GDP; poor countries would experience a loss of more than 10 percent of GDP (Stern 2007). Taking the analysis a step beyond measures of the loss in income and productivity (such as the GDP measure) and looking into the costs of climate change (by factoring in the direct health and environmental impacts and the amplifying or reinforcing feedbacks of such impacts and their outcomes), one may see that the business-as-usual costs of climate change could reduce welfare by an amount equivalent to a reduction of between 5 and 20 percent in per capita consumption. An accurate estimate is likely to be in the upper part of this range (Stern 2007). More important, the Stern Review clearly demonstrates that the poorest countries and people will suffer the impacts of climate change disproportionately and most severely. In essence, the economic, social, and environmental externalities of business-as-usual urbanization are not sustainable.

The existing and the new are a twin challenge

It is clear that if we are to absorb and sustain the powerful wave of urbanization in developing countries, while managing the existing built stock, a paradigm shift will have to occur. The fundamental questions to be addressed are the following: How can cities continue to harness effectively the opportunities for economic growth and poverty reduction offered by urbanization, while also mitigating the negative impacts? How can cities accomplish this goal given the speed and the scale at which urbanization is progressing and given their own capacity constraints? How can ecological and economic considerations be dovetailed, so that they result in a cumulative and lasting advantage for cities? How do we go from ecological cities versus economic cities (Eco versus Eco) to ecological cities as economic cities (Eco^2 cities)?

In general, cities are confronting two challenges: the challenge posed by existing urban areas and the challenge posed by rapid, new urban expansion.

In dealing with existing urban areas, cities may rely on a range of measures to enable the existing built stock to perform more effectively. Examples of retrofitting measures include implementing efficiency in the energy and water sectors; reducing, reusing, and recycling waste; and adapting the existing transportation infrastructure (roads) to make it more efficient (for instance, by designating routes for bus rapid transit and lanes for bicycles). At the same time, cities can explore cost-effective ways of remodeling the distribution, density, and use of the existing built form by increasing floor area ratios; allowing the transfer of development rights; implementing land readjustment programs; rezoning and changing land use patterns; and, more important, revising and enforcing building codes and standards. Redevelopment projects on a larger scale in certain city neighborhoods and districts have also been successful in enhancing the sustainability of existing urban areas. Retrofitting measures and redevelopment projects require holistic planning and coordination across sectors.

Meanwhile, cities are facing unprecedented rates of expansion and are in danger of becoming locked in to inefficient and unsustainable patterns of urban growth from which there is no easy escape. Initial conditions are the bedrock for urban development at every scale; they impose powerful constraints on what may be accomplished as a city matures. These initial conditions include the pattern of spatial development; the built urban form; and most of the related trunk infrastructure investments, which, because of their size and permanence, are powerful constraints on future options. This situation is typically referred to as path dependency. Such path dependency is also evident in the institutional architecture that evolves to support large complex infrastructure systems; this institutional architecture may then reinforce and perpetuate growth of a particular kind. The prospect of being able to influence new urbanization and the growth of cities is tremendous: beginning correctly is much more cost-effective than dealing with problems later on. Being proactive can provide compounded economic, social, and ecological returns. Action taken at this critical phase in the growth of cities can represent a defining opportunity to leapfrog into built-in systemic advantages in efficiency and sustainability. Timing and sequencing are crucial to ensuring the lasting impact of coordinated interventions, maximizing benefits, and reducing long-term externalities. There is an enormous opportunity cost involved in not acting at the correct time, and the correct time is now.

It is in the urgent interest of helping cities systematically capture this value, while the window of opportunity is still open to them, that the World Bank has launched the Eco² Cities Initiative.

Innovations in Urban Sustainability and Their Benefits

It has been concretely demonstrated in some innovative cities that ecological sustainability and economic sustainability can significantly reinforce each other and benefit a range of stakeholders. One role of the Eco² Cities Initiative is to reflect on these examples and find ways to transfer the lessons and successes to cities elsewhere. To begin the process, let us quickly review three case studies. Each of these cases is presented in more detail in part 3. The first case involves the implementation of a successful integrated waste management program through systematic engagement with stakeholders that led to significant environmental and economic gains. The second case involves integrated utility and resource planning and management through systematic stakeholder collaboration that led to significantly greater life-cycle benefits. The third case involves well-coordinated and comprehensive urban development, as well as social and environmental programs. The third case demonstrates that cost is not a major barrier to ecological and economic urban planning, development, and management and is an illustration of successful path dependency (spatial, institutional, and cultural) in urban development.

Yokohama: Environmental and economic benefits through stakeholder engagement

Yokohama, the largest city in Japan, initiated an action plan in 2003. The plan is known as G30 (G = garbage; 30 = a 30 percent reduction in waste generation by fiscal year 2010). The G30 Action Plan clearly identifies the responsibilities of households, businesses, and the government in achieving waste reduction through the 3Rs (reduce, reuse, and recycle) and provides a mechanism for an integrated approach to waste

reduction (City of Yokohama 2003). Environmental education and various promotional activities related to waste reduction have been undertaken to enhance the awareness and knowledge of people and the business community.

Yokohama reduced waste generation by 38.7 percent, from about 1.6 million tons in fiscal year 2001 to 1.0 million tons in fiscal year 2007, while the city's population rose by around 166,000 during the same period (City of Yokohama 2008, 2009a). This significant waste reduction allowed Yokohama to close two incinerators, which saved the city US$1.1 billion in capital costs that would have been required for their renovation. (An exchange rate of US$1 = ¥100 has been used in the cost calculation; see City of Yokohama 2006.) This reduction also led to net savings of about US$6 million in annual operation and maintenance costs (US$30 million in the operation and maintenance costs of two incinerators, minus US$24 million in the operation and maintenance costs of waste recycling operations). Yokohama has two landfill sites. When the G30 Action Plan was planned in 2003, the residual landfill capacity of these two sites was expected to be 100,000 cubic meters by 2007. However, thanks to the waste reduction, the two sites retained a capacity of 700,000 cubic meters in 2007. The value of the saved 600,000 cubic meters in disposal capacity at the two landfills is estimated at US$83 million (City of Yokohama 2006).

Calculations show that the waste reduction between fiscal years 2001 and 2007 resulted in a decline of about 840,000 tons in carbon dioxide emissions. This is equivalent to the amount that 60 million Japanese cedar trees can absorb annually. Approximately 600 km² (an area 1.4 times larger than the city) is needed to plant 60 million Japanese cedar trees (City of Yokohama 2009b). At the same time, if these emission reductions had been certified and sold, they could have provided an additional ongoing revenue stream through carbon finance.

Stockholm: Integrated planning and management through systematic stakeholder collaboration

In an ongoing redevelopment project in Hammarby Sjöstad, a district in the southern part of Stockholm, the Stockholm city council set out to improve on Swedish best practice in sustainability in 1995—the year the environmental program was adopted—by a factor of two on a range of indicators, most notably energy efficiency per square meter. In Sweden, the average annual rate of energy use in some regular new development projects is 200 kilowatt-hours per square meter; cutting-edge practice produces an efficiency of 120 kilowatt-hours per square meter (Bylund 2003). The current project is aiming for a rate of 100 kilowatt-hours per square meter. Other targets set for the project include water conservation, waste reduction and reuse, emissions reduction, reduced use of hazardous materials in construction, use of renewable sources of energy, and integrated transportation solutions. Stockholm is already a sustainable city, and the city council intended for this project to be a pathbreaking demonstration of sustainable methods of urban redevelopment. Hammarby Sjöstad is one of three designated ecocycle districts in Stockholm.

To accomplish the objectives set by the city council, the three city departments of waste, energy, and water and sewage collaboratively designed a model, the Hammarby Model. The Hammarby Model represents an attempt to turn a linear urban metabolism—consume resources on inflow and discard waste through outflow—into a cyclic urban metabolism by optimizing the use of resources and minimizing waste (figure 1.1). The model streamlines vari-

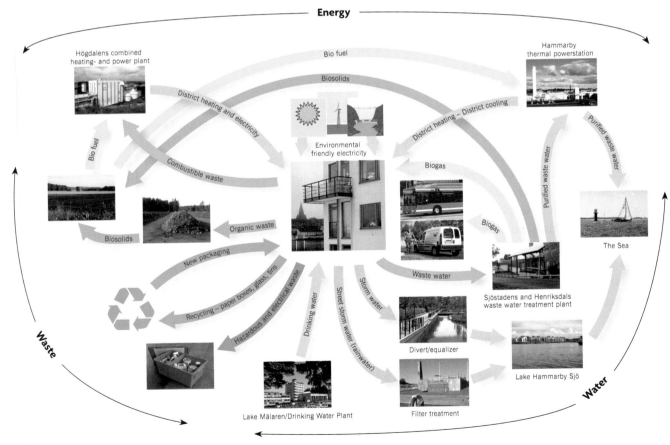

Figure 1.1 The Hammarby Model, Stockholm: An Example of Integrated Planning and Management

Source: City of Stockholm, Fortum, Stockholm Water Company.

ous systems of infrastructure and urban service delivery and provides the foundation and blueprint for achieving the sustainability objectives outlined above.

The initial findings of the preliminary evaluations of the first phase of development, Sikla Ude (SU), compared with a reference scenario (Ref) are shown in figure 1.2: a 30 percent reduction in nonrenewable energy use (NRE), a 41 percent reduction in water use, a 29 percent reduction in global warming potential (GWP), a 41 percent reduction in photochemical ozone creation production (POCP), a 36 percent reduction in acidification potential (AP), a 68 percent reduction in eutrophication potential (EP), and a 33 percent reduction in radioactive waste (RW).

Success in a project such as Hammarby Sjöstad depends on coordination among key stakeholders. To channel all efforts in a single direction, the city appointed a project team in 1997. In 1998, the project team was incorporated into the city's Department of Roads and Real Estate (now called the Development Depart-

ment). This step had several positive ramifications. First, by being housed in the city's Department of Roads and Real Estate, the project team had greater access to and control over public funds. In addition, the project team was in a much stronger position to leverage and negotiate with private interests. The structuring of the team was established as follows. Representatives of the city departments of planning, roads and real estate, water and sewage, waste, and energy were members of the team. The various departments of the city were integrated into a single fabric led by a project manager and an environmental officer who were charged by the city with the responsibility to guide and influence the public and private stakeholders toward the realization of the environmental objectives of the project.[2]

Curitiba: Cost is not a major barrier

Sustainable urban development in developing countries has also been successfully undertaken by forward-looking cities that possess relatively limited fiscal resources. Consider the case of

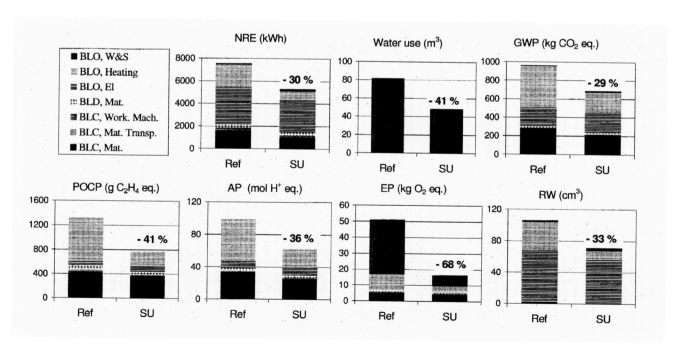

Figure 1.2 Initial First-Phase Results of Hammarby Sjöstad according to the Environmental Load Profile Life-Cycle Analysis Tool

Source: Brick (2008).

Note: The tool is described more fully in chapter 10.

Curitiba, the capital of the state of Paraná, Brazil. Since the 1960s, through its innovative approaches in urban planning, city management, and transportation planning, Curitiba has been able sustainably to absorb a population increase from 361,000 (in 1960) to 1,797,000 (in 2007) on what was initially a limited budget. It has provided critical urban services with a wider coverage and smaller ecological footprint than many cities with much greater fiscal resources at their disposal. Moreover, while doing this, Curitiba has expanded its own fiscal capacity and economic base and has gained a reputation as one of the best examples in the world of ecological and economic urban development.

The most significant planning decision made by Curitiba was to grow from the city core outward in a radial linear branching pattern, thereby opening up the city, while preserving urban density and protecting green areas. This approach contrasts with the typical concentric, ad hoc development. To encourage effective linear urban growth along major structural axes (rather than extensive sprawl), Curitiba pursued the incremental development of an integrated bus system (figure 1.3). Land use and zoning simultaneously encouraged

higher-density commercial and residential development along each structural axis, thereby providing the economic density and user base to make the system financially sustainable. The color-coded bus system is designed to provide various levels of service (interdistrict, feeder, intermunicipal, and so on) and is integrated as a single unified system within the land use plan.

As a consequence, Curitiba has the highest rate of public transportation ridership in Brazil (45 percent). This means that Curitiba has one of the lowest rates of urban air pollution in the country. Fuel loss caused by traffic congestion was valued at US$930,000 in 2002, compared with US$13.4 million in Rio de Janeiro (CNT 2002, Vassoler 2007). In contrast, in 2000, congestion in 75 metropolitan areas in the United States caused fuel and time losses valued at US$67.5 billion (Downs 2004). If these areas in the United States were planned and developed more efficiently, a major portion of the annual recurring loss and the harmful emissions could be avoided.

In the 1950s and 1960s, Curitiba suffered from persistent flooding while construction and development were proceeding at a rapid pace. Additional drainage canals would have

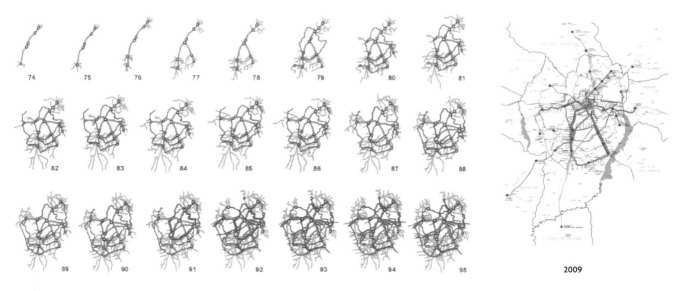

Figure 1.3 The Integrated Transportation Network, 1974–95 and 2009

Source: IPPUC (2009).

been required at an enormous cost. However, by setting aside land for drainage and putting low-lying areas off limits for development, the city managed to tackle the costly flooding problem and avoid the huge capital costs linked to flood control and drainage (Rabinovitch and Leitman 1996). The city turned these areas into parks planted with many trees and created artificial lakes to hold floodwaters. Buses and bicycle paths helped integrate the parks into the city's transportation network (Rabinovitch and Leitman 1996). This is an excellent example of how ecological assets and green infrastructure may be integrated in urban design. The cost of this strategy, including the relocation costs of slum dwellers, has been estimated at five times less than the cost of building concrete canals. Also, as a result, the property values of neighboring areas appreciated, as did tax revenues.

A special program allowed developers to transfer their development rights to land in locations the city desired to preserve to land in locations the city desired to develop and provided incentives and tax breaks for the preservation of green areas, as well as historic and cultural heritage sites. At the same time, Curitiba maintained its vibrant urban density along the axes of growth: population density rose from 1,410 to 4,161 persons per km² between 1970 and 2007, even as the green area per person rose from 1 square meter to 51.5 square meters.

Much of the success of Curitiba may be attributed to the Institute for Research and Urban Planning of Curitiba. Established in 1965, the institute is a powerful independent municipal public authority and serves as the research, planning, implementation, and supervision agency in the city. The institute has enabled coordination among the different elements in urban development and has been the most important factor in ensuring continuity and consistency in the planning process through successive city administrations. Its imaginative and integrated urban planning, development, and management solutions have significantly reduced the inefficiencies associated with piecemeal development.

Jaime Lerner, who contributed to drafting the city's 1966 master plan and worked as president of the institute in 1969 and 1970, was elected mayor of Curitiba three times (1971–75, 1979–83, and 1989–92). He is widely known as one of the most popular, creative, and successful mayors in Brazil, and his influence has spread globally. He has won many awards, including awards from the United Nations Environment Programme, the United Nations Children's Fund, the International Institute for Energy Conservation, and the Prince Claus Awards of the Netherlands.

The current administration has achieved wide-ranging success by taking on innovative new projects with a sustained focus on social, environmental, and urban planning issues through substantive consultation with public audiences. Mayor Carlos Richa, who began his administration in 2004, enjoys great popularity, which was evident in his 77 percent approval rating when he was reelected in October 2008.

Powerful Lessons from Best Practice Cities

Curitiba, Stockholm, and Yokohama, as well as the many other examples in this book, provide the positive message that change at various scales is possible, and some of the dominant myths about urban sustainability (such as the high cost) are not always based in fact. Let us examine some of these lessons a bit more.

Many solutions are affordable even if budgets are limited

One of the biggest, most prevalent misconceptions is that innovative measures are not affordable and that they do not generate significant returns. As concretely demonstrated by the cases of Curitiba and Yokohama, this is certainly not true. Many creative, practical, and cost-effective solutions simultaneously achieve greater benefits than business-as-usual scenarios.

An additional case helps underscore this lesson. The municipality of Emfuleni, South Africa initiated an energy and water conservation project that achieved savings of 7 billion liters of water and 14 million kilowatt-hours per year. At a cost of only US$1.8 million, the project saved over US$4 million each year, meaning that the project paid for itself in less than six months. Because the project contract was financed and implemented by an energy service company, the municipality not only saved large sums of money because of reduced water losses and pumping costs, but also did not have to pay for the investment up front. Meanwhile, the energy service company recouped its investment quickly by sharing in the cost savings (USAID 2005).

Ample evidence similar to Emfuleni's experience indicates that steps to improve energy and resource efficiencies can bring about strong fiscal and economic gains. Thus, while solid waste management in many medium-size cities can account for 40 to 50 percent of the total municipal budget, programs such as the one in Yokohama offer a powerful illustration of the significant returns that may be achieved and the capital costs that may be avoided (Pagiola and others 2002). For cities and utilities seeking ways to meet budget shortfalls or save municipal expenditures so as to spend more on worthy social pursuits (such as extending tap water services, waste collection and treatment, or street-lighting coverage to slums), there is no better place to look for new funds than the cost savings achieved through resource use efficiency.

Effective and well-coordinated urban planning and land policies and appropriate spatial layouts can provide strong and sustained long-term development and compound the economic, social, and environmental returns. Effective urban planning and land policies can help integrate the urban poor into the economic, social, and physical fabric of a city, thereby proving to be economically beneficial to cities, national governments, and the urban poor themselves. The financial outlays required for good planning are not significant. However, as shown by Curitiba, sustained commitment and investment in maintaining strong technical, administrative, and institutional capacity are necessary.

If they are designed and implemented appropriately, policy and regulatory measures can also generate environmental, fiscal, and economic gains. Households and businesses represent major components in energy and resource consumption and waste generation, and their savings in these areas may be translated into economic and fiscal gains for a city. Thus, since the mid-1970s, California's utility improvement programs and energy efficiency policies, which consist of standards and research and development, have held per capita electricity use constant, while per capita electricity use has risen by nearly 80 percent elsewhere in the United States. This has generated substantial savings among consumers, households, and businesses in California (California Air Resources Board 2008). Over the past three decades, consumers in California have saved more than US$50 billion through policies promoting appliance efficiency and building efficiency. Educational programs and awareness campaigns may also influence consumption patterns without significant resource expenditures.

Success is achievable through existing proven technologies and appropriate new technologies

Best practices suggest that success depends less on new technologies and more on appropriate technologies. In most cities, expensive hydrogen-fueled cars are less relevant to transportation efficiency than expanding the network of bicycle- and pedestrian-friendly pathways. Simple technology solutions such as the installation of insulation in homes or water-efficient faucets often generate more cost savings than do many new technologies (EIU 2008). Many of the new technology options are often mistakenly considered commercially unviable because an accurate and full cost-benefit analysis from a life-cycle point of view has not been carried out and because new technologies must sometimes compete with the embedded subsidies for older technologies. An example is the high implied subsidy for automobiles provided by free or inexpensive parking on public lands and by extensive highway construction. The simplest solution sometimes involves embracing appropriate new technologies through bulk procurement by cities and their partners, incubating new technologies through phased subsidy schemes until economies of scale are able to bring down local prices, or public awareness campaigns. In cases where local production may fill the need, the impact on local economic development can be a significant added value. The manufacture and installation of efficient technologies help to circulate money within the community, generating and retaining new local employment instead of exporting money for the purchase of water or energy commodities. Countries such as Denmark and Sweden have invested in the incubation of new technologies and are reaping the rewards. There is now a growing international demand for much of their technological expertise.

Developing countries should take pride in homegrown solutions

As cities in developing countries grow in size and affluence, city authorities should first look at the innovation taking place within their own city boundaries. Often, local interest groups, universities, or institutional stakeholders might already be lobbying for change and piloting innovations within the unique cultural context of the cities. Building on these homegrown initiatives, the city administration may begin to look more systematically into the successful example of cities that began their efforts in the context of similar constraints: for many cities in developing countries, Curitiba might prove a relevant example. It is as important to learn from the failures of some cities as it is to learn from the examples of best practice. Many cities in more developed western countries are locked into development patterns—such as sprawled spatial form, building heights and setbacks, parking allocations, street widths, road patterns, infrastructure systems, consumption trends, and so on—that cannot be easily changed. In many cases, performance as measured in sustainability indicators is significantly better in cities in developing countries than in cities in developed countries. It is important to remember that most energy and resources are consumed and most waste is generated in developed countries. There are additional differences in these patterns across developed countries. For instance, European cities consume much less energy and are significantly more ecological and well planned relative to North American cites, even though the level of development of the two regions is similar. This situation is partly the result of environment-friendly city, national, or European Union–wide policies that promote clean energy and energy efficiency. Europeans pay higher energy prices and have historical and cultural preferences for compact urban form and high-quality public transportation. This is also attributable to regulations requiring automakers in Europe and Japan to produce cars that are much more fuel efficient than are the cars produced in the United States. By selecting and combining actions suitable to their own capacities and needs, city authorities may adapt these lessons into their own homegrown solutions.

Many solutions benefit the poor indirectly and directly

Achieving fiscal gains in city expenditures and utility payments can free up money for social investment and, thus, indirectly benefit the poorest segments of urban populations. Furthermore, because poor people are so dependent on land policies and urban services, many planning measures taken by cities may offer other direct and substantial benefits. For example, regulatory reform and effective urban planning and land use policies have a powerful direct impact through improvements in the situation of the poor by reducing the prices for land and housing. The poor may also benefit directly from more public transportation, pedestrian access, and bicycle paths; improved water access, sanitation, and electricity connections; the provision of safe cooking fuels; and energy-efficient light-emitting diode lights in slums. Better environmental standards to address industrial pollution will strongly enhance living conditions among the poor. Innovative programs such as the Orangi Pilot Program in Karachi, Pakistan have involved the poor directly in community-based sanitation construction projects, offered jobs to families, and achieved the extremely cost-effective construction of local sanitation networks that link to the trunk lines of cities. The money and the services contribute to local economic development by generating employment and income, improving environmental conditions, increasing home values, and creating local ownership in neighborhoods.

Opportunities to Capitalize

The challenge that lies ahead is to take full advantage of the many opportunities created by rapid change and successful innovation. Best practices exist in strategic long-term planning and regional growth management, and the appearance of new tools for systems analysis and mapping offer potential for more well integrated, practical, and rigorous analysis. Because successful cities are often fundamental to successful nations, the higher levels of government should be key partners in helping cities take the initiative. There is also a growing commitment at the international level to support cities and to finance investments that enable ecological and economic sustainability in cities. New funding opportunities have emerged for city authorities in developing countries who are willing to take steps to achieve sustainable urban development, particularly measures promoting energy and resource efficiency that lead to reductions in greenhouse gas emissions. New accounting methods for estimating the full costs and benefits of various policy, planning, and investment options are also being used (for example, life-cycle costing). Channeling these opportunities toward a massive scale and accelerating the pace of urban development create the potential for tremendous impact.

An increasing number of cities have initiated actions to achieve greater ecological and economic sustainability according to their own visions, needs, and capacities. Their limited resources or capacities do not discourage these cities. Some cities demonstrate strong leadership in pioneering new approaches; some use well-established approaches and excel in the implementation of these approaches; and some work with the international community to learn from best practices and invest in technical, institutional, and administrative capacity.

The Eco^2 Cities Initiative has been developed to enable cities in developing countries to benefit from the promise of a more rewarding and sustainable growth trajectory while the window of opportunity is still open. The chapters hereafter provide a detailed framework that may be adapted by cities in developing countries to work systematically toward accomplishing many more positive results.

A case study from Dhaka, Bangladesh, illustrates the potential for improving conditions among the urban poor through interventions in the urban environment. Waste Concern, a nonprofit organization that works with the city government, succeeded in reducing emissions in Dhaka by composting solid waste instead of burning or flaring and then selling the waste to fertilizer companies. The initiative is helping to reduce 52 percent of the generated solid waste that remains uncollected in Dhaka. The city provides public land for community composting. Waste Concern coordinates with the city to undertake the house-to-house collection of solid waste in rickshaw bicycles and bring the waste to processing plants. Organic waste is separated from other rubbish. Waste is composted into enriched biofertilizers. Waste Concern arranges for fertilizer companies to purchase and nationally market the compost-based fertilizer. This approach has the potential to create 16,000 new jobs for the poor of Dhaka and 90,000 new jobs for the poor across Bangladesh.

Source: C40 Cities (2007), Enayetulla and Hashimi (2006).

Notes

1. The data vary depending on sources and methodologies. These are data of the International Energy Agency.
2. The information in this paragraph has been provided by the City of Stockholm.

References

Angel, Shlomo, Stephen C. Sheppard, and Daniel L. Civco. 2005. *The Dynamics of Global Urban Expansion.* Washington, DC: World Bank.

Brick, Karolina. 2008. "Barriers for Implementation of the Environmental Load Profile and Other LCA-Based Tools." Licentiate thesis, Royal Institute of Technology, Stockholm.

Bylund, Jonas R. 2003. "What's the Problem with Non-conventional Technology? The Stockholm Local Investment Programme and the Eco-cycling Districts." In *ECEEE 2003 Summer Study Proceedings: Time to Turn Down Energy Demand,* ed. Sophie Attali, Eliane Métreau, Mélisande Prône, and Kenya Tillerson, 853–62. Stockholm: European Council for an Energy Efficient Economy. http://www.eceee.org/conference_proceedings/eceee/2003c/Panel_4/4214bylund/.

C40 Cities. 2007. "Waste: Dhaka, Bangladesh." C40 Cities Climate Leadership Group. London, U.K. http://www.c40cities.org/bestpractices/waste/dhaka_organic.jsp.

California Air Resources Board. 2008. "Climate Change Draft Scoping Plan." Sacramento, CA. http://www.arb.ca.gov/cc/scopingplan/document/draftscopingplan.pdf

Calthorpe, Peter, and William B. Fulton. 2001. *The Regional City: Planning for the End of Sprawl.* Washington, DC: Island Press.

City of Yokohama. 2003. "Yokohama shi ippan haikibutsu shori kihon keikaku, Yokohama G30 plan." 横浜市一般廃棄物処理基本計画、横浜G30プラン [City of Yokohama, master plan for management of general waste: Yokohama G30 Plan]. City of Yokohama, Japan. http://www.city.yokohama.jp/me/pcpb/keikaku/kei1.html (accessed February 2009).

———. 2006. "Yokohama G30 plan, kenshou to kongo no tenkai ni tsuite" 横浜G30プラン「検証と今後の展開」について [Yokohama G30 Plan: verification and next steps]. Resources and Wastes Recycling Bureau, City of Yokohama, Japan. http://www.city.yokohama.jp/me/pcpb/keikaku/G30rolling/ (accessed February 2009).

———. 2008. "Heisei 20 nendo jigyou gaiyou" 平成20年度事業概要 [Operation outline for fiscal year 2008]. Resources and Wastes Recycling Bureau, City of Yokohama, Japan. http://www.city.yokohama.jp/me/pcpb/keikaku/jigyo_gaiyou/20gaiyou/ (accessed February 2009).

———. 2009a. 横浜市統計書[web版] 月別世帯数及び人口 [Yokohama statistical reports (web version): Monthly number of households and population]. Yokohama Statistics Portal, City of Yokohama, Japan. http://www.city.yokohama.jp/me/stat/toukeisho/new/#02 (accessed February 2009).

———. 2009b. ごみの分別による効果 - 二酸化炭素削減効果 [Effect of segregation of garbage—reduction of carbon dioxide]. Resources and Wastes Recycling Bureau, City of Yokohama, Japan. http://www.city.yokohama.jp/me/pcpb/shisetsu/shigenkai/lca/ (accessed February 2009).

CNT (Confederação Nacional do Transporte). 2002. "Pesquisa da Seção de Passageiros CNT, 2002; Relatório Analítico: Avaliação da Operação dos Corredores de Transporte Urbano por Ônibus no Brasil." Report. CNT, Brasília.

Downs, Anthony. 2004. *Still Stuck in Traffic: Coping with Peak-Hour Traffic Congestion,* rev. ed. Washington, DC: Brookings Institution Press.

Enayetullah, Iftekhar and Quazi Sarwar Imtiaz Hashimi. 2006. "Community Based Solid Waste Management Through Public-Private-Community Partnerships: Experience of Waste Concern in Bangladesh." Presentation at Asia 3R Conference, Tokyo, Japan. October 31. http://www.env.go.jp/recycle/3r/en/asia.html.

EIU (Economist Intelligence Unit). 2008. "Sustainable Urban Infrastructure, London Edition: A View to 2025." Siemens AG, Munich.

Florida, Richard. 2002. *Rise of the Creative Class: And How It's Transforming Work, Leisure, Community and Everyday Life.* New York: Basic Books.

Gill, Indermit, and Homi Kharas. 2007. *An East Asian Renaissance.* Washington, DC: World Bank.

IPPUC (Institute for Research and Urban Planning of Curitiba). 2009. "The City of Curitiba: Planning for Sustainability; An Approach All Cities Can Afford." Presentation at "World Bank Energy Week 2009," World Bank, Washington, DC, March 31.

Pagiola, Stefano, Roberto Martin-Hurtado, Priya Shyamsundar, Muthukumara Mani, and Patricia Silva. 2002. "Generating Public Sector Resources to Finance Sustainable Development: Revenue and Incentive Effects." Technical Paper 538, Environment Series, World Bank, Washington, DC.

Rabinovitch, Jonas and Josef Leitman. 1996. "Urban Planning in Curitiba." In *Sustainable Urban Development Reader,* eds. Stephen Wheeler and Timothy Beatley. 2008. New York: Routledge.

Rees, William E. 2001. "Global Change, Ecological Footprints and Urban Sustainability." In *How Green Is the City? Sustainability Assessments and the Management of Urban Environments,* ed. Dimitri Divuyst, 37–42. New York: Columbia University Press.

Stern, Nicholas. 2007. *The Economics of Climate Change: The Stern Review.* New York: Cambridge University Press.

Swedish Environmental Protection Agency. 2004. "Local Investment Programmes: The Way to a Sustainable Society." http://www.naturvardsverket.se/Documents/publikationer/91-620-8174-8.pdf.

UNFPA (United Nations Population Fund). 2007. *State of World Population 2007: Unleashing the Potential of Urban Growth.* New York: UNFPA.

UN-Habitat (United Nations Human Settlements Programme). 2003. *The Challenge of Slums: Global Report on Human Settlements 2003.* London: Earthscan Publications.

———. 2008. *The State of the World's Cities 2008/2009: Harmonious Cities.* London: Earthscan Publications.

USAID (U.S. Agency for International Development). 2005. "Watergy Program Pioneers Performance Contract to Save Water, Energy in S. Africa." *Energy Update* 2 (April/May): 6–7.

Vassoler, Ivani. 2007. *Urban Brazil: Visions, Afflictions, and Governance Lessons.* New York: Cambria Press.

Wheeler, Stephen M., and Timothy Beatley, eds. 2007. *The Sustainable Urban Development Reader.* Routledge Urban Reader Series. New York: Routledge.

World Bank. 1997. "Expanding the Measures of Wealth: Indicators of Environmentally Sustainable Development." Environmentally Sustainable Development Studies and Monographs Series 17, World Bank, Washington, DC.

———. 2003. "Looking Beyond Short-Term Shocks." East Asia Update, April, East Asia and Pacific Region, World Bank, Washington, DC.

———. 2007. *Cost of Pollution in China: Economic Estimates of Physical Damages.* Washington, DC: World Bank. http://siteresources.worldbank.org/INTEAPREGTOPENVIRONMENT/Resources/China_Cost_of_Pollution.pdf.

———. 2008. *World Development Report 2009: Reshaping Economic Geography.* Washington, DC: World Bank.

———. 2009. "The World Bank Urban and Local Government Strategy: Concept and Issues Note." World Bank, Washington, DC.

World Urbanization Prospects Database. Population Division, Department of Economic and Social Affairs, United Nations. http://esa.un.org/unup/index.asp (accessed May 2009).

CHAPTER 2

Eco² Cities Initiative: Principles and Pathways

Chapter 1 explores the many opportunities offered to cities as a consequence of change. It also uses examples of best practice drawn from cities worldwide to illustrate the potential environmental and economic benefits associated with innovative approaches. If the knowledge and means exist to design and implement such measures and if practical and powerful solutions have been demonstrated even among cities with limited budgets, then why are other cities not taking advantage of these opportunities? Why are these examples so rare?

Chapter 2 begins with a brief overview of the many challenges cities face in trying to adopt more well integrated approaches. Most readers are familiar with these challenges (unfortunately, they are common), and a detailed accounting is unnecessary. However, the challenges are worth highlighting because, together with the valuable ground-level lessons derived from the experience of best practice cities, they help to frame our strategic response: the key strategies and principles that define the Eco² Cities Initiative.

After reviewing challenges, the chapter describes a set of four overarching principles that provide the scope and direction for all elements of the Eco² initiative. Adopting these principles represents the first step toward the Eco² approach. Basically, the principles are proven strategies that can help cities seize new opportunities, overcome challenges, and transfer best practices to every new project.

At the end of the chapter, a summary table is provided of the Eco² approach. The principles are translated into a number of core program elements, and an example is provided for how any city might implement the program in a step-by-step fashion and create its own unique Eco² pathway.

The Many Challenges That Cities Face

Limited resources

In general, cities in developing countries face significant administrative, technical, and financial capacity constraints. These cities are also challenged by the rapid pace of urbanization. For these reasons, city staff tend to focus their attention on chronic problems and on the day-to-day and sector-to-sector problems that are piled at the front of the counter. Ask any city administrator, and you will hear the same story: no time exists to take on long-term plans or cross-cutting agendas like the Eco² Cities Initiative.

Misinformation

Another reason for the lack of initiative is the fact that the lessons described in chapter 1 have not been widely shared or understood. Instead, many local decision makers operate under a series of myths and false assumptions. Solutions like the Eco² approach are perceived as demonstration projects, rather than a permanent, alternative approach to planning, developing, and managing cities. They are assumed to be costly, dependent on advanced and complex technologies, and practical only for wealthy neighborhoods and well-resourced city administrations. This attitude is reinforced by the all-too-common assumption that the most advanced approach to city building is to import the styles and technologies used in a majority of western cities (or in the growing number of real estate ventures opportunistically branded as eco-cities) rather than relying on local culture and ecology.

Correcting misconceptions may need to be one of the first stepping stones in a city's Eco² pathway.

Institutional barriers

Inappropriate institutional structures and mind-sets are commonly cited as the greatest challenges when cities consider implementing integrated solutions. Some of the most obvious examples include the following:

- *Fragmentation of responsibilities;* separate budgets, timelines, and goals; and piecemeal solutions that serve individual interests well, but that, in combination, are precisely wrong

- *Excessive specialization* and overwhelming complexity; silos of expertise; and incomplete perspectives on urban resource use and the associated costs

- *Single-purpose funding mechanisms* that fail to address cities directly, that fail to address the urban system as a whole, or that fail to link program objectives to the priority issues in a city

- *Lengthy and challenging political processes* for allocating funds at all scales

- *Short-term and narrow accounting formats* that ignore indirect costs and benefits, separate capital costs from operating and maintenance costs, fail to capitalize the replacement of systems, do not take into account all capital assets (manufactured, ecological, human, and social) and risks, and mislead investors and the public

Locked-in relationships among networks of public and private institutions and existing technologies

Some dimensions of urban planning reflect a complex set of entrenched relationships among many different organizations, public and private. Because some groups benefit from the status quo, they actively promote more of the same and create obstacles to investment in alternatives.

A well-known example is the highway lobby, which represents everyone who makes money from roads and has been accused of promoting massive investments in road building regardless of the societal costs and alternative technologies.

Cities commonly become locked in to certain technologies as a result of past capital investments in facilities and the ongoing need to

A techno-institutional complex arises because large technological systems (such as electricity generation, distribution, and end use), cannot be understood fully as a set of discrete technological artifacts, but must be seen as complex systems of technologies embedded in a powerful conditioning social context of public and private institutions.

Such complexes are developed through a path-dependent, coevolutionary process involving positive feedbacks among technological infrastructures and the organizations and institutions that create, diffuse, and use them. Once locked in, the complexes are difficult to displace and may lock out alternative technologies for extended periods even if the alternatives demonstrate improvements on the established complex.

Source: Unruh (2000).

recover sunk costs and realize returns. If someone proposes to invest in demand-side management (DSM) or meet the need for services in other ways, the effect is to reduce the flow of revenues below projections; as a consequence, the existing facilities remain oversized and may become economically unviable. This issue may occur whenever cities or their financial partners invest in new energy plants, water factories, wastewater treatment plants, solid waste transfer stations, and incinerators. Under such circumstances, cities commonly use policy to prevent innovative approaches. If they are not developed properly, public-private partnerships may offer another example of how cities become locked in to technologies by entering into contracts that guarantee a long-term demand for services of a single type.

Human inertia

A new planning process that involves many planners and designers will certainly challenge the natural tendency of people, particularly professionals, to resist change of any kind. Without a focused effort to manage change, human inertia will invariably reproduce the same patterns of land development and exactly the same infrastructure in city after city, all in accordance with standard practice. It is difficult to change the mold. If conservative engineers are hired to consider a type of system they have never previously designed, they will invariably condemn the idea. While conservative engineers are, in most respects, the best

sort to have, the process of preliminary engineering or concept design requires a much more open and innovative mind-set, which is difficult to find without engaging specialty firms at added expense and risk.

The continuing dominance of 19th century models

Part of the difficulty with adopting a program such as Eco² is that current design and planning practices among cities are rooted in patterns established in the 19th century, when an abundance of coal, combined with new manufacturing technologies, brought unprecedented increases in wealth and improvements in the quality of life. By the beginning of the last century, millions of families in Europe and North America suddenly had access to clean water, sewage treatment, space heating, lighting, clean streets, and public transit. This wave of societal progress and modernism was achieved through single-purpose, centralized, supply-oriented utilities that operated in silos (that is, entirely independently), and capitalized on economies of scale, abundant resources, and open access to public goods such as water and the atmosphere.

Hugely successful in their time, the 19th century models are no longer the best solution and, in fact, have become part of the problem. The world is more crowded and complex and requires much more efficient, longer-term solutions for servicing urban areas. Nonetheless, the 19th century models are integral to our professional training and institutional structures. A program that encourages a more well integrated approach must overcome the inertia of past practice and the natural resistance to change within established institutions and groups of practicing professionals.

A Principled Approach That Can Overcome the Challenges

The Eco² initiative is designed on the premise that many of the opportunities and challenges described above may be addressed most effectively by adopting new principles. These principles may be used to guide the process of designing, implementing, and financing urban development. The principles function as super strategies for cities in transition. The four Eco² principles are (1) a city-based approach enabling local governments to lead a development process that takes into account specific circumstances, including the local ecology; (2) an expanded platform for collaborative design and decision making that accomplishes sustained synergy by coordinating and aligning the actions of key stakeholders; (3) a one-system approach enabling cities to realize the benefits of integration by planning, designing, and managing the whole urban system; and (4) an investment framework that values sustainability and resiliency by incorporating and accounting for life-cycle analysis, the value of all capital assets (manufactured, natural, human, and social), and a broader scope of risk assessments in decision making.

Each of these strategies has been elevated to the status of super strategy or principle because it is universally applicable, crucial to success (based on the experience of best practice cities), and frequently ignored or underappreciated.

The four principles are interrelated and mutually supportive. For example, without a strong city-based approach, it is difficult to engage key stakeholders fully through an expanded platform for collaborative design and decision making. And, without this expanded platform, it is difficult to explore creative new approaches to the design and management of integrated systems and to coordinate policies for implementing the one-system approach. Prioritization, sequencing, and the effectiveness of investments in promoting sustainability and resiliency will be greatly enhanced if one is able to appreciate the city as a single system and if one may rely on an expanded platform of collaboration.

The synergy among the principles is more apparent in other chapters. The Eco² principles are explored below.

PRINCIPLE 1: A city-based approach

A city-based approach is the first principle, and it carries two complementary messages. First, it recognizes that cities are now on the front lines in managing change and leading an integrated approach. Only at the city level is it possible to integrate the many layers of site-specific information and to work closely and rapidly with the many stakeholders who need input on an integrated solution. In addition, fiscal and administrative decentralization has brought important decision-making and management responsibilities to local governments. Second, the city-based approach serves to emphasize the importance of incorporating within any development program the unique aspects of place, especially ecological assets. Increasingly, cities depend on their natural landscapes to provide food and recreation, capture and store water and energy, absorb wastes, and satisfy many other needs. Protecting and enhancing ecological assets—the natural capital—are a priority in directing (and constraining) urban growth. A city-based approach is thus place specific and focuses on enabling local leadership and local ecologies.

We look now at each of these in turn.

Depending on their size, cities are the most influential institutions within the modern state. Not only do they represent the engines of the economy and provide homes for a majority of the population, they are also responsible for a majority of resource and energy consumption and harmful emissions. Thus, a city that works with its key sectors and stakeholders is especially well placed to explore Eco² solutions. Cities also have critical instruments at their disposal (zoning, permits, approvals, taxes, and fees) and have been empowered through fiscal and policy decentralization in many countries. It is not therefore surprising that almost all the case studies of Eco² solutions have been produced in cities that have taken leadership and applied a city-based approach.

When a city takes leadership in setting priorities and implementing solutions, two factors appear to be critical: its level of commitment and its capacity to act. Decision makers need to be convinced of the value of an Eco² approach, and they need to mobilize political support within their constituencies. A city's success will depend on how effectively and creatively it uses and develops the levers of influence within its control. These may range from the city's human and technical capacity and knowledge of local realities to its formal urban planning tools and municipal financing strategies. Often, to act effectively, a city may need technical, administrative, and financial support, including knowledge, skills, and tools.

A city's capacity to act will also depend on the levers of influence beyond its realm of control. Often its legislative, administrative, and fiscal powers are circumscribed by national or state governments, the cooperation of which is crucial. At the same time, given the growing predominance of metropolitan areas that span the jurisdiction of more than a single city, coordination is often required at the metropolitan level to implement optimal interventions within and across all sectors. Thus, leadership by cities needs to occur at many levels, including the region.

The city-based approach is not only political, but also fundamentally ecological. Cities

> The living city is not an island. Its metabolism is linked to surrounding ecosystems, and its people and culture are networked to other viable urban cells to form a living and developing tissue that is a net primary producer, not a parasitic system.
>
> *Source:* Goa 2100 Plan, 2003. For information on the Goa 2100 Plan, see Revi and others 2006.

are centers of resource consumption, and ultimately, resource efficiency will depend greatly on how well the city is integrated into the local and regional ecologies. City planning is aimed at protecting and regenerating irreplaceable natural capital, especially the natural assets and ecological services throughout the urban region in which the city is located. All cities need to be fully integrated into a viable local ecology. The integration of cities into local ecologies may occur at all scales, from food gardens and naturescaping to planning of containment boundaries that effectively separate urban areas from natural areas.

Ideally, the ecological elements mix and intersect within and stretch throughout the city as a natural blue-green web, providing multiple services to the local economy. Ecologies and open green spaces serve as a kind of green infrastructure. They might pollinate crops and orchards on behalf of agrifood systems, or recharge aquifers on behalf of the water supply system, or channel wind toward open hilltops or water basins on behalf of the local energy utility. Green infrastructure may also serve to enhance larger ecological systems.

PRINCIPLE 2: An expanded platform for collaborative design and decision making

One of the characteristics of resource-efficient and well-planned cities is their ability to capture synergies through integrated approaches and to coordinate actions among multiple stakeholders over the long term. An integrated approach and an alignment of policies are not likely to emerge by default. The process requires a platform that is suitable for the expanded scope of activity.

Cities are dynamic phenomena. They emerge from the overlapping actions of many stakeholders (the public sector, the private sector, civil society groups, and citizens), each of which has influence over the design and management of the elements composing the city. Although none of these groups of stakeholders has the mandate or capacity to address the performance of the city as a system, they all stand to benefit if the elements are well integrated. However, without a proactive effort to bring these stakeholders together and to integrate plans and policies, the likelihood exists that some policies and actions will conflict and that the costs of conflict will be borne by the economy and the environment. Even without direct conflict, the tendency for all stakeholders to act in their own immediate interests represents a barrier to the potential for positive synergies and optimum solutions.

Cities are increasingly experiencing a splintering of the responsibility for infrastructure, an overlapping in jurisdictions, and an increase in the private sector ownership of key assets. An additional constraint is the political election cycle, which may limit the capacity of cities to execute policies over the long term. The typical four-year election cycle in local government undermines sustainable decision making because the change in leadership frequently means a loss in continuity. If cities are to lead in the process of urban development, especially in the context of rapid urbanization, it is important that plans compensate for this disadvantage.

A city may lead a collaborative process on at least three tiers of an expanded platform. At the first tier, projects may be completely within the realm of control of the city administration and will require that a city get its own house in order (for example, an energy efficiency upgrade for all municipally owned buildings, or a ride-share program for employees, or energy and transportation peak-load management through the adjustment of working hours).

At the second tier, projects will involve the city in its capacity as a provider of services and include its formal planning, regulatory, and decision-making powers; this may include water provision, land use planning, or transit development. At this level, greater collaboration is warranted with other stakeholders who may influence and may be affected by the outcomes.

The third tier of the expanded platform will entail collaboration at the scale of the entire urban area or region. This may pertain to issues such as the development of new land or metropolitan management and may necessarily involve senior government officials, key private sector partners, and civil society. In collaborating at the scale of the entire urban area, the city may lack the authority to coordinate the actions of many stakeholders. Senior government officials, utilities, landowners, and private sector groups all have their own plans and agendas. At this level, it is often appropriate to develop an overarching planning framework, including a growth management strategy, which sets the context for all other plans in the urban area by all other stakeholders. At each of these scales, different levels of collaboration are necessary and different working groups are required, all participating in a city-led collaborative process.

As a city embarks on its Eco^2 pathway, many projects might be launched over a single year in which different players in the private sector, the public sector, the civil sector, and other sectors may wish to participate or may have valuable information or assistance to offer at various stages. For this reason, it is important for a city to initiate a process through which participants may develop a shared long-term planning framework that guides all projects and efforts and creates the opportunity for groups to align their policies and programs around a common set of long-term goals and strategies. The framework might also set the context for specific projects. In many cases, a primary collaborative working group may generate subgroups that meet as needed and that may also benefit from professional facilitation, research, and other support. The planning framework can be a powerful platform for collaborative design and decision making and can enable the city to steer the efforts of all stakeholders toward a commonly agreed vision. Because Eco² focuses on integrated design solutions and integrated implementation policies, projects may invariably expand to include multiple stakeholders and require a highly diverse pool of expertise.

Once the formal collaborative process is in place, it also offers the opportunity for much more intensive participation on particular projects among stakeholders in design and implementation. For example, an integrated approach to neighborhood revitalization may benefit from an iterative series of design workshops that engage a variety of experts in different groups in creative design exercises. Regular participation in such creative design workshops is much easier to arrange and approve if the groups that need to be involved are already participating in a formal collaborative process at the most senior level. The same is true during the implementation of preferred design solutions. Essentially, an expanded platform for collaboration at different scales creates a mechanism that may be used repeatedly to bring stakeholders together and to expedite the intensive and interdisciplinary process of the design and implementation of Eco² projects.

Finally, the expanded platform for collaboration, in combination with a long-term planning framework, is likely to increase the commitment by local governments to longer-term policies. It is much more difficult for a new council or mayor to reverse decisions if many other stakeholders have participated in the decisions and are cooperating through their own policy instruments. In the case of Curitiba, Brazil, for example, the creation of a separate planning institute (the Institute for Research and Urban Planning of Curitiba), provided a particularly strong basis for ongoing collaboration in long-term planning. This approach has now been followed in many other countries in Latin America. Because the platform for decision making has been extended to include planning institutes and because alignment among all stakeholders has been encouraged, the governance of a city becomes less vulnerable to the inevitable disruptions created by elections, political incidents, and the manipulation of policy by special interest groups and swing voters at election time. An expanded platform for collaboration compensates for the inherent short-term vision of the democratic process.

PRINCIPLE 3: A one-system approach

Chapter 1 offers specific examples of system integration within cities, all of which have led to sizable and lasting benefits. An integrated approach to planning and management in Stockholm helped to improve resource efficiency significantly in a large urban redevelopment project. In Yokohama, Japan, an integrated approach to waste reduction, reuse, and recycling saved over US$1 billion for the city, while allowing the city to achieve impressive environmental gains. In Curitiba, an integrated and holistic approach to urban planning, transportation planning, and socioeconomic vitalization has enabled the city to achieve tremendous results across all sectors and among many stakeholder groups. There are many more examples in this book. What distinguishes these cities from others is that they have broadened their perspective to adopt a one-system approach, which they have pursued largely through strategies of integration.

A one-system approach enables cities to plan, design, and manage the whole urban system by integrating key subsystems. The approach thereby provides the opportunity for cities to realize many benefits through greater optimization and synergy.

Systems thinking can be defined as the art of simplifying complexity, managing interdependency, and understanding choices. Once we understand something, once we see it as one system, we no longer see it as chaotic or complex.

Contrary to widely held belief, the popular notion of a multidisciplinary approach is not a systems approach. The ability to synthesize separate findings into a coherent whole seems far more critical than the ability to generate information from different perspectives.

Source: Gharajedaghi (2006).

The one-system approach aims at taking full advantage of all the opportunities for integration. Integration may apply to hard infrastructure systems and land use planning. One may integrate elements within a sector or across sectors. Integration may be applied to policies, stakeholders, plans, the sequencing of financing mechanisms, and all of these in combination. In each case, the opportunities arising from integration tend to provide greater efficiency and increased utility for a given investment and improve ecological and economic performance. By applying the one-system approach to every project, entire cities and their surrounding natural and rural areas are able to coalesce into a functional system that works well as a new whole.

The benefits of integration are especially attractive because the efficiency gains tend to be substantial and because the opportunities tend otherwise to be missed. The greatest success in best practice cities has been achieved (1) in joint land use, spatial and transportation planning, and coordinated policies; (2) through positive synergies across infrastructure sectors (such as the positive effect of increased water system efficiency on energy efficiency because of the reduction in the need for electricity to pump water); (3) in integrated utility management systems (for instance, the reuse of sludge and organic waste as biogas [methane] and fertilizer); (4) through technology solutions (such as combined heat and power plants); and (5) through the synchronization of policy, investment planning, and regulations.

Integration is a powerful concept for cities (see chapter 5). So, where does the concept originate? And where might it take us in the long run?

Integration is used here as it relates to the application of systems theory: seeing the full scope of elements that make up the city, how these different elements are connected, and how changes in one element may affect the others. This systems perspective is a way of seeing the world that has emerged from studying ecological systems, and in the end, it may help us design and manage cities so that they become efficient and adaptable in the same way that natural ecologies are efficient and adaptable.

Ecological systems are characterized by multifunctionality among elements and the looping and cascading of resources through linked, nested subsystems that greatly enhance productive utility. They also embody powerful strategies for managing change—strategies such as succession and evolution, self-organization, and adaptive management. All these strategies are part of what we mean by the integrated or one-system approach. The strategies serve two purposes: they improve the efficiency of the system as a whole, maximizing assets and information quality over time, and they help the system adapt to change at the least cost and recover quickly and fully from shocks. Many of these ideas are being applied by innovative cities that have grasped the potential of these opportunities for systemwide sustainability and resilience.

A one-system approach has many dimensions, but it is not complicated. The aim of systems thinking is to reduce complexity by understanding how parts fit into a whole. The challenging aspect is overcoming the institutional structures and inherited attitudes that prevent city leaders, investors, designers, users, suppliers, and managers from working as a team. Adopting the one-system approach as a principle for all projects is a good way to bring the team together.

PRINCIPLE 4: An investment framework that values sustainability and resiliency

Despite a rising interest in sustainability in many locations and a demonstrated capacity for urban design solutions, cities today are having difficulty investing in systems that are long term and ecological. Although many exceptions exist, our time horizons for investments appear generally to be shrinking. Perhaps the rapidly paced, deregulated global economy makes it especially difficult for corporations and political leaders to take a long view.

Whatever the explanation, the simple concept of investing in sustainability and resiliency has become extremely difficult for cities to put into action. Policies, plans, and projects are assessed on their ability to provide short-term financial returns or on economic valuations based on narrowly structured cost-benefit analyses from the perspectives of individual stakeholders. Investments are valued in monetary terms, and what cannot be monetarized is either ignored or addressed as externalities. Decisions are dominated by immediate capital costs, despite the fact that over 90 percent of the life-cycle costs for typical infrastructure are often expended on operational maintenance and rehabilitation.

Most cities worldwide have no real knowledge of the long-term impacts of new development on fiscal health. Life-cycle costs are backloaded, which means that future generations will have to bear huge costs for the repair and replacement of infrastructure without any capitalization. In many cities in developed countries, that particular future has already arrived and is creating a massive infrastructure deficit that can be addressed only through subsidies or more debt financing.

At the same time, ecological assets, the services they provide, and the economic consequences of their depletion and destruction are not accounted for in most government budgets. Because these resources are not measured, they are treated as zero value assets, and the related services they supply go unaccounted for.

For instance, city green areas are usually thought of as merely providing some sort of soft aesthetic value. But, in fact, green areas are ecological assets that provide valuable services and economic benefits in several ways: (1) they provide natural drainage (resulting in avoided infrastructure capital and maintenance costs and reducing seasonal losses related to flooding); (2) they may reduce the average temperature in cities (lowering the peak load demand on electricity, which can result in avoided capital costs for installed power and related operations and maintenance costs); (3) they absorb carbon dioxide and release oxygen, are natural air cleaners, and support overall citizen health; (4) they can be integrated into the public transportation system as a network of bicycle paths and pedestrian walkways to enhance utility; and (5) they have generally been shown to increase physical and mental well-being, while creating a sense of community and reducing crime. If all of these services were truly valued and understood in the long term, then decisions in many cities might be made in a way similar to those in the case of Curitiba.[1]

To achieve ecological and economic sustainability, decision making needs to be guided by a holistic perspective. This entails a new accounting and assessment framework that allows every city to adopt a life-cycle perspective and make investments that are fair to all stakeholders, effective at preserving all assets (manufactured, natural, human, and social), and good for our long-term fiscal health.

This framework will involve adopting a new range of indicators and benchmarks for assessing and rewarding the performance of all stakeholders. Longer-time horizons and life-cycle analysis of the implications of policies and investment options and strategies among multiple stakeholders will need to be realized to reflect a truer, more inclusive, and more complete picture. All capital assets (manufactured, natural, human, and social) and the services they provide should be appropriately valued or priced and then monitored through indicators. The combination of indicators should be viewed as a whole so that the qualitative dimensions of city life (cultural, historic, and aesthetic) are not ignored in the assessment of costs and benefits. The basis and implications of policy decisions, regulatory actions, and legislation will need to be assessed within a much broader context and understanding of value.

At the same time, investing in sustainability and resiliency will entail broadening the scope of risk assessment and risk management to include managing the many indirect, difficult-to-measure risks that threaten the viability of an investment or even the city as a whole. In reality, cities today face multiple hazards that are largely external to financial calculations. These include sudden disruptions to systems, such as epidemics, natural disasters, and socioeconomic changes. By proactively adopting the concepts of resiliency and adaptive capacity, cities will become more well positioned to absorb and respond to shocks and protect their investments.

Implementing new methodologies and a broader scope of accounting in many countries will be difficult at the beginning. But the principle behind such methods should be clearly understood and considered by decision makers. Curitiba did not undertake a detailed accounting and valuation exercise before following its development agenda. But by appreciating the broader and longer-term perspective, it managed to focus on critical interventions that continue to pay lasting and compounded benefits.

The Move from Principles to Core Elements and a Unique Eco² Pathway

The four principles define the scope of each city's unique pathway. Every aspect of a city's pathway is linked directly to one or more of the principles. Because the principles are at the core of the program, we may always fall back on the principles if complications arise.

The analytical and operational framework emerges from the principles. To begin, we derive a set of core elements from each principle.

The core elements serve to implement the principles. They provide specific information on new concepts and on the roles and responsibilities of Eco² cities and their partners. Each core element is an arena of activity and learning. (This is addressed in detail in subsequent chapters.)

Each city may translate the core elements into a series of action items or stepping stones that adapt the elements to local conditions in a logical, step-by-step sequence. The framework summarizes how each principle leads to a set of core elements and stepping stones.

Together, the stepping stones for a particular city constitute a unique pathway. The pathway should include all the essential actions needed to take leadership, collaborate, design catalyst projects, and invest in preferred solutions.

Table 1.1 provides a summary of the core elements and stepping stones. Each item is described in more detail in subsequent chapters. However, it is clear from the summary that developing an Eco² pathway is not a simple exercise, nor is it likely to be quick and easy. For this reason, this book also introduces a number of methods and tools that are designed to save time and guide decisions. The methods and tools provide practical ways for cities to take leadership, collaborate, and analyze and assess ideas for Eco² projects. The methods also address all aspects of project implementation, including the use of an expanded accounting process and a strategic approach to financing.[2] It is up to city leaders to determine whether the Eco² initiative is the kind of pathway they are seeking. The following chapters describe the step-by-step processes.

Table 1.1 The Eco² Cities: Principles and Pathways

PRINCIPLES	CORE ELEMENTS	STEPPING STONES
A city-based approach	A development program that supports cities in making good decisions and implementing these decisions using all levers of city influence and control	Review the Eco² Cities Initiative, and adapt the Eco² principles to the local context, especially current issues of concern and the local political constraints.
	A planning philosophy that recognizes the fundamental role played by local ecological assets in the health and wealth of cities and their surrounding rural communities	Identify champion(s) and the specific groups or individuals who are vital to success.
		Obtain commitments from city councils and influential groups and people.
	An action-oriented network that provides city leaders with the full support of national governments, the international development community (including the World Bank), and global best practice cities	Work closely with national governments and, where possible, dovetail the Eco² elements so they clearly fit within national priorities.
	A decision support system with methods and tools that adapt to varying levels of knowledge and skill and provide cities with the technical, administrative, and financial capacity to develop an Eco² pathway	Seek a partnership with the international development community (including the World Bank), best practice cities, and Eco² Cities Initiative partners.
		Outline a process for building capacity, and enhance the skills and knowledge of local professional staff.
		Develop fluency of concepts among local decision makers using case studies from this book and other supporting materials.
An expanded platform for collaborative design and decision making	A three-tier platform that enables a city to collaborate (1) as a model corporation, engaging all city departments; (2) as a provider of services, engaging residents, businesses, and contractors; and (3) as a leader and partner within the urban region, engaging senior government officials, utilities, rural settlements, private sector stakeholders, nongovernmental organizations, and academia	Initiate a process for collaborative decision making and integrated design to develop the Eco² approach as a corporation, as a provider of services, and as a leader within the larger urban area.
		Prepare a mandate and budget for a secretariat that can support collaborative committees through background research on cross-cutting issues and the facilitation of regular meetings, communications products, and event planning.
	A shared long-term planning framework for aligning and strengthening the policies of the city administration and key stakeholders and for guiding future work on Eco² projects	Prepare a long-term planning framework, in collaboration with others, and seek consensus on common goals and indicators of performance, an overarching growth management strategy, and an adaptive management approach.
		Select a catalyst project suitable for demonstrating the Eco² principles, aligned with the goals and strategies identified in the long-term planning framework.

Table 1.1, *continued*

PRINCIPLES	CORE ELEMENTS	STEPPING STONES
A one-system approach	Integrated infrastructure system design and management that focuses on enhancing the efficiency of resource flows in an urban area Coordinated spatial development that integrates urban forms with urban flows, combining land use, urban design, urban density, and other spatial attributes with infrastructure scenarios Integrated implementation by (1) correctly sequencing investments, (2) creating a policy environment that enables an integrated approach, (3) coordinating a full range of policy tools, (4) collaborating with stakeholders to align key policies with long-term goals, (5) targeting new policies to reflect the differing circumstances involved in urbanization in new areas and in improving existing urban areas	Provide just-in-time training and capacity building, arrange for multiple opportunities for local professionals to become comfortable with the one-system approach, and make the best use of technical support so it may be truly transformative and valuable. Conduct a series of integrated design workshops to create important opportunities for planners, designers, and engineers to come together and use new methods and information: a series of short workshops can clarify goals and set targets; and the long-term planning framework can guide, design, and stimulate creative solutions. Explore design solutions and prepare a concept plan for review: an integrated design process should be used to generate alternative proposals on ways to design, construct, and manage the project; an intensive, multiday urban systems design charrette (see part 2) can facilitate the integrated design process; and the integrated design process should culminate in a recommended concept plan for implementation, including any policy reforms. Align a full set of policy tools to ensure successful implementation, in collaboration with stakeholders, to sequence and enable a one-system approach and to coordinate actions across sectors: a strategic action plan can be prepared to clarify who is responsible for what tasks and to show how policies interact.
An investment framework that values sustainability and resiliency	Incorporation of life-cycle costing in all financial decision making Equal attention to protecting and enhancing all capital assets: manufactured capital, natural capital, social capital, and human capital Proactive attention to managing all kinds of risk: financial risk, sudden disruptions to systems, and rapid socioeconomic environmental change	Use a life-cycle costing method or tool to understand the life-cycle costs and cash flows. Develop and adopt indicators for assessing the four types of capital and for benchmarking performance. Forecast the impacts of plausible changes in climate, markets, resource availability, demographics, and technology by hosting a forecast workshop. Implement a catalyst project in ways that protect and enhance capital assets and reduce vulnerabilities: the best way to learn the accounting methods is in practice in a catalyst project, and a base case scenario may be developed as a benchmark for comparing alternative approaches. Monitor feedback results, learn, and adapt to improve performance.

Source: Author compilation.

Notes

1. In contrast, in July 2005, hundreds of people were killed, and financial losses of about US$100 million were incurred in Mumbai, largely because of the loss of natural mangrove ecosystems and an unplanned speculation-driven construction frenzy in the northern suburbs. The new investments required in drainage to compensate for bad planning will be exorbitant. Much of this cost might have been avoided.

2. The Eco² Cities Initiative will make full use of innovative financial products offered by the World Bank, such as the new Climate Investment Funds that provide clients with strong financial incentives for transformative changes in energy efficiency and clean technologies. Carbon finance will also be leveraged.

References

Gharajedaghi, Jamshid. 2006. *Systems Thinking; Managing Chaos and Complexity: A Platform for Designing Business Architecture,* 2nd ed. Burlington, MA: Butterworth-Heinemann.

Revi, Aromar, Sanjay Prakash, Rahul Mehrotra, G. K. Bhat, Kapil Gupta, and Rahul Gore. 2006. "Goa 2100: The Transition to a Sustainable RUrban Design." *Environment and Urbanization* 18 (1): 51–65.

Unruh, Gregory C. 2000. "Understanding Carbon Lock-In." *Energy Policy* 28 (12): 817–30.

CHAPTER 3

A City-Based Approach

The first step toward a city-based approach is to appreciate and apply the philosophy at all levels, from local councils to national governments to the international community. One should recognize that local governments, working in collaboration with stakeholders, are now on the front line in dealing with some of the most pressing development challenges and that, most often, they hold the key to solutions. It is this philosophy that motivates the Eco² initiative. The core elements and stepping stones of a city-based approach are designed to enable local governments to lead a development process that accounts for their specific circumstances, including local ecology.

The Core Elements of a City-Based Approach

A development program that supports cities

Cities have a wide range of powers that they may use to influence their development trajectories. In addition, many countries are now pursuing processes of fiscal and administrative decentralization. This approach has generated additional important decision-making and management responsibilities for local governments. Often, the impact of initiatives depends on the effectiveness and creativity with which city leaders cultivate and use these powers.

A development program that supports cities in their decision-making process and, more critically, in the implementation of decisions is needed to enable cities to use their powers to exercise meaningful proactive leadership more effectively.

After assessing 25 successful cases of sustainable urbanization in different European cities, Timothy Beatley (2000: 423) has concluded that the role of city leadership is crucial to success:

Government in these cities is not seen as laissez-faire or caretaking in nature, but as an entity exercising important proactive leadership; it is a pacesetter, not a follower or spectator.

A planning philosophy that recognizes the fundamental role of local ecological assets

Local ecological assets provide all kinds of services to cities, from sand and gravel for concrete to renewable sources of energy, supplies of drinking water, the assimilation of waste products, the pollination of market gardens, pleasant views, and recreational environments. The list of services for a typical city is long. These services are increasingly critical to the viability of the local economy and to the health, safety, and quality of the lives of residents. Because we lack a systems perspective and comprehensive accounting methods, the real quantity and value of such assets are rarely recognized. New accounting methods should help fill this gap. So will a new philosophy of planning that assigns priority to these assets in reaching decisions about urban form and land use.

A city-based approach alters the mind-set of the urban planner and civil engineer. Urban development moves from big-architecture industrial engineering and environmental management (coping with externalities) to the stewardship of landscapes and the integration of social and ecological values into land use planning and infrastructure design and management. This is a change from the traditional city-centric view whereby natural systems are valued only as economic inputs or amenities and wherein the rural and natural fringe of lands surrounding a city is most often ignored or treated as an urban preserve for future expansion.

The Eco² approach to planning begins with understanding the opportunities and constraints of local ecologies. How do we fit into the topography of the area so that water may be provided by gravity? How do we protect the water recharge areas and the wetlands so that water capacity and quality are sustained? How do we distribute populations so that local renewable energy—windy sites, forests, solar access—is sufficient to meet our basic needs? These types of questions are the place to start and may ultimately provide urban professionals with their most exciting design challenge: how to fit cities into the landscape in ways that respect and complement our natural capital and that ensure the availability of ecological services for present and future generations. In theory, all the constructed elements that make up a city may contribute to and benefit from the health and productivity of local ecologies and natural resources.

An action-oriented network

The city-based approach requires an action-oriented network that knits together cities, their senior or national governments, and their supporters at all levels. The composition of supporting players will vary from place to place, but should be broad enough to include local stakeholders, academic institutions, private corporations, international agencies and organizations, and best-practice cities. Each player in the network will bring different yet complementary strengths. Some will bring strong technical expertise, while others might bring financing or educational programs. It is this mix of players and resources that makes a sustainable transformation possible. However, the agenda may become confusing and problematic unless the players in the network share an understanding of their respective roles. In a city-based approach, the role for all players is primarily to support the city in a bottom-up process. Why do we look to cities to lead? Because the local level is often where the greatest opportunity exists for truly creative solutions and for maximizing the benefits of an investment across many sectors (see box 1.1). Instead of promoting a one-size-fits-all solution, the network provides the city with enabling policies, information flows, targets and guidelines, and the freedom to create and adapt. Local self-reliance is explored before investigating possible solutions at higher levels. When an action-oriented network is aligned in support of a city, it is often surprising just how much can be achieved locally.

The City-Based Approach Is Bottom-Up

In the Eco² approach, bottom-up actions are those that begin at the most local level: the city or its particular neighborhoods and buildings. Instead of investing in, for example, a remote power station or equipping a regional water utility with oversized pipelines, it is better to first explore bottom-up solutions, such as, for this example, a rooftop rain-water catchment system and a solar water heater. If the most local solutions are insufficient, move up one level and consider reclaiming water at the block or neighborhood, or a district heating system. Only when the most localized solutions are impractical, uneconomic, or undependable should the network start to focus on the next level. As we move from the local or city scale towards regional, national and international scales, the need for design and investment is progressively reduced because we are tapping into local creativity and resources. This bottom-up process is possible only if supported from the top down, especially from regional or national utilities, senior governments, and the international community.

Top-down support comes in many forms. Most important are *enabling policies*: those policies that provide cities with the authority, skills, knowledge, and financial resources to implement local solutions. This might take the form of a national group helping successful cities share their experience and lessons learned—what worked and what didn't—with other cities. Or a regional utility might agree to help set up, finance, or operate a local utility for district energy. Top-down support can also consist of clear targets and guidelines that help cities synchronize their designs with, for example, the regional economic development strategy or an international strategy for climate change mitigation. Finally, top-down support can include physical infrastructure systems that are flexible enough to allow each location to share with others its surpluses of water, energy, materials, and other services. In an action-oriented network, the top-down solutions may be diverse, but they are always city-based. They increase the capacity of cities to solve their own problems by providing them with a coordinated and complementary package of policies, targets, financial mechanisms, guidelines, knowledge, and flexible infrastructure systems.

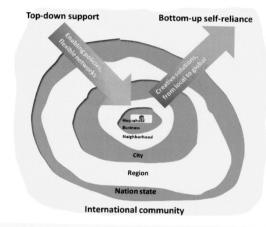

Source: Author elaboration (Sebastian Moffat).

A city-based decision support system

The Eco² approach requires that cities enhance their technical and administrative capacity, particularly with respect to leading collaborative processes and exploring integrated design solutions. Capacity building means adopting methods and tools that help simplify otherwise complex decisions. This is the role of a city-based decision support system (DSS). The city-based DSS is an evolving set of methods and tools designed primarily to help cities take leadership and make the best choices. Each city may develop its own DSS.

Coping with complexity is one of the greatest challenges in Eco². Cities represent the longest lasting, most valuable, and most complex artifact created by humanity. Even under the best of circumstances, urban planning is a complex task, and the challenge increases if there is an attempt to implement integrated solutions, which may add to the complexity. Of course, the task becomes that much more challenging in developing countries if resources are limited and if urbanization seems to be happening too quickly. Other challenges in developing countries include the lack of experience with computer-based planning tools and the inadequate performance of existing infrastructure. For all these reasons, a city-based DSS is an essential element in each city's sustainability pathway.

One of the most difficult tasks faced by someone trying to apply an integrated approach to infrastructure systems is the dynamic relationship that exists between physical flows and spatial form. Physical flows tend to be addressed through modeling and calculation, and these typically involve individuals with engineering and technical backgrounds. Spatial issues are usually addressed using mapping techniques, and these involve individuals with a planning or design background. Integrated design solutions incorporate spatial and physical flows and an understanding of interrelationships. The city-based DSS can help create a transdisciplinary platform for the purpose of involving all these people and many others. Physical and spatial effects are communicated using graphics tools, data sharing, and terms and images that may be easily understood in a multidisciplinary group. (These and other aspects of the Eco² city-based DSS are described in more detail in part 2.)

Stepping Stones for a City-Based Approach

Review and adapt the Eco² Cities Initiative

The management of change is most successful if the new ideas are clothed in familiar patterns and reflect a sensitivity to local concerns and capabilities. An assessment of local strengths and weaknesses helps tailor the Eco² initiative to local conditions and experiences. This may involve a number of different ways of customizing Eco²:

- *Traveling back to the future:* Begin by reviewing the history of the city and region, focusing on examples of cases in which city leadership has achieved positive outcomes or in which a more well integrated approach to design or a process of collaboration has already helped generate multiple benefits. Use these historical examples to explain the strengths of Eco². To obtain broad support within the city, the program should be introduced as a return to approaches that have worked in the past and as a reaffirmation of traditional values and institutions. The history of most cities is replete with stories that may be used for this purpose.

- *Talking about trigger issues:* Identify the current political issues within the community that are most likely to be addressed through an Eco² approach. All politicians want to resolve such issues, and media personnel want to report on them, not on a program or philosophy. These are the trigger issues that will build support for the Eco² approach.

- *Learning to lever influence, stand firm, and say no:* The levers of influence and control vary considerably from one location to another, and this obviously affects the potential for an Eco² approach. For example, in some countries, national governments control the financing for urban infrastructure; in other countries, the investment by cities in local renewable energy systems is prohibited by law. Cities that lack control over financing or that lack the authority to develop new policy obviously face a greater challenge. However, the biggest difficulty often revolves around using levers that influence decisions, including zoning, development approvals, hook-up requirements for infrastructure, and so on. An assessment of local strengths should clarify the full extent of influence and power available to local government. Almost always, cities have more authority than they realize, and the real challenge is learning to say no to the short-term vested interests that drive so much land development.

Identify local champions

The successful introduction of Eco² principles usually requires a strong champion who can help motivate the many groups that need to be involved, sustain the commitment over time, and provide confidence and leadership. Local champions can put forward key ideas in ways that are acceptable to various stakeholders and thereby broker solutions that are widely accepted. Local champions may also attract other influential individuals by virtue of their reputation and influence.

The champion may be drawn from any place; everyone has the potential to take leadership. However, the task is easier if the champion is someone possessing recognized authority or influence, such as a well-liked retired statesperson, the city mayor, the chief administrator, or the chair of a development committee. Sometimes, leadership may emerge from an advisory group of senior statespersons or elders who are widely respected and support the Eco² concept.

Wherever the champion is located, a support group of committed and knowledgeable individuals is also necessary. All champions depend on a support groups or change agents to develop networks and the knowledge base. In an Eco² city, support may arise from a small group of hard-working staff members or an ad hoc group of experts and community activists. Ideally, the support group should be capable of providing its champion with both administrative and technical support. In some cases, a national body may be part of the support group. For example, the support provided by national governments may include an office for supplying cities with technical and financial assistance.

Obtain a commitment from the city council

Much of the land in a city and a majority of the infrastructure may be owned by private sector groups or by the senior levels of government. Nonetheless, democratically elected local councils have a legitimate role in any effort to undertake land use planning, especially in making strategic choices that may affect the long-term health of the community. These councils are often considered appropriate lead-

ers and are able to bring together regional stakeholders and promote collaborative decision making and integrated design. If a local council is fully engaged, others join in. It is thus critical that support be forthcoming from the council and individual council members with special interests in development issues. The council needs to be engaged in the Eco² initiative from the start.

With involvement of a council, it helps if the city's own Eco² pathway is presented as a means to address the issues of greatest concern to council members. This method is not usually a problem. The integrated approach adds strength to any specific issue by providing multiple benefits and expanding the base of support for positive change. For example, affordable housing may be designed to include a project to treat wastewater for the neighborhood or to increase the space available for small shops and businesses. The multipurpose nature of catalyst projects and the more thorough analysis they represent of the impacts on the whole economy and ecology make the task of brokering them easier.

Obtaining an informed commitment from the council and sustaining this commitment may be difficult and time consuming. It is especially important to emphasize the long-term and collaborative elements and to use these features as a means to dissociate the agenda from any one political party or power group.

Work closely with the national government

National governments can play a number of complementary roles in the Eco² Cities Initiative. They can function as important centers of expertise on and networking for best practices in urban design and planning. National governments can share best practices across cities and develop new policies in support of a city-based approach. They can choose to work with cities on a locally specific planning framework (for example, a regional growth management strategy) and contribute expertise on a project-by-project basis.

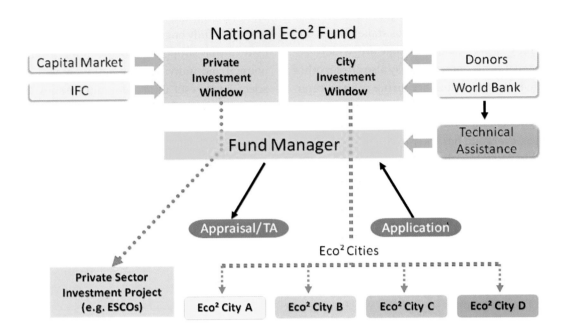

Eco² Fund Scale-Up Mechanism at National Level

Figure 1.4 A Possible Government Role: Administering a National Eco² Fund to Support Participating Cities

Source: Author elaboration.

Note: IFC = International Finance Corporation; TA = technical assistance; ESCO = energy service company.

Although the limited resources available to national government departments may constrain their ability to participate directly with cities on new initiatives, these departments should still seek ways to participate to some degree in any regional-scale collaborative working groups.

Another interesting and highly influential role for national governments involves establishing a national Eco² Fund Program, which may serve as a conduit for financing programs and disseminating knowledge on global best practices. Canada and Sweden have used similar mechanisms to support cities. In Sweden, a local investment program, which lasted from 1998 to 2002, allocated SKr 6.2 billion (€671 million) to 211 local investment programs in 161 municipalities, involving 1,814 projects (Swedish Environmental Protection Agency 2004). From municipalities, businesses, and other organizations, this national investment leveraged SKr 27.3 billion (almost €3.0 billion), of which SKr 21 billion (about €2.3 billion) represented investments directly related to sustainability and the environment. It has been estimated that 20,000 full-time short-term or permanent jobs were created through this process. (For more details on this program, refer to part 3.)

Figure 1.4 illustrates one possible model for a National Eco² Fund Program. The national government would adapt the Eco² initiative to local circumstances, working in partnership with the World Bank, other international agencies, development organizations, and the private sector. It would allocate resources among cities and administer funds.

Whatever the involvement of national government in the program, it is important for local governments to adapt their Eco² pathways to the priorities currently established at the national level. This means finding points of commonality and adopting terms and language similar to the terms and language used by the national government. In this way, the national government automatically becomes an ally and a potential partner.

Engage the international community, best practice cities, and the World Bank in the Eco² Cities Initiative

Engaging the World Bank and other partners directly in the Eco² pathway is an option for every city. The Eco² initiative can offer cities a variety of literature, including guidelines and technical reports, to support every stage of the Eco² pathway. On a case-by-case basis, the World Bank, in concert with national governments and global development partners, may be in a position to assist in financing Eco² integrated solutions. For example, the World Bank can help cities integrate and consolidate a variety of financial mechanisms given that Eco² projects will tend to need different types of financing at each stage and may qualify for multiple types of funding. The World Bank's various financial instruments are examined in chapter 7 (part 1) and in part 3.

Other global development partners may also be willing to provide support to cities in cases where special expertise is needed and resources can be found to cover costs. Best practice cities, for example, are often glad to share information and may provide additional assistance and support to cities directly or through a joint capacity-building initiative with the World Bank.

Outline a process for building capacity

Capacity building involves the familiar process of professional development and demonstration projects. The DSS, described in this chapter and in part 2, should be a key element in any capacity-building plans. The city-based DSS includes methods and tools without which it is almost impossible to adopt an integrated approach to design and policy. Most methods and tools within the city-based DSS are well support-ed by tool developers and may be accompanied by useful handbooks, guides, and tutorials.

As cities engage in outlining a process for capacity building, it is important to recognize that the Eco² initiative represents a significant departure from standard urban planning, development, and management. The examples of integrated infrastructure cited in chapter 1 are not yet commonplace. The large majority of growing cities, including those located in the developed world, are still unable or unwilling to contain urban sprawl, optimize land use and infrastructure, adopt life-cycle costing, or apply many of the alternative designs and policies used in best practice cities. For these reasons, an Eco² pathway must incorporate a carefully planned process to manage change and pay special attention to adopting new ideas for leadership, visioning, collaboration, and analysis.

Develop Eco² fluency

Another challenge associated with any change in standard practice is the task of familiarizing the city leadership group with the key concepts and helping officials to understand what is truly different about the new approach and why it may be especially beneficial. Individuals need quiet time to absorb new ideas. The Eco² Cities Initiative provides resources, including this book, that can help introduce key concepts and terms. Case studies are an excellent place to start. Talking with best practice cities or viewing video testimonials from other experienced decision makers may also be helpful in providing leaders with the confidence to adopt and promote a new approach.

Developing fluency with the concepts might require special sessions for local politicians and executives that allow these officials to explore new concepts and practice defending new approaches. Within the Eco² initiative, for example, the concept of collaboration involves consensus decision making at different tiers and a formal commitment by stakeholders to attend regular meetings and align their policies in cases where consensus exists. These distinctions need to be made clear and to be accepted because they expand on the traditional view of best practice in governance.

Fluency is also important in realizing the concepts of ecological design. The looping and cascading of resource flows within a city may be explained using graphical case studies. It may be helpful to gather key decision makers together for several hours in a comfortable environment to discuss the case studies and the key lessons learned or even to participate in mock workshops and design exercises. The fluency campaign is aimed at helping decision makers become comfortable in using a new language for design and investment.

References

Beatley, Timothy. 2000. *Green Urbanism: Learning from European Cities*. Washington, DC: Island Press.

Swedish Environmental Protection Agency. 2004. "Local Investment Programmes: The Way to a Sustainable Society." http://www.naturvardsverket.se/Documents/publikationer/91-620-8174-8.pdf.

CHAPTER 4

An Expanded Platform for Collaborative Design and Decision Making

The principle of the expanded platform speaks to the importance of adopting a design and decision-making process that is more well integrated, adaptable, and lasting. If we want to improve economic and ecological performance through integrated, fine-scaled, flexible, and long-lasting solutions, we must also pursue shifts in the institutional arrangements that enable design and decision making. In many ways, the constructed environment is a mirror of the way we think and relate.

The solution is twofold: (1) engage stakeholders at all scales in a collaborative process as part of every major project, and (2) develop an overarching planning framework for sustainability and resiliency that includes goals, targets, and strategies. Each of these elements is discussed in this chapter. The elements are mutually supportive. The collaboration at all scales generates the skills, goodwill, and creative interchange needed to adopt new business models. The shared planning framework provides the context for integrated project design and also aligns everyone's plans and policies with a common set of community goals.

Collaboration is also a new form of governance. By engaging stakeholders at all scales, the city creates a planning forum that is more appropriate to mixed economies in which private sector groups often control a majority of the infrastructure systems. Because the process is driven by long-term goals and strategies, it can help cities compensate for the impacts of frequent election cycles, which tend to focus attention on short-term agendas and crisis issues.

The single greatest difficulty in adopting collaborative arrangements is the lack of any institutional champion to lead and guide the process. Almost by definition, no department, group, or government has the mandate, funds, or independence to undertake such a broad, cross-cutting process. Without a sponsor or host, the process never gets started. This is one reason so few collaborative models exist at the city scale. It is also a key reason the Eco² Cities Initiative proposes that cities assume leadership in creating a platform for ongoing collaboration. Methods and tools to help cities organize an expanded platform for collaboration and to help cities use this platform to develop effective planning frameworks, including regional growth strategies, are included in the city-based decision support system (part 2).

The Core Elements of a Platform for Collaboration

A triple-tier platform

The city can lead a collaborative process on at least three levels or tiers (see figure 1.5). Each tier affects the others, and in an ideal world, every city should lead a collaborative working group at every tier. In practice, the process may be incremental or periodic. However, it is still important to differentiate the options. The tiers reflect the varying levels of control and influence.

Inner tier: The house in order (corporate operations)

The first and most fundamental tier is the collaboration that may occur within and among the departments of the city. At this innermost tier, the city has a great measure of control. Here, the city government may address how well it functions as a corporation and how well it works as a team to put its house in order.

Various departments may routinely collaborate to make decisions that are more well integrated and more effective. Cross-cutting goals and targets may be adopted and incorporated into the strategic plan. A reporting and monitoring process may be implemented that informs the wider community of how well the city is looking after its various assets, including employees, equipment, capital, publically owned buildings, and so on. Special internal programs

may be warranted. For example, the city might reduce transportation costs for employees by means of ride sharing, bicycle storage, a new parking policy, the purchase of efficient vehicles, tele-work, and so on. Such a project might require changes in the facilities in buildings and employee benefits, changes that may only be possible through a collaborative process involving many city departments. Other internal initiatives might include improving the efficiency of building operations, procurement processes, waste management systems, and energy use.

Whatever the program or project, this inner-tier collaboration provides the city with an immediate opportunity to learn how to lead an effective process and demonstrate the benefits of the process. The city may use the collaborative process as a model for all efficient and sustainable corporate operations. In almost every sector worldwide, the leaders in sustainability not only provide sustainable products and services, but also take pride in corporate performance (for example, their green headquarter offices). The same logic applies to cities. There is no excuse for failing to collaborate internally because the city may initiate the process unilaterally. The benefits extend well beyond internal operations. It is always easier for cities to lead a collaboration process externally with stakeholders and partners if the city has already succeeded internally.

Middle tier: The city as a provider of services

The middle tier of the collaborative platform may be focused on municipal services—the various public services delivered by city government to residents and businesses within city boundaries. Although the services and associated investments may be largely or completely

The future seems to lead quite clearly to a consensus-led approach where everything is discussed and the participation of all interested groups spans all phases of the development plan process.

Source: Lahti (2006).

within the control of the city, they nonetheless affect many other stakeholders at all levels. Collaboration at this tier may assist in policy development in many areas. For example, the choice of a transit system may be a city responsibility, but it has major long-term impacts on land values and development potential, the competitiveness of local businesses, local job creation, street safety and livability, and the development of neighborhoods. Ideally, a local transit system needs to be integrated with land use planning, parking policies, energy supply systems, street profiles, neighborhood planning, regional transportation connections, and much more. Without a well-structured collaborative process, it is difficult for any city to understand the full implications of alternative policies. Moreover, the impacts of new investments may be uneven, and it may become necessary to manage the political agenda. Rather than debates and autocratic predict-and-provide models, a meaningful dialogue is required on the best long-term strategies. All complex system designs benefit from a process that encourages creative solutions and allows for consensus decision making by key stakeholders.

Collaboration is necessarily more complex in the middle tier than in the inner tier. A larger number of groups must commit to the process and share information with other stakeholders, such as businesses and households, and also with their respective constituencies. Larger financial investments may be required to launch citywide programs and implement capital projects, and this may require collaboration with the financial community.

Outer tier: The urban region

The outer tier of collaboration focuses on the urban region. In a metropolitan area, this may mean focusing on the city composed of cities. In almost all locations, it means expanding beyond the strict boundaries of the municipality to include adjacent towns, cities, rural lands, and natural areas that are part of the economic region and the bioregion. This scale is the most

The Danger of Predict-and-Provide Models

In the middle of the 20th century, "a new 'scientific' and professional endeavor was born through the transport planning and traffic engineering disciplines. The basic philosophy of the [urban transport planning] process was to plan for infrastructure supply to meet projected traffic growth: a 'predict and provide' approach. This approach became characterized by self-fulfilling prophecies of spiraling traffic growth, congestion, and road building.

This method of transport planning has proven damaging to cities around the world. Freeways have been punched through neighborhoods, demolishing large sections of urban fabric, severing communities, and destroying natural environments and food-producing areas. Roads have been built and widened to accommodate more traffic, reduce congestion, save fuel, and reduce emissions, despite evidence that this approach fails. Public transport and particularly non-motorized modes have been big losers in a planning process optimized for the automobile.

Source: Kenworthy (2006: 81).

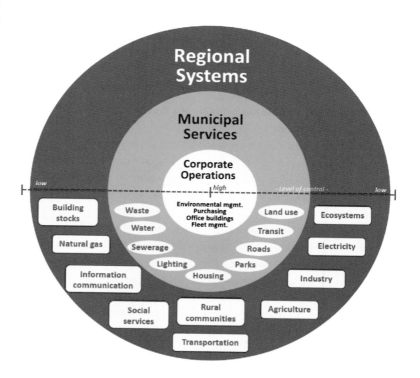

Figure 1.5 The City's Collaborative Working Group at Three Tiers: Corporate, Municipal, and Regional

Source: Author elaboration (Sebastian Moffatt).

Note: Moving from the inner tier to the outer tier increases the number of stakeholders and the complexity and scope of the potential benefits.

challenging for cities, but potentially the most rewarding. At the outer tier, the city is merely one player among many. It is not immediately clear why or how the city becomes a leader. It is also difficult to find (except in the case of island states) any one definition of boundaries for a region because the ideal boundaries will change with each issue. The urban region is always a fuzzy concept. However, many examples now exist of cities that have risen to the challenge and, in so doing, greatly enhanced the capacity of their communities to articulate and achieve economic and ecological goals. To a large extent, the sustainability of a city depends on the city's capacity to provide leadership and collaborate at the scale of the urban region in which it is immersed.

Stakeholders at the outer tier may resist attempts to develop a formal platform for collaboration. For an electrical utility, for example, the service territory may form a logical planning unit, not a particular urban region. For adjacent towns and cities, the habitual mode may be competition for land rents and a tax base or access to development funding. The focus of collaboration needs to be long term to find common purposes. Absent a collaborative process, the regional stakeholders will almost certainly be working at cross-purposes. Collaboration provides an unusual and important opportunity for such groups to meet, develop personal relationships, agree on long-term directions, and discuss current plans. For example, electricity companies might meet with natural gas companies and begin a conversation about the best long-term uses for scarce energy resources within the city. The owners of buildings might likewise discuss with city departments the appropriate level of investment to be made to upgrade existing building stock for resource efficiency. These are crucial issues for Eco^2 cities, and they can be resolved only through a continuous, well-managed dialogue and collaborative decision making.

The outer tier collaborative platform requires a strong structure. It may include senior statespersons at the core; team leaders selected from private firms, knowledge institutions, and public bodies; and experts and champions from a variety of sectors. The structure may build on existing partnerships and committees if these exist and if they are consistent with the collaborative process. A collaborative working group does not need to be time limited. Ad hoc subgroups may be formed to meet regularly on specific issues as appropriate. (Part 2 provides

---

The Emergence of the Regional City as a Crucial Scale for Long-Term Planning

Peter Calthorpe and William B. Fulton (2001) describe the resurgence of a regional approach to city building. They argue that the economic, social, and ecological patterns of cities now seem to be more well understood and planned at the regional scale. As cities mature, the traditional combination of urban sprawl and satellite or edge cities is transformed into a structure that is more accurately described as polycentric, that is, more like a cluster of grapes than a single fruit with a dense core. The polycentric forms are complex; instead of the focus on a single center, we see layers of networks—economic, open space, resources, and connections—with many more centers or nodes nested within other nodes. The challenge is to fit these complex forms into the landscape in ways that suit the ecology of the region and its resource base and also to limit and contain the nodes so they are at a human scale and walkable. "The regional city," write Calthorpe and Fulton (2001: 10), "must be viewed as a cohesive unit—economically, ecologically, and socially—made up of coherent neighborhoods and communities, all of which play a vital role in creating the metropolitan region as a whole."

"Coordination and collaboration between national, provincial, and local authorities can achieve harmonious regional and urban development, provided they share a common vision and demonstrate sufficient political will. . . . local authorities, working with regional authorities, need to develop clear visions and strategies that articulate short- and medium-term responses to enhance economic and social conditions in their cities."

Source: UN-Habitat (2008: xvi).

more detail on the potential makeup and activities of collaborative working groups.)

From 2003 to 2009, the urban region of Auckland, New Zealand, undertook a collaborative process, including the preparation of a shared long-term (100-year) planning framework. The process of developing a framework was highly inclusive, with many conversations feeding into the framework and into the emerging responses. The regional growth strategy, for example, facilitated regionwide discussions and a reference group of council members to provide direction and support. Similarly, local authorities and the central government formed a working group to ensure representative influence, enable shared responsibility for funding the Auckland Sustainability Framework, and ensure that staff would be actively involved. The process was neither linear nor predictable, and its messiness may be seen as an inherent quality of the positive outcome. A key collaborative element was the relationship between central and local governments aligned with common governance elements, including a joint commitment to developing a shared long-term view of a sustainable Auckland. (Part 3 includes a full case study of the Auckland collaborative process and the sustainability framework the city created.)

A new approach to governance and, perhaps, a new way of living together

Collaboration is a process that may evolve from a simple working group for interdepartmental planning to a new forum for governance for the urban region as a whole and to a new culture of cooperation and flexible teamwork that is adopted as a matter of course. Whatever the scale of collaboration, the capacity to lead a collaborative process can greatly enhance the potential for integrated design and policy and for sustainable development. The first step toward success is to understand how a city may organize and support a collaborative process. (More detail is provided in the city-based decision support system, part 2.)

A shared long-term planning framework for the urban region

A second step in the creation of an expanded platform for collaboration is the adoption of a shared long-term planning framework. The framework ensures that all public decisions, including capital investments, are supported by a logical, transparent rationale. An Eco² framework needs to combine two perspectives on the future: achieving goals for sustainability and managing risk for greater resiliency. Box 1.2 summarizes how these two perspectives become integrated into a strategic plan for the region. The framework needs to be developed through a collaborative process if it is to be influential across the region. Once in place, the framework becomes a tool that supports collaborative efforts at all levels.

Not everyone will immediately see the point of developing a broad framework that transcends the urgent issues of the day and that transcends the authority of any one group. To introduce the concept, we explore how frameworks function.

A framework is a structure for connecting visions to actions, a kind of mental map or way-finding system that provides us with a sense of how elements fit together and relate to each other. All of us use some kind of framework to help us make decisions. Most frameworks rely on a hierarchical structure to reduce complexity, moving from the big ideas or categories to the details and specifics. The Aalborg Charter, which has been adopted by 2,500 European communities, is an example of a comprehensive framework for long-term planning by cities. The outline of planning steps in the Aalborg Charter helps each city address the key steps in the planning process, from problem identification and visioning to implementation and monitoring (figure 1.6). The framework also provides a common language and a standard sequence for planning.

A shared framework is useful for all aspects of planning and design. Local governments may use the framework to organize and align

Combining Forecasts and Backcasts to Achieve Resiliency and Sustainability

FORECASTS: Projecting the impacts of forces and planning for mitigation and adaptation

Forecasts (or narratives) explore the likely impacts on infrastructure of changes in population, climate, economics, and technology. The impacts may be presented visually using chains of cause and effect that help tell stories about the future of complex systems. Forecasting with diagrams can sensitize all design teams and decision makers about the types of futures that may be encountered by the city and its systems. The forecasts may also be used as a mind-map or decision tree to help groups to brainstorm about the most appropriate interventions to mitigate threats or adapt to change. A primer for cities on climate change that has been published by the World Bank provides many examples of the ways climate change might affect the various parts of a city and how the city community might respond (see Prasad and others 2009). Similar kinds of exercises are needed to address other external forces such as technological and population changes.

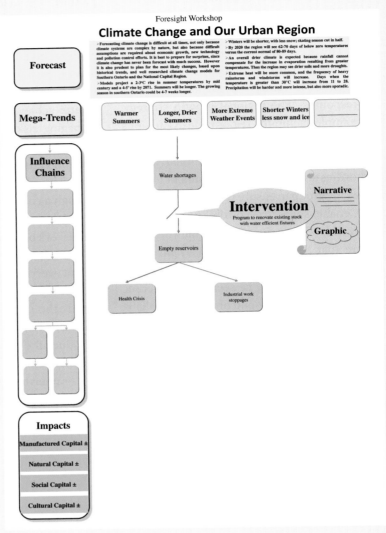

Source: Author elaboration (Sebastian Moffatt).

their strategic plans, master plans, concept plans, transportation plans, and economic development plans. Integrated design teams may use the framework to direct each stage of their designs and to remind designers about the full scope of community goals and priorities. At every stage, a shared framework helps us communicate and work together in a coordinated fashion. Because everyone shares the framework, it is clear how each activity fits into the whole, and there is less need to micromanage

the myriad city departments and stakeholders who might be involved in project planning and implementation.

As cities undertake a more well integrated approach to system design and develop an extended platform for collaboration, the shared framework may help to solve the problems in organizing and communicating complexities. It puts first things first and cross-links every concept and action into an easily understood argument. This creates an easy-to-follow men-

BACKCASTS: Managing the transition to end-state goals

Backcasts involve making changes in those areas where a city has real influence and control. Backcast refers to the process of working backward from a goal set for a point in the future to the current situation and creating a critical pathway for managing change. Interim targets may help to set the pace of change to fit the ambitions and priorities of the city. Moving too rapidly may be as destructive as moving too slowly. The biggest problem in backcasting occurs if trends are taking the city in the wrong direction altogether. The use of automobiles for commuting is increasing in most cities, for example, but is not sustainable. Such a trend must be counteracted through interventions that accelerate and leverage the preferred alternatives.

INTEGRATION: Creating a proactive strategy that addresses sustainability and resiliency

By responding to forces that cannot be controlled and by managing what can be controlled, one may create the potential for managing the transition to a more resilient and sustainable city. A city must define a solution space that avoids moving too rapidly or too slowly and that provides the room required to recover from the inevitable shocks and surprises that will be encountered over the next few decades. If the trend is in the wrong direction, a focused policy intervention or a catalyst project may help redirect it. By mitigating threats and adapting to changes, one reduces the number of surprises or dislocations. In this way, one may sustain a managed transition to end-state goals.

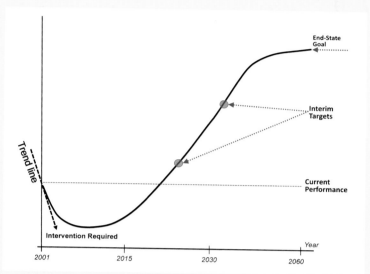

Source: Author elaboration (Sebastian Moffatt).

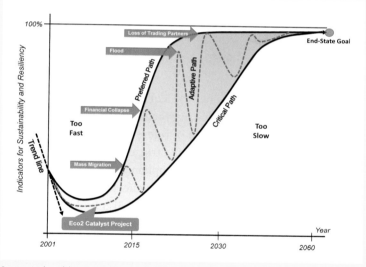

Source: Author elaboration (Sebastian Moffatt).

tal map for establishing and justifying specific recommendations. Everyone involved in the planning is able to follow the transparent, logical connections between, on the one hand, intended goals and overall vision, and, on the other, detailed actions and results. This allows all agencies and stakeholders to understand how their work fits within and contributes to the long-term vision and goals. Ideally, the framework serves to align the various initiatives a city may undertake.

"Automobiles are often conveniently tagged as the villains responsible for the ills of cities and the disappointments and futilities of city planning. But the destructive effects of automobiles are much less a cause than a symptom of our incompetence at city building."

Source: Jacobs (1961: 16).

Aalborg Charter for Local Action Planning
(est. 1994)

1. Recognise existing frameworks, other plans and programs
2. Systematic identification, by means of extensive public consultation, of problems and their causes;
3. Prioritisation of tasks to address identified problems;
4. Creation of a vision for a sustainable community through a participatory process involving all sectors of the community;
5. Consideration and assessment of alternative strategic options;
6. Establishment of a long-term local action plan towards sustainably which includes measurable targets;
7. Programming of the implementation of the plan including the preparation of a timetable and statement of allocation of responsibility among the partners;
8. Establishment of systems and procedures for monitoring and reporting on the implementation of the plan.

Figure 1.6. Aalborg Charter

Source: Author compilation based on EU (1994).

Over time, the collaboration around goals and strategies may begin to coalesce around a common frame of reference throughout a city. This framework is a triggering factor leading to innovation and compliance across all sectors. Such positive spillover effects are the ideal result, creating a powerful local culture of sustainability. In Curitiba, for instance, the overarching vision of sustainability inspired citizens voluntarily to plant 1.5 million trees along the streets (ICLEI 2002).

A shared long-term planning framework is most effectively focused at the urban region as a whole, even if this area exceeds the jurisdictional boundaries of the city. In this context, *urban region* refers not only to the city or group of towns and cities that make up the existing urban area, but also to the combination of rural and natural areas immediately surrounding the city. A lot of city planning is city centric and treats areas outside the city limits as simply the responsibility of another jurisdiction. However, without regional planning, it is impossible to address long-term goals and to benefit from an ecological and economic perspective, the Eco² approach.

Part of the reason a regional context is important is the unregulated and unplanned urban expansion that is occurring worldwide, in some cases even where populations are shrinking. This threatens the long-term health and prosperity of cities and countries. Increasingly, cities depend on the rural and natural areas in which they are immersed. These areas provide ecological functions, capturing and cleaning water; cooling, slowing, and filtering air; growing fresh produce for food security and public health; and providing energy resources that are renewable and secure. A shared regional strategy becomes an umbrella plan that defines how to direct city growth in ways that protect and enhance many ecological functions. This type of umbrella plan is sometimes called a regional growth strategy.

The urban region is also a vital scale for economic planning. Almost all economic patterns are formed at the regional scale, and efforts to intervene and control economic development must also be executed at this critical scale.

Defining the boundaries of an urban region may be difficult. In fact, the actual dimensions

"In many cases, the decline—and possible renewal—of cities cannot be divorced from their wider regional contexts. Declining cities are almost always concentrated in declining regions."

Source: UN-Habitat (2008: 44).

of a region may be kept deliberately flexible so that borders may adapt to reflect the concerns of stakeholders. For example, the regional boundaries used for growth strategies might need to include watershed planning, commuter sheds, air sheds, utility service territories, market gardens, local energy generation, ecological systems, and economic development planning, each of which requires a different delineation. Regardless of the label and scope, a regional strategic plan must help everyone to understand how the city will fit into its ecological surroundings and how the pace and direction of growth will be consistent with near-term targets and long-term goals. The city-based decision support system in part 2 includes information on long-term planning frameworks and the creation of a regional growth strategy.

Stepping Stones for an Expanded Platform for Collaboration

Initiate a process for collaborative decision making

The process of creating collaborative committees begins with an invitation to key stakeholders to discuss the collaborative process and to consider the benefits of participating in an Eco² pathway. It is usually necessary for the Eco² champion to meet individually with key stakeholders and establish a common basis of good will and interest prior to a group meeting. Each stakeholder needs to see the benefits of participation from the stakeholder's own position. For example, land developers have a chance to affect the regulations under which they must work and, ultimately, to improve business by influencing policy. Utilities and landowners can become more well informed about the opportunities for new business and improved customer relations. For second- and third-tier committees, it is especially important to clarify the role of the city as initiator and secretariat, but not as a group in control of decisions. Sometimes, it is necessary for the city to explain to everyone that the integrated approach means it must temporarily remove its regulator hat and join with others in the search for integrated solutions.

Prepare a mandate, and budget for a secretariat

A secretariat needs to support the collaborative committee, which means that it needs to be distinct from other city departments, even if it shares city offices to reduce costs. The size of the secretariat may be adjusted to fit the pace and scope of collaborative processes. If only one person is involved, this person must have skills in communications (facilitation, writing), research, and data collection. Finding a budget for a secretariat may be challenging because collaborative committees are not normally budget items. One option is to include collaboration under the costs of strategic planning. Regardless of the funding source, the secretariat needs at least three years of secure budget to prove its worth.

Prepare a long-term planning framework for sustainability and resiliency

Part 2 provides the detailed methods and tools that may assist in preparing a framework. If time or money is limited, a rapid process is possible, using predefined goals and strategies from appropriate best practice cities. In this context, the case study reports are helpful in providing examples of goals and strategies. Software tools are available on the Web that may help in developing a framework that connects visions and goals to specific strategies and projects; the tools also allow the public and other stakeholders to explore the content of the framework. The framework requires a locally specific set of external forces (for example, climate change in the surrounding location or the demographics of each city). An extensive collaborative effort may be required to complete the framework, supported by tools such as visioning workshops and foresight workshops (part 2).

Select a catalyst project

A catalyst project is a key part of managing change. Catalyst projects should be projects that offer substantial benefits to the most influential stakeholders and that may be completed relatively quickly at low risk to the city. With luck, the catalyst project will contribute to a rising spiral of goodwill and acceptability for the Eco² pathway. Choose carefully; first impressions count for a lot. The creation of positive expectations among participating stakeholders and the public is crucial in successful change management.

References

Calthorpe, Peter, and William B. Fulton. 2001. *The Regional City: Planning for the End of Sprawl.* Washington, DC: Island Press.

EU (European Union). 1994. "Charter of European Cities & Towns Towards Sustainability." http://ec.europa.eu/environment/urban/pdf/aalborg_charter.pdf.

ICLEI (ICLEI—Local Governments for Sustainability). 2002. "Curitiba: Orienting Urban Planning to Sustainability." Case Study 77. ICLEI, Toronto, Canada.

Jacobs, Jane. 1961. *The Death and Life of Great American Cities.* New York: Random House.

Kenworthy, Jeffrey R. 2006. "The Eco-City: Ten Key Transport and Planning Dimensions for Sustainable City Development." *Environment and Urbanization* 18 (1): 67–85.

Lahti, Pekka, ed. 2006. *Towards Sustainable Urban Infrastructure: Assessment, Tools and Good Practice.* Helsinki: European Science Foundation.

Prasad, Neeraj, Federica Ranghieri, Fatima Shah, Zoe Trohanis, Earl Kessler, and Ravi Sinha. 2009. *Climate Resilient Cities: A Primer on Reducing Vulnerabilities to Disasters.* Washington, DC: World Bank.

UN-Habitat (United Nations Human Settlements Programme). 2008. *The State of the World's Cities 2008/2009: Harmonious Cities.* London: Earthscan Publications.

CHAPTER 5

A One-System Approach

A one-system approach enables cities to plan, design, and manage the entire urban system by integrating or optimizing key subsystems. In doing this, it provides opportunities for cities to realize many benefits through synergy.

As we explore the possibilities for a one-system approach, we first address the enhancement of the efficiency of resource flows in an urban area through integrated infrastructure system design and management. The approaches apply to most urban infrastructure sectors, such as transportation, energy, water, and waste management, and may be applicable within each sector and across sectors.

Next, we look at the possibilities for applying a one-system approach to integrate urban form and urban flows. We consider spatial planning, land use, density, connectivity, proximity, and other attributes of urban form, and we examine the large extent to which overall system efficiency depends on the integration and coordination of these attributes with infrastructure systems. There is a fundamental relationship between a city's infrastructure systems and its urban form. Urban form and spatial development establish the location, concentration, distribution, and nature of the demand nodes for the design of infrastructure system networks. Urban form establishes the physical and economic constraints and parameters for infrastructure system designs, their capacity thresholds, and technology choices, and the economic viabilities of the various options. These have tremendous implications for resource use efficiency. At the same time, infrastructure system investments (transportation, water, energy, and so on) typically enable and induce particular spatial patterns on the basis of the market response to the investments.

The final section of the chapter explores ways to implement projects using a more well integrated approach to implementation. This means that investments are sequenced so that the city sets the correct foundation by addressing long-lasting, cross-cutting issues first. This also means creating a policy environment that enables an integrated approach, coordinating a full range of policy tools, collaborating with stakeholders to align key policies, and targeting new policies to reflect the different circumstances involved in urbanization in new areas and the improvement of existing urban areas.

It is critical for cities as they strive for greater ecological and economic sustainability to develop a systems perspective and apply the one-system approach. A review of this chapter reveals a more complete picture of the opportunities and the possibilities for new development paths. In addition, methods and tools introduced in part 2 may help planners, engineers, and designers visualize system dynamics; model the systemwide impacts of different design and policy options at varying scales; and, generally, think outside of the silos created by professional training, institutional structures, and historical practice. As outlined in box 1.3, this will include the use of material flow analysis and the layering of information on maps to create a transdisciplinary platform for integrated design.

The Core Elements of a One-System Approach

Integrating flows: Infrastructure system design and management

We first address the issue of enhancing the efficiency of resource flows in an urban area through integrated infrastructure system design and management. These approaches apply to most urban infrastructure sectors, such as transportation, energy, water, and waste management, and may be applicable within each sector and across sectors.

Integration of demand and supply: Addressing efficiency and conservation before supply-side investments

An integration of supply and demand must always begin by asking why one should bother about new infrastructure if investments in demand reduction and the more efficient use of existing infrastructure are more economical and beneficial. The integration of supply and demand is a strategic approach that needs to be supported by careful investment planning. For any given investment in services, an optimum balance exists between investments in systemwide and end use efficiency and investments in new supply systems. In an ideal scenario, supply- and demand-side investments are considered on a level playing field, and money is placed where the returns to society, the economy, and the environment are greatest. In most utilities, proper tariff structures based on full cost recovery principles, together with progressive block tariffs with precisely targeted subsidies (where needed for social considerations), are an effective mechanism to reduce demand. This is because tariffs that do not reflect the true economic cost may send the wrong signal to consumers and lead to waste or the overuse of resources. It is widely recognized that, historically, too much has been invested too quickly in supply solutions as opposed to reducing demand through resource efficiency standards; building retrofits; and the replacement of lighting, fixtures, vehicles, and appliances. In every sector, significant gains have been realized by demand-side management (DSM); examples are the cases of Yokohama, Japan, in the waste sector (a capital expenditure of US$1.1 billion was avoided) and Emfuleni, in the energy and water sector (where singular investments of US$1.8 million led to annual savings of US$4.0 million). Not only do the net economic returns tend to be higher for DSM, but so, too, do the many indirect benefits for a city, including improved living environments and reduced vulnerability to future price fluctuations or interruptions in resource supply.

While DSM may be easy to implement and may quickly pay dividends in some instances, in other instances, it is difficult to implement because of the incentives of various stakeholders. Consider the case of housing and commercial buildings. On the one hand, they represent a tremendous potential for DSM because most buildings have not been constructed to energy or water efficiency standards and may quickly generate high returns on relatively small

BOX 1.3

Combining Flows and Forms to Create a Transdisciplinary Platform

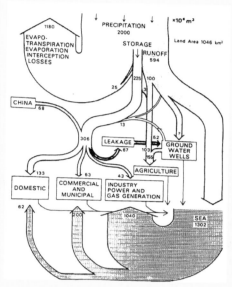

This flow diagram summarizes all the water flow through Hong Kong (China) and is one of the first illustrations of an urban metabolism.

Source: Boyden, Millar, and Newcombe (1981).

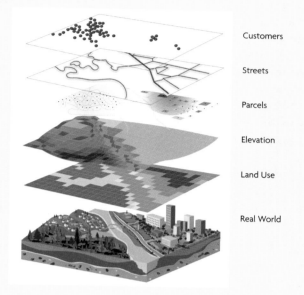

Source: Copyright © ESRI, used by permission, http://www. esri.com/.

FLOWS: Material flow analysis and Sankey diagrams

Material flow analysis and Sankey diagrams are a method for calculating and illustrating the flow of resources through an urban area of any size. Inputs and outputs are determined as resources are extracted from nature; processed by infrastructure; consumed by homes and businesses; treated by infrastructure; and, finally, returned for reuse or delivered back to nature as waste. Colorful, but simple, diagrams are used to educate everyone on the resource flows and the effectiveness of their use, all on a single page.

FORMS: Layering of information on maps

Maps are especially useful in collaboration because they speak so well to so many. (A picture is worth a thousand words.) The layers of information make it possible immediately to interrelate the various features and qualities of the landscape and also easily to quantify important spatial relationships. Layering is an old technique that has become more powerful as a result of computer technology and satellite imagery.

INTEGRATING FORMS AND FLOWS:
A transdisciplinary platform

Because diagrams and maps may be easily understood and shared by a broad range of professionals and decision makers, they help to bring stakeholders and experts together, facilitating a common understanding of integrated approaches to design and decision making. Forms and flows should be analyzed and understood for current and future scenarios. In combination, the methods represent a transdisciplinary platform for understanding the spatial dynamics of a city and its physical resource flows—elements that are interdependent, but difficult to integrate because they involve such different skills and stakeholders.

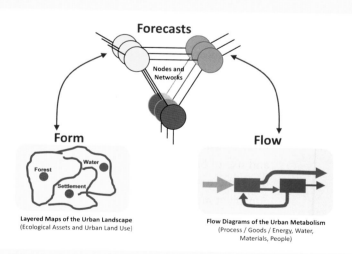

Forecasts

Nodes and Networks

Form

Flow

Layered Maps of the Urban Landscape
(Ecological Assets and Urban Land Use)

Flow Diagrams of the Urban Metabolism
(Process / Goods / Energy, Water, Materials, People)

A platform is needed to integrate the design concepts for urban form with the corresponding resource flows.

Source: Redrawn and adapted from Baccini and Oswald (1998).

investments. On the other hand, changes to existing buildings require collaboration among decision makers, and the benefits do not always accrue to those who must make the investments, thereby fracturing the incentive structure. For example, if owners cannot capture the benefits of energy efficiency savings, they will not invest in retrofitting; renters who have a short time horizon also have no incentive to invest in retrofitting. In addition, the standards for products, including building codes, are often established by senior government entities and are therefore difficult to align with local goals and strategies. For all these reasons, a well-planned collaborative process is necessary if the benefits of integrating supply and demand are to be captured.

DSM measures may apply to all sectors and may include investments in more efficient technology. Typical examples include energy retrofits of building envelopes; resource-efficient lighting and appliances; low-flow water fixtures; waste reduction, reuse, and recycling; and use of bus rapid transit instead of cars on the same roads (thereby avoiding the need to construct more roads). It may also imply a culture of doing more with less and living lightly on the Earth by voluntarily limiting consumption and waste. DSM may also be achieved by improving designs at multiple levels and through regular audits; commissioning, process enhancement; and improved training among the personnel who install, operate, and manage systems. The proliferation in the number of energy service companies (the end product of which is energy efficiency) is evidence of the growing and still untapped potential of the energy efficiency market.

Often, DSM in one sector may lead to benefits in another sector. For this reason, integrated approaches across sectors are critical. For instance, the significant urban energy benefits of the water sector's DSM led to a program launched by the Alliance to Save Energy. The program is called Watergy. The alliance has achieved significant benefits in developing-country cities by increasing access to clean water, while reducing energy costs and water losses. In Fortaleza (northeast Brazil), the alliance worked with the local utility, the Companhia de Água e Esgoto do Ceara, to develop and implement measures to improve the distribution of water and the access to sanitation services, while reducing operating costs and environmental impacts. The utility invested about US$1.1 million, including in the installation of an automatic control system, and saved US$2.5 million over four years. The efficiency gains were so great that 88,000 new households were connected to the water system, without the need to increase supply (Barry 2007).

DSM can even apply to spatial systems. For example, the demand for land may be reduced through a review of regulations (including adjusting minimum lot sizes, increasing floor area ratios, revising zoning, and adjusting land subdivision parameters), layering of different land use functions on the same site, or a reduction of parking spaces. (Additional analysis of the management of the spatial structure of cities may be found in part 3.) In all cases, the demand-supply relationship needs to be replaced with an approach that enables and encourages demand management.

Peak load management: Managing the demand for services to minimize the requirements for peak capacity

Energy, water, and transportation systems all tend to suffer from daily and seasonal peak loads that force utilities to use oversize systems to meet peak demand at a particular time or period. This may be significantly inefficient from an economic and resource point of view. Peak loads also force utilities to supplement supply using backup or imported resources and services that are especially costly. Spatial systems likewise suffer because of the highly uneven demand for spaces that are dedicated to uses such as parking, roads, and restaurants.

The effort to reduce the need for greater overall system capacity through the management of daily and seasonal peak loads is known as peak load management. The object of peak load management is to even out the demands throughout the system and to distribute demand across time to avoid investment in new permanent capacity. In some cases, peak load management may also help avoid the high cost of topping up capacity if the primary system has reached maximum output.

By delaying or avoiding costly capital investments and costly backup strategies, peak load management may be extremely economical. It may also reduce resource consumption requirements and can make more optimal use of existing capacity. However, recognizing the best locations for intervention at each stage of the system requires a systems perspective.

For instance, in Europe, heat demand varies significantly during the heating season. To deliver all district heating through a combined heat and power plant would require the utility to size a plant in accordance with the maximum heat load, which would mean greater investments. A strategy is therefore sometimes used whereby only the base load is supplied by the combined heat and power plant, while the peak load is supplied by a simple boiler plant (figure 1.7).

Peak load management is often applied in public transportation systems and highway

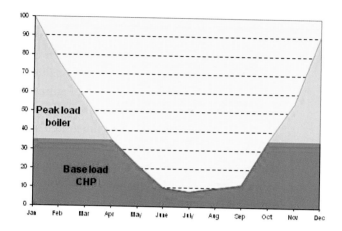

Figure 1.7 The Load Curve of a District Heating System

Source: Author compilation (Bernd Kalkum).

Note: The system is designed for 35 units of base load instead of 100 units of peak load, thus representing significant savings. CHP = combined heat and power plant.

systems to reduce overcrowding or congestion during rush hour. In Japan, most commuter railway systems adopt an off-peak-hour tariff (lower tariff) to induce passengers to take trains during off-peak hours. The Tokyo Metropolitan Highway Authority also uses an off-peak-hour tariff for highway tolls. The highway authority adjusts tariff levels across different highway routes to divert traffic from one route to another to reduce congestion.

Peak load management may also benefit from a more collaborative approach because demand profiles are influenced by many factors that are sometimes difficult for cities to control unilaterally: land uses, time-of-day pricing structures, metering technology, control technology, business and school operating hours, daylight savings time, and the determination of the size of distribution and storage facilities at each level. Meanwhile, simple alterations to business and school operating hours may have a significant impact on peak loads in transportation.

Cascading resource use: Matching resource quality to the requirements of each user

Cascading resource use is another option for integrating flow pathways. Cascading is achieved by matching the quality of a resource to the requirements of the end user. As the quality deteriorates, the resource is directed to uses with lower quality requirements. In this way, water, energy, and materials may achieve two or more functions in sequence. Figure 1.8 illustrates the transition from a once-through flow water supply system to an integrated system that matches quality to needs. It cascades water flows from drinking and cooking and sanitation to toilet flushing and the subsoil irrigation of gardens. The chief benefit of cascading is efficiency gains (satisfying many demands with the same unit of supply); however, an added advantage is the capacity to direct scarce resources to essential needs during difficult times. Resources may be cascaded through multiple uses and then, through processing, may be looped back to the original point of use.

Considered a water-scarce city-state, Singapore adopted an integrated water resource management strategy that includes many integration strategies, including the cascading and looping of water resources (figure 1.9). The approach successfully lowered annual water demand in the city from 454 million tons in 2000 to 440 million tons in 2004 (Tortajada 2006), while the city's population and GDP per capita grew by 3.4 percent and 10.3 percent, respectively. Cascading and looping represented a welcome departure from conventional supply-driven investment approaches (often based on business-as-usual scenarios) to a new resource management approach, including effective demand-management control.

Looping resource use: Reclaiming the secondary resource values

Looping refers to the closed loop systems that ultimately return water and materials to their points of origin. Returnable drink containers are an obvious example, but the same concept may apply to the much larger flows of organic material and water that are carried within drinking containers.

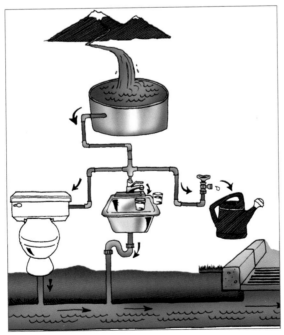
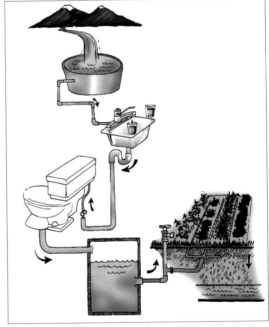

Figure 1.8 Cascading Water Use

Source: Author elaboration (Sebastian Moffatt).
Note: As the resource cascades through the system, its quality is matched to the needs of successive uses.

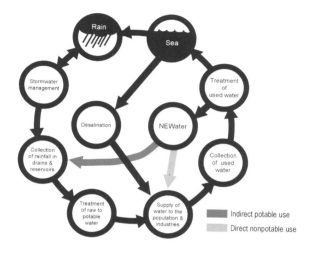

Figure 1.9 Cascading and Looping Water in Singapore

Source: Singapore Public Utilities Board, http://www.pub.gov.sg/about/Pages/default.aspx (accessed January 2009).

The City's Life Support Systems

The overall aims of environmental technologies are to maximize the possibility that cities can meet their needs from the natural capital of their own bioregions in a renewable way and to move to closed loop infrastructure systems that recycle and reuse their own wastes, so that the absorptive capacities of natural systems are not overwhelmed with the waste loads from urban areas. (Kenworthy 2006: 76).

Figure 1.10 Looping Resources

Source: Author elaboration (Sebastian Moffatt).

Note: In the example on the left, a factory consumes resources and generates waste. On the right, an urban ecology has emerged, where the waste heat, water, and materials are reused by other land uses, looping within the city to lower costs and reduce the negative impacts on the environment.

Looping is common in natural ecologies, where it is manifest in the water cycle, the carbon cycle, and the nitrogen cycle. City infrastructure is most successful if loops are closed. This might mean recharging aquifers during rainy periods or converting organic waste into soil supplements for local parks, gardens, and farms (figure 1.10). Looping close to home is especially effective because it reduces transportation costs and creates many potential benefits such as jobs close to home and local stewardship.

An example of cascading and looping across infrastructure sectors is the case of Hammarby Sjöstad in Stockholm. Energy, water, and waste are looped many times to enhance and optimize the utility derived from resources. (See the case study on Stockholm in part 3.)

Looping also provides an opportunity to invest strategically in the weakest link. Once the connections in the loop are understood, it is possible to retrofit existing infrastructure based on greater knowledge of the most effective investments in each sector. For example, for water, wastewater, and gas distribution systems, leakage reduction in existing pipelines represents an effective investment for improving water and energy use efficiency.

Distributed systems for omnidirectional flows: Achieving greater functionality for nodes and networks

The integration of nodes and networks is achieved through distributed systems. In a traditional supply-oriented approach, the number of nodes is few; a single supply facility might be the only supply node, for example, and the distribution network may be a simple one-way hierarchy from a big facility node directly to users. A fully distributed system actually works in both directions and enables omnidirectional flows. The supply system may begin at or near the home, office, or shop where the demand for services originates. Local and renewable options may be explored for on-site supply, storage, or treatment; roof-mounted technology may capture and store water, for example, or capture and convert sunshine. It is conceivable that public utilities may still own and manage technologies, but they are located on-site. If on-site facilities are not practical, sufficient, or economical, the next choice is to examine options for the cluster, block, or neighborhood.

Economically, it is often viable to locate a significant capacity in supply and treatment facilities at the neighborhood or district level or at the center of small clusters of mixed use buildings so that equipment may be well managed and used continuously (figure 1.11). Combined heat and power plants in Europe often operate at this scale and thereby provide power and heating across city districts. Another example might be septic tanks attached to buildings, which may be interconnected to a small in-ground wastewater treatment facility at a local park or to a high-rate composting vessel at the nearest recycling depot or community garden. Distributed systems make greater use of networking. Local networks for capturing water or generating power may allow nearby sites to share surpluses with others, creating a two-way micronetwork. Surplus energy (for instance) generated by a cluster of users may be stored for later use or sold back to a smart grid. Local networks may be nested within larger networks. In this way, the pattern changes to a system with many nodes serving clusters of users and connected through a complex network with omnidirectional flows. Distributed systems may cover large areas of the city, but the nodes are more numerous, and the networks are more adaptable.

The Rocky Mountain Institute's pathbreaking and comprehensive work on the viability of distributed energy systems catalogues over

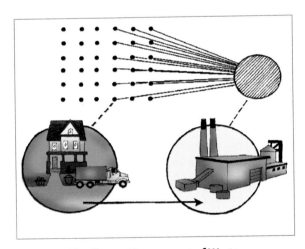 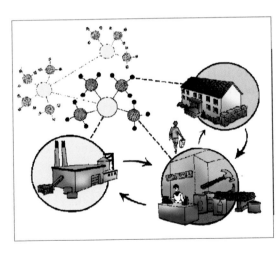

Figure 1.11 The Cluster Management of Waste

Source: Author elaboration (Sebastian Moffatt).
Note: The scenario shown on the left is the common supply model, whereby solid waste is collected from many sources by a centralized trucking system and then processed at a large, remote facility. On the right, a two-way network evolves to eliminate waste within the cluster.

200 benefits (see Lovins and others 2002). The most significant benefits relate to system modularity, which contributes to a reduction of economic and financial risk by several orders of magnitude (figure 1.12). Other potential benefits from the integration of nodes and networks might include the reduced costs for land dedications and reductions in the characteristically large transmission and conversion losses. For many cities, a growing proportion of utility resources are being used in unproductive generation and distribution, especially because DSM has reduced the demand for services at each node. A distributed system not only helps to avoid these costs, but also may offload the costs of new facilities from taxpayers to developers and give developers a long-term interest in the efficiency of neighborhoods. Other benefits from more distributed systems include the more incremental pace of investment that is shaped by demand, the more effective matching of capacity to existing load, and the lower vulnerability to whole-system collapses. As infrastructure facilities move close to buildings, so do the jobs, and the city becomes more efficient and walkable. The proximity of facilities also increases the potential for almost all other types of integration (for example, recycling, looping of resource flows, multipurpose uses, and culturally distinct structures).

As demonstrated in a Worldwatch Institute case study (Bai 2006) on the City of Rizhao in Shandong Province, China, distributed solar water heating systems can be effective urban energy solutions, while also helping address social equity issues.

Spatial planning may also benefit from distributed systems that make nodes of the population more self-reliant. This is the philosophy behind smart land uses such as mixed use walkable communities that provide easy access to transit, services, shops, and parks, rather than forcing everyone to travel to the city center or mall with the attendant costs of time, energy, and emissions.

Multifunctionality: Serving different ends by using common spaces and structures

The integration of infrastructure facilities across sectors is achieved through multipurpose elements that serve different sectors simultaneously or at different times. A common example is the integration of energy and water systems. In many towns and cities, the largest single energy account in the community is for pumping municipal water from wells or water bodies. Sewage digesters also require large motors and energy expenses. Thus, saving water automatically means saving the energy required for water supply and sewage treatment. An integrated approach is the logical option.

The integration of energy and water can involve more than simply the shared efficiency gains. The water system for the Olympic Village in Vancouver, Canada, for example, is closely integrated with the energy supply systems in the city. As the water travels down from the city's mountain reservoirs, it turns a turbine within the pipes. The turbine creates electrici-

Solar Energy Systems in Rizhao, China

Rizhao, a city of about 350,000 people in northern China, is using solar energy to provide lighting and water heating. Starting in the early 1990s under a municipal government retrofit program, the city required all buildings to install solar water heaters. After 15 years of effort, 99 percent of the households in the central district had obtained solar water heaters. Solar water heating now makes economic sense. The city has over a half-million square meters of solar water heating panels, the equivalent of about 0.5 megawatts produced by electric water heaters. Most traffic signals and street and public park lights are powered by solar cells, reducing the city's carbon emissions and local pollution. Using a solar water heater for 15 years costs about US$1,934 (Y 15,000), less than the cost of running a conventional electric heater, which equates to saving US$120 per household per year in an area where per capita incomes are lower than the national average. This achievement is the result of a convergence of four key factors: a regional government policy that promotes and financially supports research; the development and deployment of solar water heating technologies; a new industry that takes the opportunity in stride; and city leadership that not only has a vision, but also leads in action and brings along other stakeholders.

Source: Bai (2006).

Figure 1.12 Distributed Systems

Source: Author elaboration (Sebastian Moffatt).

Note: Centralized, remote facilities with one-way networks may be transformed into distributed systems as shown in these two extreme examples of energy systems. In the centralized example, a remote facility services all end users in a one-way distribution network. In the distributed case, all buildings within a 5-kilometer radius are connected to a local heating and cooling plant, using low-temperature water to move heat or cooling from one location to another. Excess heat may be captured from local industrial processes, sewage, or large buildings such as the hospital and then shared at low cost. Local power generation is an option through the creation of a small electrical utility that offers waste heat for use in buildings or for the operation of a cooling system. Typically, such a combined system is able to raise overall efficiencies from 55 percent to 80 percent. The on-site power may be used for local transit year-round. Flexibility is also enhanced because energy sources may be mixed to take advantage of market rates, local waste products, weather, new technology, and so on. Any excess electricity from the local utility may be offered to the regional grid and used for more efficient load management and backup.

ty. After the water is used in the village, a heat pump draws thermal energy from the sewage and returns the heat to buildings that require space and water heating. When the sewage is eventually treated, the methane gas that is released is used to power the treatment facility. Is this a water system? A hydroelectric system? A gas-fired electrical system? A district heating system? A sewage treatment system? Answer: all the above.

The photographs in figure 1.13, from a West Coast Environmental Law study, describe the integration of a trail system and other forms of infrastructure. Many possibilities exist for such multipurpose facilities and amenities (see figures 1.14–1.19). At some point, the integration of systems is most successful if it is, in fact, difficult to isolate any particular system from the others. The functional components of urban services are tightly woven into the fabric of the community at the most local scale.

Integrating forms with flows: Spatial planning and urban design

We now look at the possibilities for the application of a one-system approach in integrating urban form with urban flows. We consider land use, density, connectivity, proximity, green infrastructure, and other attributes of urban form and examine how a large portion of overall system efficiency depends on integrating and coordinating these attributes with infrastructure systems.

Urban form, land use mix, density, connectivity, and proximity

The integration of spatial planning and infrastructure system design represents the most significant opportunity to enhance overall system performance. Urban form, land use mix, density, connectivity, and proximity all have effects on infrastructure performance. Yet, few land use plans are evaluated from this perspective. Planners and engineers sit in different meetings at different times and ask different questions. Seldom do infrastructure concerns influence land use plans or vice versa. Despite this disconnect, the best time to consider ways to minimize infrastructure costs is during the early stages in land development processes.

In principle, spatial planning may contribute to lower infrastructure costs by increasing density and compactness and by locating development sites in close proximity to key facilities (box 1.4). The amount of linear infrastructure required for low-density, single-family

Figure 1.13 Uses of a Pedestrian Pathway

Source: Rutherford (2007).

Note: A pleasant pedestrian pathway (an example in each photograph) serves the transportation needs of a walkable community, providing a quiet, safe, and cool option for moving around. At the same time, it also functions as an element in other infrastructure systems. The garden strip on both sides of the pathway is used to grow plants and flowers that help keep the city cool, reducing energy requirements for air-conditioning. The pathway is bordered by a gentle swale or depression in the earth that functions as an infiltration trench, intercepting and slowing storm water flows. The soil in the trench is enriched with composted organic waste, avoiding the need to truck such waste out of the community. The enriched organic soil is highly absorbent and, thus, needs little irrigation to stay green, helping reduce the city water budget. The subbase for the pathway is composed of ground glass and rubble from returned bottles and from industrial waste. In essence, the pedestrian pathway is a transportation facility that also serves to manage and treat storm water flows, recycle organic and inorganic waste, cool the city, and provide a water-efficient garden amenity.

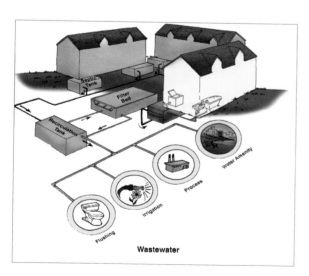

Figure 1.14 A Distributed System for Wastewater Treatment

Source: Author elaboration (Sebastian Moffatt).

Note: In this example, a distributed system for wastewater treatment incorporates low-flow fixtures in buildings, primary treatment in septic tanks attached to each building, and an advanced secondary treatment system in a courtyard to serve a cluster of nearby buildings. The water is decanted from septic tanks and sprayed over a gravel bed in the recirculation tank. The reclaimed water exiting the tank is safe for all uses other than drinking. This water may be used for flushing toilets in a two-pipe system or for irrigating and fertilizing gardens. It may also be used as an input into local industrial processes or as a way to augment water in streams, fire prevention reservoirs, and fish ponds.

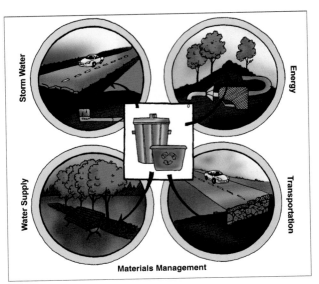

Figure 1.15 Integrated Materials and Waste Management

Source: Author elaboration (Sebastian Moffatt).

Note: This simple illustration shows how the solid waste stream from a city neighborhood (center) is diverted to other sectors: crushed glass provides a base for roads; composted organics provide nutrients; compost provides soil additives for parks and public green spaces; and coarse organics are used to create drains next to highways that capture and clean storm water flows and runoff water from roads. Finally, a facility converts organic matter to biogas for use in generating heat and power.

Figure 1.16 Innovative Energy Infrastructure

Source: Author elaboration (Sebastian Moffatt).

Note: Energy systems may take advantage of flows from other sectors. For example, a sound barrier along a highway has photovoltaic panels that generate electricity to be reused for domestic hot water; a small turbine in the water supply system harnesses excess water pressure to generate electricity; and methane from composting facilities is used to generate heat and power.

Figure 1.17 Integrated Storm Water Management

Source: Author elaboration (Sebastian Moffatt).

Note: Storm-water management systems may realize synergies with other urban systems. Here, a bicycle path doubles as an infiltration trench; shady trees and green roofs reduce energy use and also keep precipitation away from drainage systems and slow rainwater flows; roof rainwater catchment and storage systems provide water for gardens and lawns; and storm water is directed into finishing ponds for reclaimed wastewater, helping to treat sewage and maintain amenities.

Figure 1.18 Traditional Dwelling Supply Systems

Source: Author elaboration (Sebastian Moffatt).

Note: The high demands of this dwelling are satisfied through disconnected infrastructure supply systems.

Figure 1.19 Combined Trenching for Infrastructure Systems

Source: Author elaboration (Sebastian Moffatt).
Note: This dwelling is much more resource efficient by design. The infrastructure application combines trenching and more varied flows to facilitate the sharing and cascading of resources within the housing cluster.

Form and Flows

The sustainability of infrastructure systems depends on the path chosen for spatial development. To appreciate impacts, one must compare the spatial layout of Atlanta and Barcelona at the same level. The comparison of these spatial distributions of population illustrates differences in urban spatial structure and the consequences for the operation of transit and other infrastructure. Imagine the differences in capital costs to service these populations of similar size. Remember that distribution network cost is a large share of the total cost; for example, pipes account for about 70 percent of the cost of a water supply system. Also, imagine the differences for these cities in the operation and maintenance of water systems (the pumping of water and the collection and treatment of waste) and transportation systems. Keep in mind that about 30 percent of urban energy bills often goes for pumping water and wastewater.

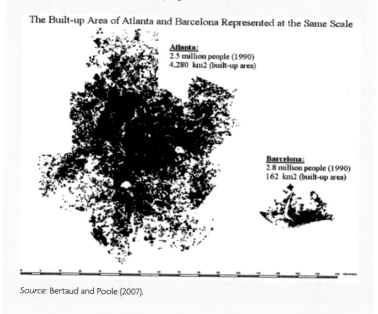

Source: Bertaud and Poole (2007).

housing can be 17 times greater than the amount required for multiunit dwellings in dense urban developments (the cost of sprawl). The capital cost savings are roughly proportional to the average length of the system per serviced unit. In low-density, single-use developments, local governments often generate less in development fees and property taxes than they spend in services and infrastructure costs, such as roads, water mains, and sewerage. An analysis found that, for every U.S. dollar raised in development fees and property taxes in southwestern Ontario, Canada, US$1.40 needs to be spent on services. The rest of the city subsidizes lower-density development.

Urban form and density lock in some of the most significant physical and economic parameters of supply-side infrastructure investments. Public transportation and district heating and cooling are examples of efficient technologies that become financially viable only at certain threshold urban densities.

As cities sprawl and splinter, the energy consumption involved in transporting people can increase by a full order of magnitude, and cities may decisively exclude pedestrians, as shown in photographs of Houston, Texas (figure 1.20). Houston has a population of 2.2 million, and its area is 1,600 square kilometers.

Figure 1.21 illustrates this relationship across a range of cities, underscoring that urban form and density significantly affect energy consumption in transportation.

Proximity and connectivity to key facilities are other factors because scattered pockets of development are likely to be far from supply and processing systems and require relatively greater investments in trunk lines, major roads,

Figure 1.20 A Broad View of the City Center of Houston

Source: Houston-Galveston Area Council, Google Earth.
Note: The expanded view shows a parcel of land that is theoretically within walking distance of the city center.

pumping stations, and so on. The additional capital costs for remote connections are typically shared by all users and translate into cost premiums. If higher-density, transit-oriented developments are instead located downstream from water reservoirs and close to existing mains, the capital costs of development will be much lower, and the city can avoid the otherwise significant costs—often 30 percent or more of total energy bills—associated with

pumping water and wastewater to and from new households and businesses.

Because resource efficiency and emissions are influenced directly and permanently by urban form and density, intelligent spatial planning is the first proactive step toward DSM in infrastructure (figure 1.22). The mixing of land uses at the neighborhood level can reduce system costs by evening out the demand for services and reducing the peak loads that directly

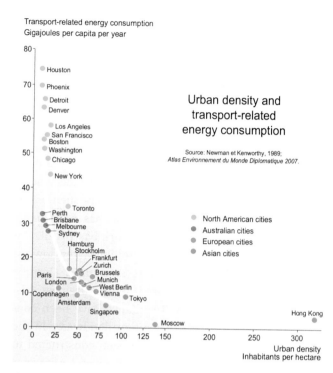

Transport-related energy consumption
Gigajoules per capita per year

Urban density and
transport-related
energy consumption

Source: Newman et Kenworthy, 1989;
Atlas Environnement du Monde Diplomatique 2007.

- North American cities
- Australian cities
- European cities
- Asian cities

Urban density
Inhabitants per hectare

Figure 1.21 Urban Density and Transport-Related Energy Consumption
Source: Kirby (2008).

affect the design capacity and capital costs of infrastructure systems.

Land use plans also need to consider existing and planned infrastructure capacity and direct growth plans accordingly. Certain locations within a city may be especially suitable for infill or new development precisely because they have excess capacity for power, roads, and water. In other locations, capacity may be nonexistent, and land development may require major new investment for one or more systems. Ideally, infrastructure capacity needs to be analyzed and mapped on a fine scale and included in the overlay analysis that guides land use planning.

At the same time, spatial development (and its coordination with broader investment strategies and plans) has significant implications for economic competitiveness, and it affects land and real estate markets. Spatial development and infrastructure investments set in place and shape the contours of these larger economic dynamics. Spatial development is also influenced by these dynamics. (For more detailed information on how spatial form and land use regulations affect mobility and affordability, see part 3.)

Poor spatial planning can fragment labor markets and make cities unaffordable for people who cannot buy cars. They render the cities vulnerable to fluctuations in oil prices. For instance, the spike in gasoline prices in the United States in spring 2008 resulted, over a four-month period, in a 6 percent drop in vehicle kilometers traveled. By moving people out of cars and onto public transit, the area required for roads and parking may be drastically reduced, as suggested by the photographs of Houston. Cheap or free parking subsidizes car use, as do massive land investments in roads. In cities, parking should be market priced and compete for real estate.

Policies that affect the viability of transit also affect the cost and performance of transportation infrastructure. For example, densities of around 50 people per hectare are required to provide convenient alternatives to

Figure 1.22 A Different Paradigm for Urban Design

Source: Author elaboration (Sebastian Moffatt).
Note: Big roads, lengthy pipes, big wires, and larger pumps are replaced by a mixed use, compact, pedestrian-friendly design whereby public funds are used for parks and social and local services.

the car. These two issues are impossible to separate. Thus, a key strategy for improving overall system performance is to organize land use, densities, connectivity, and access to ensure viable public transit and other infrastructure.

Green infrastructure: Integrating natural systems with built systems

The integration of natural systems with infrastructure is possible through green infrastructure and ecological engineering. Green infrastructure refers to the city's naturescape, that is, the mix of trees, shrubs, hedgerows, gardens, green roofs, lawns, parkland, and waterways. These natural elements may be effective in providing a variety of services for other sectors (figure 1.23). For example, when the mayor of Los Angeles, California, faced brownouts and severe energy shortages in 2004, his response was to invest in the planting of thousands of trees along the streets of the city. Urban forests save energy by reducing temperatures, shading buildings, cooling the air, and reflecting sunshine. The trees in Los Angeles are part of the city's energy infrastructure.

The most common examples of green infrastructure are the ribbons of green riparian areas along streams and rivers. These green strips act as filters, preventing silt and nutri-

ents from entering streams. Storm water permeates the soil, or is retained on leaves or captured by roots, and the result is less damage to aquatic environments or reduced requirements for investing in treatment systems.

Such natural systems may be more or less engineered to suit the city's needs. For instance, being surrounded by rivers such as the Iguaçu, flooding has been a big problem in Curitiba, Brazil. Instead of controlling water flow using concrete structures, Curitiba has created natural drainage systems. Riverbanks have been converted to parks that retain floodwater in the soil, and lakes have also been constructed to hold floodwaters. River water and rainwater that lead to flooding may be contained naturally in the lakes and in the parks surrounding the lakes. The ecosystem is preserved in a natural way. As floodwater to the park area is released from the ground to the river naturally (rather than being drained at high speeds through straight concrete drains), downstream flooding may be avoided. People are less exposed to environmental hazards and the diseases caused by flooding. The cost of building parks, including the cost of the relocation of slum dwellers, is estimated to be five times lower than the cost of building concrete canals.

Land use plans may also be used to incorporate green infrastructure if they are associated with policies that manage the demand for services. In Freiburg, Germany, for example, land use plans address storm water runoff by taxing land differently based on the permeability of the surfaces. As a consequence, developers are careful to minimize hard surfaces on parcels, using crushed stone for pathways, paving stones for parking, and so on. The result is less cost for taxpayers because the city avoids investment in infrastructure for capturing, transporting, and treating storm water.

Layering: Integrating different uses for a common space over time

The layering of uses may also occur over time. A school and its playgrounds might serve to educate children during the day. However, these areas might become sites for after-school programs in the afternoon and schools for adults in the evening and, on weekends, serve as coffeehouses, theaters, or open-air craft and farmers markets (figure 1.24). The schoolyard might also be an overflow flood control basin during the monsoon season. Smart cities do not build schools; they build multipurpose civic facilities that change their use by the hour, day, weekend, and season. It is the community (not the school board), that controls usage, and the building continues forever as a community asset, even if the need for schooling decreases.

The multiple functioning of elements in a system creates a layered approach to design whereby each location serves many purposes. This is part of a slow evolution away from the highly segregated land use patterns that are typical of many modern cities. More attention to design helps mitigate or eliminate the negative impacts of diverse uses on surrounding parcels. Industries are not necessarily dirty affairs that need to be isolated from the homes of workers. In fact, their processed wastes and emissions are now recognized as valuable resources, providing feedstock for new industry. Mixing shops and residential areas enhances livability and sustainability and creates jobs close to home.

Colocation: Using the advantageous siting and placement of new structures and rights-of-way

The more efficient use of facilities may be achieved through the strategic and cooperative siting and placement of new structures and rights-of-way. One common example is the mounting of photovoltaic and solar water heating panels on rooftops where they take advantage of unobstructed sunshine (and possibly provide useful shading for buildings). Rights-

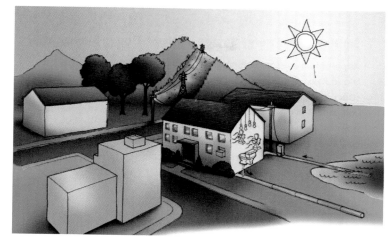

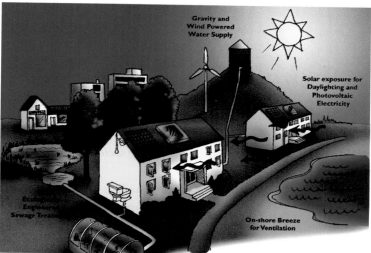

Figure 1.23 Integrating the Benefits of Natural Systems in Communities

Source: Author elaboration (Sebastian Moffatt).

Note: The settlement in the upper illustration is not blended into its surrounding ecological systems; it neither benefits from nor uses these systems efficiently. By contrast, the settlement in the lower illustration harnesses ecological attributes to its advantage, including wind, elevation, sunlight, and ecological sewage treatment options. This reduces the settlement's footprint and ongoing costs.

Figure 1.24 The Multiple Uses of a Public School

Source: Author elaboration (Sebastian Moffatt).

of-way may be shared by many different services. A wet-waste composting facility might be colocated with community gardens to facilitate easier looping and to manage the noise, odor, and activity impacts more effectively. Although the relevant structures and activities may be planned by different groups, their integration benefits everyone.

Place making: Creating social amenities as intrinsic attributes

Hard infrastructure facilities may be designed to contribute to the community in many social and aesthetic ways. It is no longer necessary to hide wastewater reclamation plants if the treatment basins have become pleasant nature ponds with landscaped shorelines served by quiet trails. The reclamation system in Irvine, California, is frequently used by residents as a park area because the connected water bodies and trails provide a unique and enjoyable experience. Water storage tanks can become sculptures and way-finding landmarks. Recycling depots can become community gathering places. The opportunities are unlimited if the mandate for design involves integration.

Employing Integrated implementation

We now examine ways to implement projects using a more well integrated approach. This means sequencing investments so that the city sets the correct foundation and addresses long-lasting, cross-cutting issues first. This also means creating a policy environment that enables an integrated approach, coordinating a full range of policy tools, collaborating with stakeholders to align key policies, and targeting new policies to reflect the different circumstances involved in urbanization in new areas and existing areas.

Sequencing: Use the phasing of investments to capture whole-system synergies

Sequencing refers to the ordering of integration strategies so that decisions in one sector do not preclude integration in another. For instance, a city's location and its route of growth are primary factors determining the city's spatial advantages and constraints. Location determines the physical and environmental conditions of the city: altitude, topography, and climate. Location has implications for a city's urban form and density, its infrastructure systems (demand and supply), and its built environment requirements and possibilities. Location also determines the access and proximity to natural resources (such as renewable energy resources) and the access and links to the economic geography of a region.

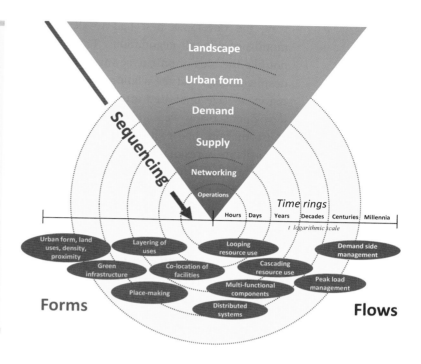

Like people, cities function most efficiently if they have good bones, the strong structural elements that are able to provide the proper context for the shorter-lived elements. Within an urban region, the sequence typically progresses from slow-moving elements such as local ecologies and natural assets, land use patterns (including rights-of-way), and building stocks to the more rapidly moving elements such as management policies and consumer behavior.

The longer-lasting elements are given priority because they may be changed only slowly and at great cost and will constrain the possibilities in other sectors. If significant opportunities for integration at this level are missed, it may take a long time to rearrange and correct problems.

Figure 1.25 provides a rough guide for sequencing integration opportunities during the diagnostics stage of a project. Moving from the outside in, we see how specific integration opportunities tend to line up. Harmonizing infrastructure with the surrounding ecology and resource base is a good first step. Integration with urban form and land use is next. Demand reduction is next. All this is common sense. There is no point in investing in large remote supply or processing systems if a similar in-

Figure 1.25 Time Rings

Source: Author elaboration (Sebastian Moffatt).

Note: Time rings help to sequence investments for optimum returns. A strategic approach to infrastructure planning examines all opportunities for integration, but moves in sequence from the most slowly changing elements (such as the integration of infrastructure with natural systems and land use plans), to the most rapidly changing (such as integrating management systems, providing incentives for consumers, or monitoring and adaptation).

vestment in local ecologies, or smart growth, or demand reduction is able to provide a more sustainable solution.

Enabling: Develop policies that enable implementation of the different types of integration strategies

Despite the best intentions, cities often stumble in the attempt to implement sustainable infrastructure and land use. Outdated polices prevent new approaches and artificially freeze the technology. Policies developed for one goal inadvertently influence design solutions in unforeseen areas. Any developer who has tried to apply ecological design will have plenty of such stories. Consider a proposal for a new underground onsite sewage treatment system for Dockside in Victoria, Canada's premier example of sustainable mixed use development. Although the system was eventually built and is now working well, the developer first had to deal with months of difficult negotiations. The city did not like the idea of on-site sewage treatment, despite the fact that the whole city was dumping its sewage untreated into the ocean. The developer's plans were rejected initially because the city had regulations against treatment plants in residential areas. Other regulations of the health department forbade any use of reclaimed water for toilet flushing and gardening, making the advanced technology in the new Dockside neighborhood much less ecological. How could they benefit from reusing wastewater? Yet another obstacle was the city's property tax structure, which forced the residents in Dockside to pay a share of the city's upcoming sewage system, although they would not be using it.

The reality is that every city has many policies that conflict with a new one-system planning framework and the ecological design and management of new projects. One of the most important outcomes of an Eco² catalyst project is the exposure of such policy conflicts. A collaborative framework may help quickly resolve such issues in catalyst projects and lead to new policy in the process.

In general, enabling policies have impacts well beyond conflict resolution. Ideally, the policy environment that evolves in a city reinforces the goal-oriented framework and stipulates performance requirements rather than prescribing specific solutions. While the city needs to articulate constraints and targets clearly for the community, the most creative design solutions are likely to be achieved at the most local scale, such as the building, parcel, or neighborhood. Primary responsibility needs to remain with local actors and decision makers. They are the first to explore design options and, thus, have the greatest freedom to innovate. Only those service or performance requirements that cannot be successfully satisfied locally because of technical, economic, or other practical reasons should be passed on to the scale above. At the region-

al level, there may be less need for policy or investment in infrastructure, except as required for regional integration (such as in regionwide transportation systems). From this ideal perspective, implementation policies in the city may begin to emulate the self-organizing and self-reliant properties of natural ecologies.

Coordination: Offer instruments in at least five flavors

Local governments have many policy tools for the implementation of a one-system approach. Too often, the focus is exclusively on legislation and enforcement. An integrated approach to implementation requires that cities take full advantage of all instruments available to the city and to the stakeholders who collaborate with the city. Every Eco² project benefits from the integration of at least five different categories of instruments. Financial instruments may include incentives, subsidies, pricing, taxing policies, fee structures, market reforms, purchasing policies, and much more. Special planning initiatives may include new plans, new institutions, institutional restructuring, special reporting, and special events. Research and demonstration may focus on innovative technology applications, tours, fact-finding missions, surveys and assessments, conferences, policy research centers, and forecasting. Education and inspiration may include professional training, visioning exercises, cotraining, communities of practice, curriculum reform, special publications, communications, social networking, and investments in social capital. Legislation and enforcement may include a wide variety of regulations, codes and standards, specific fines, and policing policies.

In some cases, the capacity of a city to apply particular policies will be restricted by national governments and by statute. However, through collaborative working groups composed of senior government officials and stakeholders, such limitations may be overcome. The best approach to implementation is always the application of the full set of instruments in concert as time and resources allow.

For instance, if a city wants to reduce water consumption at the lowest cost, it might explore an integrated approach on a collaborative basis. This might involve (1) using a public awareness and education campaign to convince households and businesses of the need for and benefits of water savings and to seek their support in designing tariff increases (stakeholder engagement); (2) adjusting the structure of water tariffs, fees, and pricing (a policy and regulatory issue and demand management); (3) promoting the use of water-saving faucets and toilets (a regulation and building code issue and public awareness); (4) designing guidelines and standards for new residences and businesses to encourage investment in the best performing water-saving faucets and toilets and procurement policies for private sector suppliers so that the best technology is supplied at high-volume market prices (engagement with private sector stakeholders); (5) providing incentives for capturing rainwater and reusing treated wastewater (resource management and market reform); (6) reducing peak load demand by creating incentives for distributing use across time or by integrating water storage into the delivery system in areas where capacity is at its peak; and (7) reducing water leaks by upgrading the system.

All these measures reduce water consumption; the energy requirements for pumping; and the load-bearing requirements (and, therefore, design specifications) of pipes and pumps, a major component of water system costs. On the supply side, if water system investments are being planned, then the design and layout of the pipes and distribution network and the location of the treatment plant should be undertaken with a view to energy and spatial efficiency. (For instance, topography and its relationship to the location of demand are often examined to foster the efficient use of gravity in water and wastewater networks.)

Collaboration: Synchronize policies among all the stakeholders

The best approach is to help everyone row in the same direction. All stakeholders and project partners bring a unique combination of policy tools based on their mandates, skills, and resources. Part of the challenge in cities implementing new projects is ensuring that all stakeholders have aligned their existing policies and programs and are using their particular strengths to support the project goals and strategies. By collaborating with senior levels of government, local utilities, private sector corporations, and nongovernmental organizations, one creates the potential for a broad and diverse suite of policy tools. A collaborative process may be able to identify potential actions for the public at large and for individuals with special talents or interests.

Alignment: Develop consistent policies aligned with goals and strategies in the planning framework

All new policy should be based on the relevant goals and strategies identified in the long-term planning framework and should use them as a rationale. The appropriate references may be included directly in the policy document. Sometimes, policy changes must occur within the natural rhythm of policy review, and this may result in delays. However, proposed changes may be worked out earlier in the cycle and placed in queue in the review process. For an increase in alignment, institutional reform may also be necessary, especially if patterns of development are locked in to networks of public and private groups (see chapter 2).

We have seen that the shape of urban spatial form is significant. The interaction of government action (transportation investments, land and tenure regulations, and taxation) and market forces is complex, and this interaction shapes cities spatially. Table 1.2 represents an attempt to summarize the complex interaction between government action and the shape of a city. Of course, much depends on the specifics of a particular case, and the table might not be applicable in all cases. For instance, the preservation of sensitive areas through urban growth boundaries may be combined with an increase in the floor area ratio and in the permissible transfer of development rights in a particular context so as not to cause necessarily a spike in land prices.

Most important in the one-system perspective is the fact that most government actions listed in table 1.2 have limited objectives and do not reflect any consideration for the impacts on land supply and demand, the shape of the city in the long term, and the attendant implications for economic and resource inefficiency. For instance, in the construction of ring roads, the objective is usually to alleviate congestion by allowing through traffic to bypass the city center. Little thought is given to the impact on the supply and the price of land.

Because the objectives of urban regulations and investments do not consider the one-system approach, it is not surprising that many government actions are contradictory. For instance, in Bangalore, India, the local government finances a bus rapid transit system that tends to concentrate jobs in the center of the city. At the same time, the floor area ratio has been kept lower in the central business district than in the suburbs, thereby preventing the concentration of jobs in the central business district, which would have been the justification and rationale of the bus system in the first place.

Table 1.2 Impacts of Government Actions on Land Markets, the Size of the Informal Sector, and the Spatial Structure of Cities

| Government action | Market reaction | | | | Impact on size of informal sector | Spatial Impact | | | |
| | Land supply | | Land price | | | Dispersion | | Concentration | |
Sector	Center	Suburbs	Center	Suburbs		Population	Jobs	Population	Jobs
Transport infrastructure									
Improving or/ and building radial roads		(+)	(+)	(-)	(-)			(+)	(+)
Building ring roads		(+ +)	(-)	(-)	(-)	(+)	(+)		
Building transit in radial pattern		(+)	(+ +)	(-)	(-)			(+)	(+ +)
Building transit in grid pattern		(+)	(-)	(-)	(-)	(+)	(+)		
Land use regulations									
Low floor area ratio			(+ +)	(+ +)	(+ +)	(+)	(+)		
High minimum plot area		(-)		(+)	(+ +)	(+)			
High standard of land development		(-)		(+)	(+ +)	(+)			
Long approval process for building permits	(-)	(- -)	(+ +)	(+ +)	(+ +)			(+)	(+)
Restrictive zoning practices	(-)	(- -)		(+ +)	(+ + +)				
Setting up an urban growth boundary (UGB)		(- -)		(+ +)	(+ +)	(?)		(?)	
Land tenure									
Large government land holding	(- -)	(- -)	(+ +)	(+ +)	(+ +)	(+)	(+)		
Rent control	(-)		(+ +)		(+ + +)				
Restriction on land transactions in periphery		(- -)	(+ +)	(+ +)	(+ + +)			(+)	(+)
High stamp duty on land transactions		(-)			(+ + +)				

Source: Bertaud (2009).
Increase = +; decrease = -; (?) = not known.

This type of contradictory action between two branches of local government—transportation and land use planning in this case—is rather typical. Transportation engineers want high densities along transit routes to ensure a large number of passengers for the transit they design. Planners, faced with congestion in the city center, find it easier to regulate a decrease in densities to alleviate congestion. This is where a planning framework may be valuable. A framework helps ensure that the misalignment of actions is drastically reduced.

Targeting of policies: Recognize the different needs of existing urban areas and new development

One of the biggest factors influencing the sequencing of investments and their capital costs is the focus of the development, whether a newly urbanizing area or an existing part of the city. Most cities include both types of situations, and it is important to adjust and target the strategies accordingly.

NEW DEVELOPMENT

In newly urbanizing areas, the extent of one-system integration is wide open. The major constraints may be the financial resources and the capacity of the design team. The clear advantage of new urbanization is the opportunity to apply the best land use practices and spatial design principles and to integrate land use planning and the design of infrastructure systems. The stage may be set for cost-effective, incremental urbanization through optimal sequencing. Reserving rights-of-way for roads and services is easier, as is the allocation and designation of land for key government and utility functions and open spaces.

An example in Freiburg, Germany, is the alignment of transit services with land development planning. Because occupancy permits are not granted for new residences until light rail transit services have begun operations in a block, newcomers are discouraged from using cars for commuting. The road-building requirements in developments are thus kept to a minimum.

However, the pace of change may be a complex issue governed by the individual agendas and financing capacities of various landowners and government actors. In most cities, one of the biggest roadblocks to the implementation of well-thought-through spatial plans in new areas is the ground-level realities of landownership and the limitations of the city's influence on land and of the city's finances. Special policies may be required to help unorganized landowners cooperate and to avoid incremental and largely unplanned expansion into new areas. An example of such a policy is urban land pooling and land readjustment. This method is particularly interesting because it tackles two problems at once: land and finance. It is briefly described in box 1.5.

THE RETROFITTING AND REDEVELOPMENT OF EXISTING AREAS

One of the difficulties we all face when we are confronted by urban problems is the illusion of permanence. The physical reality of buildings, roads, and trees conveys a strong message that only through superhuman effort will radical changes occur. But, of course, the reality is almost the reverse. Maintaining neighborhoods in their current form, delaying the deterioration of buildings and roads, and providing services to all residents and businesses typically require vast amounts of energy and time on a day-to-day basis. In fact, the operating and maintenance costs of many city neighborhoods are often so high that it is possible to justify a complete retrofitting of neighborhoods and, in some cases, even redevelopment if the disruption to people's lives and businesses is not an issue.

Badly planned cities represent a constant drain on resources. In dealing with existing urban areas, cities may rely on a range of measures to enable the existing built form to perform much more effectively. The measures

Urban Land Pooling and Land Readjustment

Urban land pooling and land readjustment are innovative techniques for managing and financing urban land development. Local and central governments are applying such techniques to assemble and convert rural land parcels in selected urban fringe areas into planned road layouts, public utility lines, public open spaces, and serviced building plots. Some of the plots are sold for cost recovery, and the other plots are distributed to landowners in exchange for rural land parcels. For viability, the value of the urban plots distributed to landowners after subdivision needs to be significantly higher than the value of the plots before the project begins.

In a typical project, the authorized land pooling and readjustment agency selects and designates the urban fringe area to be developed and identifies the land parcels and owners to be included. A draft scheme is then prepared to plan, define, and explain the project and demonstrate financial viability.

Majority landowner support for each proposed project is a key requirement for the successful application of the technique and is therefore an important consideration in selecting project sites. Although the emphasis is on landowner agreement and support for each proposed project, the land pooling and readjustment agency must also be able and willing to use the government power of compulsory purchase against any minority holdout landowners in the designated project area if this becomes necessary.

The sharing of project costs and benefits among the landowners, such as increased land values, is based on their contributions of land to the project. The calculation of each landowner's share may be based on the area of his or her land parcel relative to the total land area or on the estimated market value of the land relative to the estimated market value of the total area.

There is an important legal difference between land pooling and land readjustment in landownership. In a land pooling project, land is legally consolidated by transferring the ownership of the separate land parcels to the land pooling agency. Later, the ownership of most new building plots is transferred back to the landowners. In a land readjustment project, the land parcels are only notionally consolidated, and the land readjustment agency has the right to design services and subdivide them on a unified basis. Then, at the end of the project, the landowners exchange their land parcel title documents for the corresponding documents of the new building plots.

There are many successful examples of such projects, for instance, in Indonesia, Japan, and the Republic of Korea. A similar process of land pooling and land readjustment is practiced in the state of Gujarat in India, where the projects are known as town planning schemes. (See figures 1.26 and 1.27, which illustrate the before and after scenarios of land readjustment in Gujarat.)

Source: Mehta and Dastur (2008).

---

usually fall into two categories: retrofitting and redevelopment. Retrofitting existing city areas entails working with the existing built stock and infrastructure and making improvements to enhance performance, without redeveloping the entire area. Examples of retrofitting measures include implementing end use efficiency in the energy and water sector; reducing, reusing, and recycling waste; and adapting existing transportation infrastructure (roads) to more efficient uses (for instance, by designating routes for bus rapid transit and lanes for bicycles).

Redevelopment entails demolishing and rebuilding certain areas of the city and is typically more complicated. Redevelopment is challenging because of the political, social, and economic costs of making changes in existing land uses and structures. New zoning or transportation corridors cannot be imposed unilaterally or quickly. Nor is it easy to upgrade the systems serving so many unconnected buildings. Many stakeholders must participate in decision making. Projects require longer time frames so that communities may adjust. An incremental approach may be required, which makes the sequencing of strategies difficult. Development may need to include, for example, complex arrangements for slum upgrading and arrangements for new utilities and rights-of-way. The pace of change may need to evolve incrementally in sync with the natural turnover rate for the stocks, or it may be necessary to wait until the service quality and operating costs justify large-scale urban redevelopment.

However, cities may explore creative and cost-effective ways of remodeling the distribu-

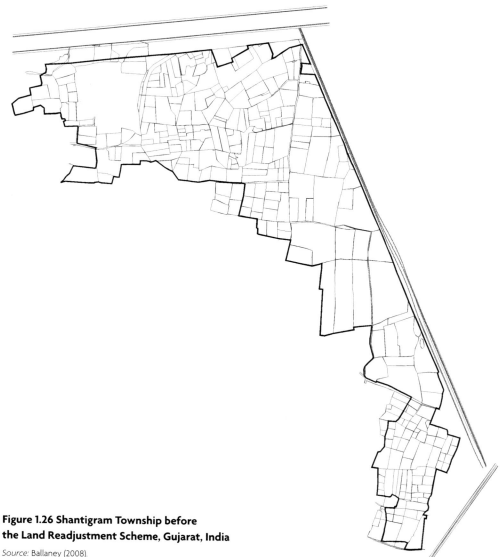

**Figure 1.26 Shantigram Township before
the Land Readjustment Scheme, Gujarat, India**

Source: Ballaney (2008).
Note: This figure and the following illustrates the 'before' and 'after' scenarios of Land Readjustment in Gujarat, India.

tion, density, and use of the existing built form by increasing the floor area ratio; allowing the transfer of development rights (see the Curitiba case in part 3); rezoning and changing land use patterns; and, more important, revising and enforcing building codes and standards. These steps might create incentives for private redevelopment efforts. In some cases, land readjustment may be used, though it is much harder to convince stakeholders in an existing urban community to demolish their properties for the purpose of the city's redevelopment if they have already built up urban structures. However, if accompanied by a sizable increase in the floor area ratio or, in the case of slums, by formal recognition or the introduction of basic services such as drainage, water, and sanitation, the returns may make economic sense. Redevelopment projects at a larger scale for certain areas and districts of a city have also been successful in enhancing the sustainability of existing areas. Such is the case of the redevelopment of old manufacturing sites as waterfront residences. Because the old sites are not

Figure 1.27 Shantigram Township: Final Serviced Land Parcels for Sale, Gujarat, India

Source: Ballaney (2008).

being used, it is easier to coordinate projects and gather consensus. The redevelopment of existing residential neighborhoods is significantly more disruptive and less likely to be supported by consensus. In such cases, retrofitting existing structures or creating incentives by increasing the floor area ratio in exchange for greater compliance with new resource-efficient constructions is often more realistic.

Stepping Stones for the One-System Approach

Provide just-in-time training and capacity building

The city leadership must provide multiple opportunities for local professionals to become comfortable with the one-system approach. An Eco² catalyst project, for example, represents a concrete opportunity to train professional staff in new procedures and methods. Ideally, the training occurs in a timely fashion because new skills need to be immediately applied or they may be lost.

A special effort is warranted to ensure that relevant institutions and people become well informed, supportive, and capable. Training may include invitations to local consultants and firms who would benefit from exposure to the catalyst project and any new approaches. Without training, these local experts may tend to obstruct projects and refuse support. Cultivating local expertise is an investment and will ultimately determine what may be achieved citywide and in other cities in the country.

A training program in the one-system approach may benefit from a variety of resources:

Other cities: Interested cities may access the critical expertise of other cities and planning institutes or agencies. It may be especially helpful to learn from other cities that have successfully implemented the approach and created the institutional framework for sustaining such efforts.

Part 2, methods: To apply the one-system approach in the design, analysis, formulation, and implementation of the options outlined in this chapter, it is necessary to develop capacity and proficiency in the use of key methods and tools, some of which are introduced in the city-based decision support system (see part 2). The full scope of integration options may be explored using methods and tools to develop and assess the performance of integrated solutions. Familiarity is required, especially in material flow analysis and overlay mapping.

Part 3, case studies and sector notes: The sector notes in the Field Resource Guide provide more information on individual sectors and more specific and detailed ideas. The case studies on best practice cities that are also featured in the Field Resource Guide may introduce staff and consultants to examples of approaches in the real world and also reveal critical lessons learned.

Conduct a series of preparatory integrated design workshops

Integrated design workshops create important opportunities for planners, designers, and engineers to come together and use new methods and information. The number and the scope of the workshops vary with the situation. Sometimes, it is best to plan one or two short workshops to clarify goals, set targets, and share information among stakeholders. The directions and priorities may be refined and aligned with the city's shared planning framework. The framework (if already established) may be used within workshops to orient discussions or stimulate creative thinking and then to evaluate preferred strategies and actions. Workshops may also examine analytical methods, producing, for example, a business-as-usual scenario for benchmarking purposes. This scenario may include a material flow analysis and meta diagrams, overlay mapping, risk assessment, and other analytical exercises. Workshops may also be used to review and finalize a design brief in preparation for more intensive design work.

Explore design solutions, and prepare a concept plan for review

An integrated design process should be used to generate alternative proposals for designing, constructing, and managing the project. A multiday urban-systems design charrette (a type of intensive workshop discussed in part 2) is a tool that may facilitate the integrated design process, helping to generate creative and effective proposals in the shortest time. A well-planned systems design charrette often produces a final concept plan that is more than 90 percent complete. A charrette involving regulatory and management personnel may help reveal existing policies that may need to be revised or removed to enable innovation. A design charrette may benefit the project indirectly by generating goodwill among stakeholders and by helping experts become familiar with new concepts and technologies. The integrated design process should culminate in a recommended concept plan for implementation, including any policy reforms.

Align policy tools among all stakeholders to ensure successful implementation

Use the procedures outlined in this chapter to implement the project in an integrated fashion. This may help sequence investments, enable contributions from partners and residents, coordinate strategies among stakeholders, and align and target policies to match the planning framework. A collaborative exercise helps all interested parties explore how to use complementary policy tools to implement the concept plan and achieve the intended outcomes. A strategic action plan may be prepared to clarify who is responsible for each of the various tasks and to indicate how policies interact. Where appropriate, a feasibility plan and a detailed master plan may be prepared with specifications and guidelines for each element and for each phase of the work.

References

Baccini, Peter, and Franz Oswald. 1998. *Netzstadt: Transdisziplinäre Methoden zum Umbau urbaner Systeme.* Zurich: vdf Hochschulverlag.

Bai Xuemei. 2006. "Solar-Powered City: Rizhao, China." In *State of the World 2007: Our Urban Future,* ed. Worldwatch Institute, 108–9. Washington, DC: Worldwatch Institute.

Ballaney, Shirley. 2008. "The Town Planning Mechanism in Gujarat, India." World Bank, Washington, DC.

Barry, Judith A. 2007. "Watergy: Energy and Water Efficiency in Municipal Water Supply and Wastewater Treatment; Cost-Effective Savings of Water and Energy." Handbook. Alliance to Save Energy, Washington, DC. http://www.watergy.net/resources/publications/watergy.pdf.

Bertaud, Alain. 2009. "Urban Spatial Structures, Mobility, and the Environment." Presentation at "World Bank Urban Week 2009," World Bank, Washington, DC, March 11.

Bertaud, Alain, and Robert W. Poole, Jr. 2007. "Density in Atlanta: Implications for Traffic and Transit." Policy Brief 61, Reason Foundation, Los Angeles.

Boyden, Stephen, Sheelagh Millar, and Ken Newcombe. 1981. *The Ecology of a City and Its People: The Case of Hong Kong.* Canberra: Australian National University Press.

Kenworthy, Jeffrey R. 2006. "The Eco-City: Ten Key Transport and Planning Dimensions for Sustainable City Development." *Environment and Urbanization* 18 (1): 67–85.

Kirby, Alex. 2008. *Kick the Habit: A UN Guide to Climate Neutrality.* Nairobi: United Nations Environment Programme.

Lahti, Pekka, ed. 2006. *Towards Sustainable Urban Infrastructure: Assessment, Tools and Good Practice.* Helsinki: European Science Foundation.

Lovins, Amory B., E. Kyle Datta, Thomas Feiler, Karl R. Rábago, Joel N. Swisher, André Lehmann, and Ken Wicker. 2002. *Small Is Profitable: The Hidden Economic Benefits of Making Electrical Resources the Right Size.* Snowmass, CO: Rocky Mountain Institute.

Mehta, Barjor, and Arish Dastur, eds. 2008. "Approaches to Urban Slums: A Multimedia Sourcebook on Adaptive and Proactive Strategies." World Bank, Washington, DC.

Motloch, John L. 2001. *Introduction to Landscape Design,* 2nd ed. New York: John Wiley and Sons.

Revi, Aromar, Sanjay Prakash, Rahul Mehrotra, G. K. Bhat, Kapil Gupta, and Rahul Gore. 2006. "Goa 2100: The Transition to a Sustainable RUrban Design." *Environment and Urbanization* 18 (1): 51–65.

Rutherford, Susan. 2007. "The Green Infrastructure: Issues, Implementation Strategies and Success Stories." West Coast Environmental Law Research Foundation, Vancouver, Canada. http://www.wcel.org/wcelpub/2007/14255.pdf.

Tortajada, Cecilia. 2006. "Singapore: An Exemplary Case for Urban Water Management." Additional Paper, Human Development Report. United Nations Development Programme, New York.

An Investment Framework That Values Sustainability and Resiliency

Chapter 6 introduces the accounting method and framework that are needed to understand the full costs and benefits of projects and policies. It begins with an introduction to the basics of life-cycle costing for cities and the policies and methods that make this possible. Next, the chapter explores the need for an expanded framework for economic accounting by Eco² cities. The framework gives equal consideration to various categories of assets: manufactured capital, natural capital, social capital, and human capital. The chapter explores an expanded framework for risk assessment that incorporates foresight methods, including long-term forecasts for all sectors, and a design philosophy that increases the resiliency and adaptive capacity of city lands and infrastructure. The chapter concludes with suggestions for key actions or stepping stones that might direct the city as it learns to invest in sustainability and resiliency.

The Core Elements of Investment in Sustainability and Resiliency

Incorporation of life-cycle costing

Life-cycle costing (LCC) is a decision support method that helps cities improve project cost-benefit accounting measures and derive more accurate estimates of the financial and economic costs and benefits associated with any development project. Life-cycle costs include all the costs incurred by a project throughout its life cycle, including construction, operation, maintenance, rehabilitation, disposal, and replacement.[1] Part of the challenge faced by all cities today is the integration of cash flows over time. This includes optimizing capital and operating costs, ensuring adequate cash flows over the longer term, and recapitalizing investments so that funds are available for the replacement of assets at the end of a project's life cycle.

LCC is especially important for the long-lived investments that are a large part of city infrastructure and land development. LCC is important for decisions regarding fleets, which are decisive in determining new vehicle acquisitions; infrastructure, which is especially relevant for water, transportation, and energy systems; land use planning as it pertains to infrastructure costs; civic buildings, which are relevant for premium efficiency targets for new and existing stock; and residential and commercial buildings.

LCC requires that life expectancy and rate of deterioration be estimated for each type of asset. It then becomes possible to quantify maintenance and rehabilitation requirements. The maintenance of city infrastructure systems—pipes, facilities, pumps, and roads—may be extremely costly and may have significant impacts on the cash flow and financial sustainability of any project. It also affects the fiscal health of a city; in fact, the lack of policies based on LCC has left many cities essentially bankrupt and unable to manage assets.

Operating and maintenance costs for long-lived elements such as buildings and pipelines can represent over 90 percent of life-cycle costs. The City of Hamilton, Canada, has estimated that initial construction accounts for only 8 percent of a civic building's cost over its 30–40 year life, whereas operation and maintenance account for 92 percent. It is obviously dangerous to place too much emphasis on initial capital costs in making large public investments in city infrastructure and buildings. Nonetheless, it is still common worldwide for cities to have separate capital and operating budgets and to make investment decisions based on the initial capital investment costs, without considering the net present value of the future flows of the associated operating and maintenance costs. If, however, life-cycle costs are well quantified for a variety of development scenarios, they may be minimized at the design and implementation levels of land use and infrastructure planning.

LCC makes possible a more prudent and responsible approach to the long-term financing of investments. The calculations may be rapid and comprehensive. For example, a new neighborhood development project may be analyzed for a variety of densities and configurations, and then each scenario may be compared in terms of the capital and operating costs for utilities and services, including roads, water, sewerage, garbage, schools, recreation facilities, public transit, private vehicle use, fire protection, and policing. The interest rates for borrowing, tax rates, and service revenues may be calculated for different development plans and fiscal policies.

Life-cycle costs are typically annualized (converted into an annual cost) over a long period (75 years in the case of the neighborhood construction project in Hamilton) allowing for the operation, maintenance, and replacement of all utilities. All costs may be allocated on a per household basis for residential developments or normalized for standard office space.

Part 2 includes details on how the LCC method may be applied to Eco² cities and information on simple spreadsheet-based computer tools that make LCC easy and rapid. The tools include a preset list of many life-cycle cost categories that are worth considering in development projects, but that are typically ignored. All the default values may be adjusted to match the historical costs for any specific country and community.

Included in part 2 is an example of the help provided by an LCC tool to the City of Fort St. John, Canada, in assessing the potential costs and benefits of a proposed concept plan for a sustainable neighborhood. A design workshop had proposed smaller lots; narrower streets; more tightly packed buildings; greater diversity in building types; more open public space between buildings; and more well integrated, multiuse designs for open spaces (including greenways, green infrastructure for storm water management, community gardens, all season pathways, and a large commons around a school and community center). The proposed design represented a significant departure from conventional neighborhoods in the city. It was thus necessary to move beyond debate and opinion into a comprehensive analysis of costs and benefits.

Authorities in Fort St. John compared the new approach with a base case scenario that had been modeled on existing adjacent neighborhoods. Capital costs were estimated and allocated to each household. Operating costs were calculated, including the cost of water, roads, sewerage, school transit, recreation facilities, and police and fire departments. In the final analysis, the LCC assessment helped clarify the potential gains of the new approach. Per household capital costs averaged US$35,000 less compared with the base case; annualized operating cost savings were estimated at US$6,053, a reduction of more than 25 percent relative to the base case. Of course, the sustainable neighborhood plan had potential benefits unrelated to capital and operating costs, including improvements in livability, streetscapes, social interaction, and amenities. However, the comprehensive financial analysis helped win over the community and provided the city council with a stronger argument for defending changes in standard practices. All politicians find it easier to make the right decisions and to stand strong in the face of vested interests or institutional inertia if they are provided with simple, transparent arguments about ways to save taxpayers money and reduce liabilities. This is an important function of LCC.

Reserve funding

One of the most effective tools in sustainable financing is the reserve fund. The aim of a reserve fund is to set aside money incrementally and gradually so that sufficient funds are available to finance upgrades and replacement at the end of a project's life cycle. Such an approach not only helps ensure the viability of an investment and its various components, but also avoids the dumping of huge liabilities and a potential financial crisis onto future generations. The inadequate capitalization of infrastructure systems also unfairly shifts maintenance and replacement costs toward the end of a system's life. Reserve funds make good economic and ethical sense.

The biggest challenge is to keep the reserve fund truly reserved. The funds are subject to raids by those who see opportunities to spend the funds elsewhere. Consequently, reserve funds must be earmarked and legally protected.

A reserve fund is particularly necessary in non-revenue-generating projects. It is important to retain an appropriate amount in reserve as determined by the overall investment plan. A larger reserve is not necessarily better given that the fund is exposed to inflation risk. To reduce the amount of such a fund, similar assets should be pooled, and as far as possible, the annual investment level should be maintained.

How much is sufficient for the reserve fund? In the case of a reserve fund for educational fa-

---

Tokyo Waterworks: How to Finance a Water Pipeline Replacement Project

Fees and charges are important to revenue-generating enterprises, such as water companies, in considering an appropriate level of reserve funding. The Tokyo Waterworks, which serves 12.5 million people in metropolitan Tokyo, has been financing operating expenses and capital expenditures by relying on water tariff revenues. Various reserve funds have been set aside to cover fluctuations in these costs. Currently, the utility is facing the daunting task of replacing old water pipes beginning in 10 years time. The total investment is estimated at around ¥1 trillion (US$10 billion), which represents 40 percent of the utility's total assets of ¥2.5 trillion (US$25 billion) in current yen. To meet this challenge, Tokyo Waterworks started identifying ways to level out the ¥1 trillion planned investment over a reasonable period by planning for maintenance and rehabilitation well ahead of the project and establishing a detailed construction plan. Meanwhile, the utility has already started accelerating debt repayments so that outstanding debt may be maintained at the current level of ¥0.5 trillion even after project financing has been undertaken. The accelerated repayments are being covered by water tariff revenues even though the Tokyo metropolitan government lowered the water tariff on January 1, 2005. The utility plans to finance the ¥1 trillion replacement project by implementing a reasonable tariff adjustment.

cilities in Tokyo's Chuo Ward, the fund covers the total investment cost that will be required in a few years. However, even if the fund does not cover the total required investment cost entirely, it may be considered sufficient if Chuo Ward is able to mobilize additional funds from other sources. Important sources of external funds for cities are municipal bonds and bank borrowings. To raise these external funds in a timely manner, cities should keep the terms and amounts of their debt within borrowing capacity. They should also level the investment requirement over long investment periods so that annual capital funding requirements are minimized. The LCC method provides a useful base for long-term investment planning.

Equal attention to all capital assets: An expanded framework for accounting

A persistent challenge in accounting for the cost of urban development projects is the measurement and valuation of the many indirect costs and benefits. Economic analysis has evolved over the past few years because of attempts to understand these indirect costs and to provide decision makers with an assessment that more accurately reflects the true costs and benefits of any particular option. For example, cost-benefit analysis, the primary method for assessing economic viability, has been expanded to incorporate many indirect

effects into monetary values. Cost-effectiveness, the other standard method currently used to assess the economic viability of a project, has also been expanded to include the examination of additional indirect benefits. Despite the efforts toward fuller cost-benefit accounting, most development projects are undertaken without a firm grounding on the real nature of the impacts on people, ecologies, and social systems. Many of the indirect costs of concern to communities cannot easily be measured or explained, nor can they be easily converted into credible monetary values. The proper techniques for converting impacts into monetary values have been debated for many years, and appropriate solutions continue to be sought.

A more comprehensive economic analysis requires that greater attention be paid to environmental accounting as a separate rigorous method. Every project needs a standard protocol for assessing environmental effects by category on the basis of well-defined methods such as input-output analysis, life-cycle analysis, and material flow analysis. One example of an expanded approach to the quantification of impacts is the environment load profile adopted in Hammarby Sjöstad, Stockholm. (See part 2 for more on the environment load profile.) A separate set of indicators may be used to express each category of effect in parallel with the economic analysis. The effects may sometimes be added together to advantage; air quality, for example, is commonly addressed in terms of an air quality index that bundles multiple factors such as the quantities of particles, organic compounds, and nitrogen oxides.

A number of techniques have been developed in an attempt to evaluate a wider range of environmental and ecological effects so as to arrive at one or several overarching measurements of natural capital. A notable example is the ecological footprint, which converts energy and material use into the total area of productive land that would be required to sustain such flows indefinitely. Officials and experts in many cities and new neighborhood developments

> "For too long now ministries of finance and planning have paid scant attention to the exploitation of the natural resource base or the damaging effects of environmental pollution, while countries have been developing National Environmental Action Plans that read as if they were written *by* the environment ministry *for* the environment ministry, with no links to the economic ministries."
>
> *Source:* World Bank (1997: 7).

have appreciated the usefulness of a single rating of this type and have calculated ecological footprints as an indicator of the overall impact on natural capital. For example, London has calculated that people in the city require, on average, 6.6 green hectares of land per person to support their lifestyles, which is more than three times the amount available per person on a planetary scale (figure 1.28). London discovered that its combined ecological footprint is 293 times the land area of the city, mostly as a consequence of the high rates of food and material consumption.

All techniques for adding up or aggregating ecological impacts into a simple metric suffer from a number of significant problems (Mcmanus and Haughton 2006). For example, the ecological footprint fails to address well the important issue of water flows, which vary so much in value depending on location. A city that annexes agricultural land, thereby increasing its administrative boundary, suddenly appears much less of a burden on the planet, though the reverse may be true. Multifunctional land use—encouraged by Eco² —is ignored if all land is divided into discrete categories for ecological footprint analysis. By using a single unit, such as hectare of ecological land, the footprint ignores the major differences in ecological system values, including factors such as biodiversity, species scarcity, and habitat uniqueness. In fact, all indicators that aggregate impacts tend to ignore the many local

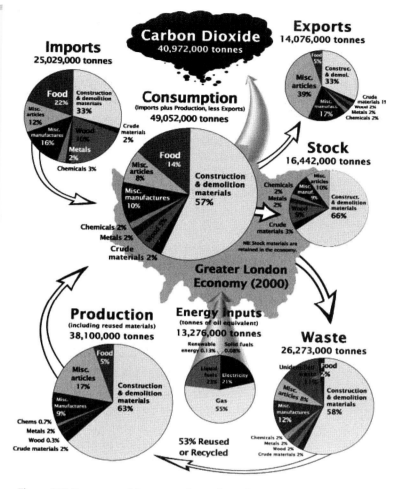

Figure 1.28 Summary of Resource Flows through London, 2000

Source: Best Foot Forward Ltd. (2002).
Note: This summary for greater London reveals all inputs and outputs and helps explain why the city's ecological footprint is approximately 300 times the size of the city's land.

factors that relate to the quality of ecosystems, the sensitivity of local environments to emissions and wastes, and the differences from place to place in the value of natural capital.

Despite these types of methodological problems, one should seek a method that allows a quick summary of the range of impacts arising from any development scenario. The method should rely on standardized measurement protocols for comparability, as well as simple graphical tools, so that one may rapidly communicate the basics to interdisciplinary teams of designers and decision makers. European Cooperation in Science and Technology, a European intergovernmental framework program, has struggled with the challenge of assessing environmental effects during a multi-year effort to analyze and describe sustainable infrastructure projects in cities throughout Europe. After reviewing all the options for assessing impacts, the program experts chose a simple matrix to summarize key effects (table 1.3). The final publication, *Towards Sustainable Urban infrastructure: Assessment, Tools and Good Practice*, describes 44 sustainable infrastructure projects and a matrix for each project (see Lahti 2006). The publication concludes that a holistic assessment of sustainability with many dimensions and numerous impacts requires a technique and tool capable of reviewing all relevant aspects in a compact space, that is, hopefully even in one page through a visually effective presentation.

Eco² cities need a framework for evaluating the costs of projects that is sufficiently flexible to accommodate a wide range of measurements and, yet, is sufficiently balanced to ensure that the trade-offs and impacts on critical thresholds and targets are well understood. The emphasis on integration at many levels means that

a broad and balanced assessment is much more important than might otherwise be the case. Eco² requires a framework that is designed to reveal not only who benefits and who pays specific costs, but also how well a project has maximized benefits of all types. The framework must be transparent, allowing a mix of professionals and residents to easily follow what is actually being measured, why it is being measured, and how the numbers relate. The framework needs to combine categories of benefits and costs so that they may be tracked as a whole and so that indicators on ecological health, for example, may be given equal consideration with indicators on economic wealth. Fortunately, many economists and communities have been experimenting with frameworks over the last 10 years, and it is now possible to learn from best practice and adopt a framework for accounting that is suitable for Eco² cities.

Protection and enhancement of capital assets

An appropriate method for use with Eco² cities is the four-capitals approach outlined by Ekins, Dresner, and Dahlström (2008). The method

Table 1.3 A Design Assessment Matrix

ECOLOGY		ECONOMY		SOCIAL ASPECTS	
Are emissions to air, water, and soil within the restrictions set locally and internationally? Are emissions decreasing?	↑	Are the cost-effectiveness and cost-benefits of the system reasonable compared with other systems? Are they reasonable compared with other needs in the city and to political goals?	↗	Have the planning and decision making for the infrasystem been carried out in a democratic and participatory manner?	↘
Is the use of natural resources reasonable relative to other, comparable systems? Is the use decreasing (for example, fossil fuels, water, phosphorus, or potassium)?	↘	Are citizens willing to pay for the services offered? Are the services affordable for all citizens?		Are the function and consequences of the system transparent for and accepted by citizens? Is the system promoting responsible behavior by citizens?	↗
Is the system allowing a reasonable level of biodiversity for the area studied? Is the biodiversity increasing?	↗	Are the organizations that finance, maintain, and operate the system effective?	↗	Is the system safe for citizens to use (hazards, health, and well-being)?	↗
Is the system more or less ecologically sustainable than a conventional system?	→	Is the system more or less economically sustainable than a conventional system?		Is the system more or less socially sustainable than a conventional system?	→

Source: Based on Lahti (2006).

Note: The matrix presented in the figure has been used in many case studies of sustainable infrastructure in Europe. It is intended to provide decision makers with instant and reliable insight into the sustainability of any design option. The arrows indicate performance in a sample project.

has evolved by combining a new approach to environmental economics, developed by David Pearce (2006), with a number of assessment tools that have been used in urban development. It is sufficiently flexible to include any type of measurement, and yet, it is well balanced. It has been successfully applied in a number of sustainable planning projects in Europe.

Most economic analysis incorporates an inventory and valuation of capital assets; however, the focus is primarily on manufactured goods and systems that produce or facilitate the delivery of goods and services. This kind of capital is referred to as manufactured capital and includes the hard infrastructure of cities.

The four capitals method begins by recognizing that benefits may flow from many sources other than manufactured capital. We need to account for the quality of labor (human capital), the networks through which labor is organized and that create the context for economic activity (social capital), and the natural resources and ecological systems that provide inputs into the economic process and maintain life on Earth (natural capital). A more detailed definition of these four capitals is provided by Ekins and Medhurst (2003):

1. **Manufactured (or human-made) capital** is traditional capital, that is, produced assets that are used to produce other goods and services. Examples are machines, tools, buildings, and infrastructure.

2. **Natural capital** includes, in addition to traditional natural resources (such as timber, water, energy, and mineral reserves), natural assets that are not easily valued monetarily, such as biodiversity, endangered species, and the ecological services provided by healthy ecosystems (for example, air and water filtration). Natural capital may be considered as components of nature that may be linked directly or indirectly to human welfare.

3. **Social capital,** like human capital, is related to human well-being, but on a societal, rather than individual, level. It consists of the social networks that support an efficient, cohesive society and that facilitate social and intellectual interactions among society's members. Social capital refers to those stocks of social trust, norms, and networks that people may draw upon to solve common problems and create social cohesion. Examples of social capital include neighborhood associations, civic organizations, and cooperatives. The political and legal structures that promote political stability, democracy, government efficiency, and social justice (all of which are good for productivity, as well as desirable in themselves) are also part of social capital.

4. **Human (cultural)** capital generally refers to the health, well-being, and productive potential of individual people. Types of human capital include mental and physical health, education, motivation, and work skills. These elements not only contribute to a happy, healthy society, but also improve the opportunities for economic development through a productive workforce.

All four types of capital are defined and identified through the flow of the benefits they provide. Sustainable development mostly revolves around the maintenance or increase of the four capitals so that the flow of benefits is sustained indefinitely. Some trade-offs may be considered acceptable; for example, a reduction in the net area of ecological systems may be offset by increases in the net productivity of ecologies resulting from good design and management practices. However, many systems (such as ecologies) and assets require that critical thresholds be respected or the system will begin to break down. For example, smaller green spaces may be more productive, but they may fail to provide sufficient habitat for some species, and as a result, biodiversity declines.

The four capitals method is a good choice for Eco^2 cities for the following reasons:

1. It incorporates critical intangible assets into the decision-making framework.

2. It considers externalities (indirect costs and benefits) in a more comprehensive fashion than other options now available.

3. It allows for easy comparisons of different categories of costs and benefits and allows cities to focus on critical thresholds (for example, limits that should not be crossed) and to recognize the trade-offs that frequently arise between one type of asset and another.

4. It fits well into the economic accounting already in place in many cities because it uses an inventory of capital assets, and it makes use of much of the data that are already collected by cities on a regular basis.

5. It reinforces the important concept that assets need to be conserved and enhanced because they provide the flows of goods

and services that ultimately contribute to human well-being.

Use of indicators to set targets and monitor impacts

Monitoring the capital assets of a city and balancing the trade-offs among types of capital require standardized measurements or indicators that correspond to the capacity of assets to provide goods and services. Indicators that cover all four capitals are referred to as indicators of sustainable development. They include monetary values, when these are available and appropriate, and also many physical dimensions.

Table 1.4 provides an example of a few of the indicators used by various cities participating in a European project for sustainable development planning. Based on the European experience, the quality of indicators tends to vary by capital. Manufactured capital tends to

Table 1.4 Sample Indicators in the Four-Capitals Approach

Manufactured capital	• GDP per capita • Gross fixed capital formation • Employment (by sector) • Change in real income	• Travel times and average speeds • % population connected to internet • Agricultural produce • Inflation rate
Natural capital	• CO_2 emissions • Air quality • Stocks of endangered species • Value per drop of water	• Quantity of collected waste • Green areas (km^2) • Energy use per capita • Resource efficiency
Social capital	• Wage differentials and poverty • Disparity between income of average, highest and lowest deciles • Male/female wage differentials • Number of social welfare recipients	• Districts with special development needs • Out migration of young people • Number of cooperative, inter-municipal projects and strategies • Crime rates
Human capital	• Employment growth and rates • Creation of new high skill jobs • Levels of education and vocational training • Public and private R&D expenditures	• Numbers of patent applications • Number of business start-ups • Improvement in human health • Participation rate in education and training

Source: GHK (2002).
Note: The sample indicators were used in 19 urban regions in Europe as part of a sustainable development assessment. CO_2=carbon dioxide; km^2=square kilometer; R&D=research and development.

be oversimplified because of the use of only GDP measurements. Social capital, however, is measured using too many different indicators. Human capital is difficult to measure directly. Natural capital indictors are often difficult to calculate.

The precise choice of indicators for a city and a specific project will vary with circumstances. In general, indicators need to be affordable so that they may be measured on a regular basis. Otherwise, what is the point? They also need to be relevant; so, they need to measure the larger changes that cities are trying to effect. The relevance varies depending on who is going to be using the indicator. For a city council and its city partners, performance indicators are required that help clarify the intended long-term results or performance. A common performance indicator for manufactured capital is GDP per person; another might be the asset value of city-owned infrastructure. The other capitals tend to be more difficult to capture. In the case of natural capital, performance indicators need to address at least the different types of ecological services: sinks (capacity to absorb wastes), sources (capacity to provide useful products and services), and life support (capacity to cycle resources and regulate environments so they support life). In addition to broad performance indicators that measure how well key targets or goals are being achieved, it is helpful to develop a set of indicators for monitoring progress at the strategic level and at the operations level.

Figure 1.29 illustrates how three different levels of indicators correspond to the scope and responsibility of city personnel. As scope narrows, so, too, does the indicator. For example, a new distributed electricity system for a city might need feedback at three levels of detail:

- *Performance:* percentage of residents in the service territory receiving power from the new system

- *Strategic:* percentage of buildings retrofitted according to new energy efficiency standards

What Makes a Good Indicator?

Affordability and practicality: Can the data be collected easily at little or no cost? Is the analysis simple and easily automated?

Relevancy: Do the indicators actually measure the key issues of concern? Do they respond sufficiently to show that progress is being achieved?

Clear explanations and measurement protocols: Is it easy to define what is actually being measured and how?

Comparability: Is this a standard measurement from which other measurements may be derived to provide comparisons and benchmarks of performance?

Aligned with objectives: Is the effort to measure appropriate given the priorities established in the planning framework?

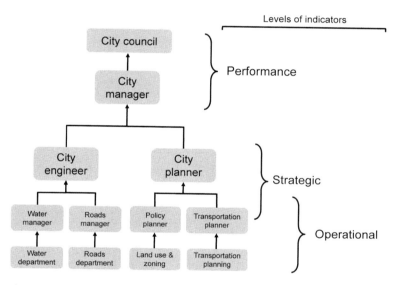

Figure 1.29 Targeted Indicator Type, by Level of City Personnel
Source: Lahti (2006).

- *Operational:* average time required to repair outages

Each project may require a family of indicators because decision makers will be interested in different time frames and levels of detail.

Proactive risk management for all threats

Standard practice in financial risk management involves an analysis of any investment in terms of sensitivity to changes in the key factors used for determining costs and benefits. Each factor has a certain probability of changing over time, with consequences for the financial bottom line. This assessment of risk based on the known probabilities for change in the direct economic factors is referred to as sensitivity analysis. It is the principal risk-assessment method used in urban development projects, and it is an important and necessary part of due diligence. If a 15 percent drop in ridership is sufficient to undermine the financial viability of a new transit system, city leaders will want to know the odds of such an occurrence. Sensitivity analysis is not a replacement for good judgment, but it is a good way to educate decision makers about the variables that might undermine the critical viability of an investment. The other well-known method for risk assessment is the Monte Carlo assessment, which expands the analysis to include the possible correlations between changing variables essentially by making many random changes to variables in combination.

What is missing from these standard risk-assessment methods is the many indirect, difficult-to-measure risks that threaten the viability of an investment. Also missing is the assessment of uncertainties, the factors that cannot be assessed statistically, but that represent significant threats. In a similar fashion to economic analysis, the risk assessment needs to be coupled with methods that expand the scope of the issues or elements examined and rated. In reality, cities today face many threats and hazards that are largely external to financial calculations, but that may nonetheless influence the viability of projects. These include sudden disruptions to systems, such as natural disasters (earthquakes, hurricanes, tsunamis, and so on), and the possibility of rapid socioeconomic-environmental change, such as the recent global financial crisis. Over the next 30 years, for example, it is highly likely that we will witness fundamental changes in energy, communication, transportation technologies, climate, demographics, global markets, and environmental regulations. The onset of epidemics is probable, and the availability of critical resources such as water, food, and fossil fuels will likely be problematic. For a city, 30 years is the blink of an eye. The infrastructure investments planned for the near future will need to perform for much longer than 30 years. But will they? How might Eco² cities assess and improve the overall *resiliency* of development projects?

Expansion of risk assessment to include resiliency and adaptive capacity

Resiliency is a concept traditionally used to describe two characteristics: the robustness of a system (that is, its ability to continue to perform by resisting changing conditions), and the adaptability of a system, (that is, its ability to continue to perform by responding appropriately to changing conditions). Resiliency may be used as a potential design criterion for all urban systems, including built infrastructure, culture, and governance.

The basic idea is that it is possible to manage risk more productively by forecasting the impacts of external forces on urban areas and by designing and operating urban land uses and infrastructure in ways that are inherently more resilient. This means including in any assessment indicators that help designers, managers, and decision makers understand the relative capacity of systems to survive and recover from shocks and rapid change. The World Bank's primer on climate resilient cities provides information on how cities may effectively assess and manage the risks associated with climate change (see Prasad and others 2009).

Elements of resilient design appear to reinforce a number of the ecological design strategies that are so effective at improving efficiency. Remote generating plants, incinerators, treatment plants, and communications facilities are far more vulnerable to catastrophic failure than a network of modular, distributed systems closely integrated into the fabric of the city. Thus, urban security helps reinforce distributed systems, a design strategy already proposed as a way to improve urban resource efficiency and environmental sustainability. The positive synergy between security and efficiency (or resiliency and sustainability) is an important outcome of integrated design solutions.

One measurement of resiliency might be redundancy, a strategy that is an example of ecological design (see chapter 5). Redundancy in urban systems may mean that critical resources are supplied by a variety of systems, each capable of drawing resources from as wide a geographic area as possible. In the eventuality that droughts, floods, or other disasters affect any one area, an alternative source of supply is already in place to meet at least minimum requirements. For each type of critical resource, the region may develop redundancy through a diversity of supply options or through a contingency plan. Redundancy may also need to include the entire supply chain for each critical system, all the way back to the ecological resource. Redundancy may then be provided for the weakest links in the chain. Links are the processes or nodes that provide essential services, wherever they may be located. If we discover nodes that are essential, but not duplicated elsewhere in the system, we have found a weak link.

Redundancy and self-reliance work on different levels. Even links within a region may benefit from contingencies. For energy sys-

tems, for example, this may mean a mix of sources, some local and some renewable. For potable water, this may mean distributed reservoirs and multiple sources of water.

Localized and distributed infrastructure systems may be more flexible and responsive in the face of external threats. To the extent that systems are self-organizing, they do not require lots of external regulation or direction to function or adapt to opportunities or constraints. Such systems may operate by a set of rules, similar to those in the market place, rather than a mechanistic, top-down approach that imposes a final solution from start to finish.

Adaptability and durability

Adaptability may be broken down into a number of simple strategies that are familiar to most engineers and designers:

- Flexibility, or enabling minor shifts in the way systems function or spaces are used

- Convertibility, or allowing for changes in use for parcels of land or buildings or changes in inputs for infrastructure systems

- Expandability, or facilitating additions (or deletions) to the quantity of land or space dedicated to particular uses

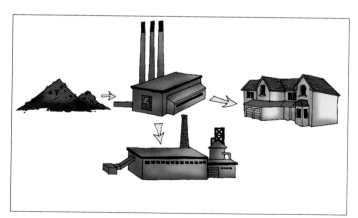

Figure 1.30 An Inflexible Energy System

Source: Author elaboration (Sebastian Moffatt).
Note: This coal-fired energy system is brittle because it cannot adapt, expand, or convert.

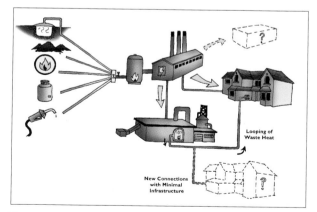

Figure 1.31 An Adaptable Energy System

Source: Author elaboration (Sebastian Moffatt).
Note: This system is resilient because it is more adaptable by design.

Infrastructure that is designed to adapt easily at low cost is likely to survive longer and to operate more efficiently throughout its lifetime (figures 1.30 and 1.31). An example might be combined trenching (utilidors) that allows easy access to pipes and wires.

Durability is a concept that may extend the useful lifetime of materials and technology; it is complementary to adaptability. In practice, adaptability and durability may be achieved through changes in design and the use of alternative zoning, materials, and technologies. For good performance, adaptable designs might begin with the concept of a fixed investment cost. The object is then to achieve maximum durability by means of flexibility and adaptable design features, while, at the same time, minimizing the running costs for energy, cleaning, maintenance, and operation. Part of a durable design strategy might involve setting minimums; for example, no secondary components may last less than 30 years. In other cases, the solution may be to minimize the maintenance and service costs for components.

Monitoring of performance, learning based on results, adaptation and improvement of the system

An integrated approach to monitoring has two dimensions: first, it considers performance objectives from the beginning of project design and uses these targets as a basis for comparing actual performance against intended results; second, it involves integrating monitoring into a feedback and accountability process that ensures adjustments in policy and systems to achieve or exceed the intended results. Both of these dimensions need to be addressed in each project.

Establishing performance targets at the commencement of a design project may be a positive experience that helps focus and inspire the project design team. The selection of targets requires a means of measuring performance easily and affordably. The choice of performance indicators should be based on best practices in other locations and on analytical methods used for system design (such as material flow analysis). Measurements have value only if a basis exists for comparison; so, it helps greatly to use well-established indicators based on standardized data collection and calculations. Ideally, the targets for performance are set only after a review of precedents and case studies, including the experiences of sister cities and best practice cities.

After project completion, it is important to integrate the monitoring program into regular reporting, staff evaluations, and management philosophy. If monitoring is used to guide continuous learning and improvement, it is referred to as adaptive management. Adaptive management originated with fisheries and forestry biologists who discovered that natural ecosystems are so complex and interconnected that all management efforts fail. It became necessary to assume things would not work and, therefore, to plan for failures. As urban environments become more complex and as we consider a broader range of goals for environmental, social, and economic sustainability, it helps to adopt the adaptive management solution discovered by ecologists.

From this perspective, all policy and practice are considered experimental and have lasting value only if they are proven over time. Policy may become a problem if it cannot be easily adjusted to accommodate new knowledge. If monitoring programs are integrated into an adaptive management process, the Eco² long-term planning framework must be included. The framework provides a transparent context for target setting and evaluation. On the one hand, the framework keeps the targets connected to end-state goals, and on the other hand, the framework connects the targets to project strategies and actions.

Stepping Stones for Investing in Sustainability and Resiliency

Use the LCC method to understand costs and cash flows

An Eco² catalyst project represents an opportunity to make LCC a standard part of project planning. Every city may develop this capacity. (Suitable methods and tools are introduced in part 2.)

Develop and adopt indicators to assess the four capitals and to benchmark performance

Indicators may be selected from lists provided by knowledge institutions and industry coalitions. A good place to start is the long list of sustainable development indicators used by cities in the countries of the Organisation for Economic Co-operation and Development (OECD) or by progressive cities in developing countries. The choices need to be guided by the selection criteria listed elsewhere above. An indicator is not successful unless it is regularly measured and reported.

Forecast the impacts of plausible changes

Forecast the impacts of plausible changes in climate, markets, resource availability, demographics, and technology. Forecasting the impact of external forces helps to begin the process of proactively incorporating resiliency and adaptive capacity into the management of risks. Foresight workshops may assist in clarifying the various chains of cause and effect that lead to significant impacts on urban infrastructure systems and the city. Some of the external forces that may be examined through such workshops, in addition to climate change, include changes in global markets, resource availability, demographics, and technology. (These are discussed in part 2.) The World Bank's primer for cities on climate change is a good starting point for understanding climate-related risks (see Prasad and others 2009).

Implement an Eco² catalyst project so as to protect and enhance capital assets and reduce vulnerabilities

The best way to understand the accounting methods in practice is to use them in a catalyst project. This will require a multicriteria assessment of projects using the methods and tools described in part 2. Generally, a base case scenario should be developed using business-as-usual assumptions, and then this base case should be applied as a benchmark for the evaluation of any alternatives that have been proposed during project design exercises. Eventually, the accounting methods should provide a sound basis for making recommendations on a preferred investment strategy.

Monitor results, provide feedback, learn, and adapt to improve performance

Monitoring requires indicators adapted to the city, the project, and the budget. It is most important that indicators be reported over time. There must be a budget allocation for data collection, analysis, and publication. The collection of measurements over time adds strength to the process of urban development. The feedback on key indicators makes it easy to see trends and patterns, educate decision makers on the performance of the city, provide benchmarks, set targets for upgrading future projects, and provide a solid basis for employee and contractor accountability. The key to evaluation and learning is consistency and perseverance.

Note

1. In some applications, the LCC method also attempts to include the embodied or upstream costs that are associated with the use of construction materials such as the energy inputs and emissions that result from the extraction, processing, fabrication, and transport of these materials. However, in most projects, this information is not examined in the applied methodology because the data are difficult to collect and the impacts tend to be most relevant to procurement policy rather than design concepts.

References

Best Foot Forward Ltd. 2002. "City Limits: A Resource Flow and Ecological Footprint Analysis of Greater London." Chartered Institution of Wastes Management (Environmental Body), Northampton, U.K. http://www.citylimitslondon.com/downloads/Complete%20report.pdf.

Ekins, Paul, Simon Dresner, and Kristina Dahlström. 2008. "The Four-Capital Method of Sustainable Development Evaluation." *European Environment* 18 (2): 63–80.

Ekins, Paul, and James Medhurst. 2003. "Evaluating the Contribution of the European Structural Funds to Sustainable Development: Methodology, Indicators and Results." Paper presented at the "Fifth European Conference on Evaluation of Structural Funds," Budapest, June 26–27.

GHK. 2002. "Annexes to Volume 1: Synthesis Report." In *The Thematic Evaluation on the Contribution of Structural Funds to Sustainable Development.* Brussels: European Commission. http://ec.europa.eu/regional_policy/sources/docgener/evaluation/doc/sustainable_annexes_rev1.pdf.

Lahti, Pekka, ed. 2006. *Towards Sustainable Urban Infrastructure: Assessment, Tools and Good Practice.* Helsinki: European Science Foundation.

Mcmanus, Phil, and Graham Haughton. 2006. "Planning with Ecological Footprints: A Sympathetic Critique of Theory and Practice." *Environment and Urbanization* 18 (1): 113–27.

Pearce, David. 2006. "Is the Construction Sector Sustainable? Definitions and Reflections." *Building Research & Information* 34 (3): 201–7.

Prasad, Neeraj, Federica Ranghieri, Fatima Shah, Zoe Trohanis, Earl Kessler, and Ravi Sinha. 2009. *Climate Resilient Cities: A Primer on Reducing Vulnerabilities to Disasters.* Washington, DC: World Bank.

World Bank. 1997. *Expanding the Measures of Wealth: Indicators of Environmentally Sustainable Development.* Environmentally Sustainable Development Studies and Monographs Series 17. Washington, DC: World Bank.

CHAPTER 7

Moving Forward Together

The Eco² Cities Initiative is a collaborative exercise that requires close working relationships among all stakeholders and a willingness to consider and apply new concepts and methods together. Of course, cities are in the driver's seat. This book is designed to explain the key principles of Eco²—how they translate into core elements and stepping stones—and to introduce cities to some of the methods and tools that will enable them to develop their own Eco² pathways. The opportunities for positive change are great at this time. We strongly encourage cities to take the first step toward ecological and economic sustainability, while the window of opportunity to achieve lasting impact is still open.

For forward-looking cities in developing countries that intend to adopt the Eco² approach, support may be available from best practice cities worldwide; the international community, including donor agencies; and academia. Cities are encouraged to tap the unique resources of each of these partners. In this context, the World Bank Group, together with other development partners, is in a position to provide technical assistance, capacity-building support, and financial support to cities that demonstrate a strong commitment to the implementation of the Eco² initiative.

Knowledge Sharing, Technical Assistance, and Capacity Building

One of the most effective methods of knowledge sharing, technical assistance, and capacity building is peer-to-peer engagement with best practice cities. It is conceivable that such engagement might be supported through donor funding. At the same time, the international community has a wide range of programs that provide technical assistance and capacity building. Academic institutions may become engaged in the process, as in the case of the environment load profile tool that has been used by the City of Stockholm and that was jointly developed by Stockholm, the Royal Institute of Technology, and Grontmij AB (a private consultancy firm). Other options for technical assistance include the World Bank Group's technical assistance and capacity-building support, which may be available to cities through a project or through stand-alone funding.[1]

Technical assistance and capacity building can provide support to cities on many stepping stones of the Eco² pathway. They may also help through more detailed applications of the core methods and tools. Some examples of possible support include (1) adapting Eco² to suit a city's unique demands and priorities; (2) conducting diagnostic analysis using the Eco² methods and tools; (3) developing Eco² pathways and plans (including investment and financial plans to realize the vision and strategies); (4) enhancing institutional capacities to implement Eco² projects, with particular attention given to the key principles; (5) equipping local institutions with the technical requirements (a GIS [geographic information system], for instance) to use Eco² methods and tools; (6) designing a national strategy to institutionalize the Eco² initiative through a national financing mechanism; (7) implementing an integrated design workshop or a forecasting workshop; and (8) focused study tours or secondment opportunities in Eco² best practice cities.

Ultimately, knowledge sharing, technical assistance, and capacity-building agendas will be based on the specific needs of each city.

Financial Resources

In general, cities may access a range of financial resources available from the international community and donor agencies. Many of these financial resources may be used to fund technical assistance. Larger donor agencies such as international financial institutions and multilateral development banks (the Asian Development Bank, the World Bank, and so on) may also provide financial resources for infrastructure investment through projects. From an Eco² perspective, it is most important that the number and diversity of financing tools are increasing and that it is possible to combine instruments to fit the dimensions or phases of a project. We consider the case of the World Bank.

In most cases, cities seeking financial support from the World Bank Group need to submit requests through their respective national governments to ensure that the provision of limited loans, credits, or grants is consistent with national priorities and strategies.[2] The World Bank Group has diversified financial tools that may be used in combination to finance Eco² projects. The tools are listed below, along with other donor financial instruments. Unlike a conventional one-project financial instrument approach, the World Bank Group may package these instruments to facilitate an integrated approach that is critical to the success of the Eco² initiative and specific investment projects.

1. Development policy loans provide quick, disbursable financing to support policy and institutional reforms at national and subnational government levels.

2. Specific investment loans finance a broad range of specific infrastructure investments (water supply, wastewater management, power generation and distribution, solid waste management, roads, public transportation, and so on).

3. If policy and regulatory reform leads to a significant reduction of greenhouse gas emissions in specific components based on the Clean Development Mechanism methodology or if direct investments accomplish the same (for instance, through solid waste management), then the World Bank's Carbon Finance Unit may enable the purchase of emission reductions. This may increase the bankability of projects by adding an additional hard revenue stream.

4. The International Finance Corporation, also part of the World Bank Group, is able to finance corresponding private sector investments (for instance, energy-efficient buildings or technologies).

5. The Global Environment Facility is a global partnership that provides grants to address global environmental issues in projects in six focal areas: biodiversity, climate change, international waters, land degradation, the ozone layer, and persistent organic pollutants. An Eco² project may qualify for a Global Environment Facility grant if it focuses on one or more of these areas.

6. Climate Investment Funds, which provide concessional financing, may be made available if projects contribute to the demonstration, deployment, and transfer of low carbon technologies, with significant potential for long-term greenhouse gas savings.

7. By insuring investments against political risks, the World Bank's Multilateral Investment Guarantee Agency may help certain developing countries attract private investment.

By integrating, sequencing, and linking these financial instruments, the World Bank may enable an integrated approach to the sequenced implementation of a city's Eco²-related financing needs. Of course, all these instruments are not required in every case. Figure 1.32 provides a sample of how instruments might be mixed. The World Bank Group may also help national governments and Eco² cities mobilize cofinancing resources from other donors, as indicated on the far right of the figure. (The features of these financial instruments are explained in part 3.)

Financial resources are important. They enable many of the initiatives discussed in this book. However, the reader should keep in mind that some of the most remarkable innovations and approaches profiled here have been implemented without the luxury of these complex external financial resources. The true test of the Eco² Cities Initiative will not be its ability to link cities with financing, but to facilitate a process whereby cities may adapt and apply the four Eco² principles to unlock their own full potential.

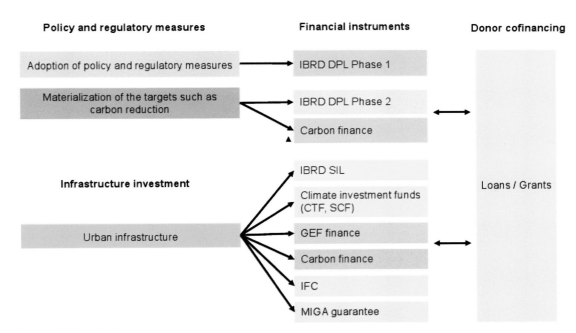

Figure 1.32 Financial Instruments

Source: Author compilation.
Note: The World Bank Group's financial instruments and the instruments of multidonor facilities administered by the World Bank may be packaged and sequenced to support a more well integrated approach to the financing of Eco² projects. CTF=Clean Technology Fund; DPL=Development Policy Lending; GEF=Global Environment Facility; IFC=International Finance Corporation; MIGA=Multilateral Investment Guarantee Agency; SCF=Strategic Climate Fund; SIL=Specific Investment Loan.

Notes

1. The World Bank Group consists of five institutions: the International Bank for Reconstruction and Development (IBRD), the International Development Association (IDA), the International Finance Corporation (IFC), the Multilateral Investment Guarantee Agency (MIGA), and the International Centre for Settlement of Investment Disputes (ICSID).

2. International Bank for Reconstruction and Development loans and International Development Association credit must also be covered by sovereign guarantees.

PART 2

A City-Based Decision Support System

Methods and Tools for Eco² Cities

Part 2 is aimed at everyone who wishes to become familiar with the core methods that, together, provide cities with a decision support system. It explains the role of these methods in assisting cities in implementing more strategic and long-term management and decision making. The decision support system is part of a city-based approach because it enables cities to develop their capacity to render operational the core elements of the Eco^2 initiative. Even if one does not expect to work with the methods directly, understanding what they accomplish adds to one's understanding of the overall framework.

Each chapter deals with a different category of the methods and tools. Chapter 8, titled "Methods for Collaborative Design and Decision Making," is an overview of the operational and process methods that help cities undertake leadership and collaboration. Chapter 9, titled "Methods for Analyzing Flows and Forms," provides an overview of the most practical analytical methods. The combination of analytical methods helps cities to develop the transdisciplinary platform described in part 1 by revealing the important relationships between the spatial attributes of cities (forms) and the physical resource consumption and emissions of cities (flows). Chapter 10, titled "Methods for Investment Planning Assessment," is an overview of accounting methods; it includes details on ways to apply life-cycle costing and proactive risk mitigation and adaptation.

The methods support the typical planning process at different times and in different ways. Some methods may be used repeatedly. For example, the meta diagrams that summarize resource flows may be used initially to provide a base line for how a location is currently performing and, later, to help in diagnosing, target setting, scenario development, and cost assessment.

All the methods represent proven approaches able to accomplish the task. They are expected to remain relevant for many years. The fundamental purpose of the methods is to simplify the process of analysis, assessment, and decision making. They provide practical ways for cities to take leadership, collaborate, and analyze and assess various ideas for the Eco^2 projects of the cities.

Wherever possible, the methods are accompanied by tools. The tools are instruments such as templates, checklists, diagrams, maps, and specialized software applications, that is, anything that is convenient to use and helps to effectively and quickly render a method operational. The tools referenced here are examples and are indicative of some of the practical options available to cities.

Part 2 is a good place to begin if cities are planning a process of capacity building with the methods and tools to achieve urban sustainability. It is mainly an introduction to the issues, and many cities may wish to follow through in more detail on suitable methods, obtain more information, acquire specific tools, deepen and broaden their capacity, and apply the new methods through catalyst projects.

Capacity-building plans are usually developed in stages, beginning with the most simple tools and applications. The benefits are substantial. For example, sophisticated computer printouts are not necessarily more effective than maps drawn on transparencies by individuals with extensive ground knowledge (community mapping). Sometimes, computers and fancy presentations may actually get in the way.

All capacity-building plans should focus on tools that may accommodate varying levels of data and skill and allow for the capacity to evolve over time. Tools with the following characteristics may assist in this evolution:

1. **Transparent:** analytical tools must be easy to understand and adjust, so that even beginners may follow the rationale and the flow of information. Complex black box computer models are inappropriate.

2. **Scalable:** tools easily adapt to the level of effort warranted by the project and to the level of knowledge and skill of the user. As conditions change, the same tool should accommodate a larger scope or more precise inputs.

3. **Web-friendly:** by designing most tools so they may take full advantage of the Internet, one may more easily train people, update tools, share results, interchange data and results, and use the tools to enhance stakeholder and public participation.

4. **Modular:** experience with the use of tools for city planning suggests that it is a mistake to adopt models and tools that are overly general and all-inclusive in purpose (Lee 1973). Models work best if they are limited to specific tasks and sufficiently flexible to be applied on their own or in combination with other tools. A modular approach based on strong theoretical foundations, but allowing for changes in key assumptions, may be more easily adapted to the complexities of the real world and to changing user needs.

Acquiring capacity in particular methods and tools may appear challenging. Training seminars and user-friendly software may make the process more manageable. Despite the challenges, however, most cities in developing countries will need to adopt new methods and invest in capacity building. Problems in cities in the developing countries are often more complex and demanding than those faced by wealthier cities in developed countries, and, thus, the need for effective decision support systems is greater. The investment will yield compounded benefits.

Reference

Lee, Douglass B., Jr. 1973. "Requiem for Large-Scale Models." *Journal of the American Institute of Planners* 39 (3): 163–78.

CHAPTER 8

Methods for Collaborative Design and Decision Making

Organizing and Managing Collaborative Working Groups

Adopting basic rules for collaboration

Collaboration is a method by which diverse groups join together for a common purpose without necessarily altering their mandates, relinquishing their authority, or sharing their budgets. Power structures are retained. In fact, the reason collaboration works is that nobody is forced to give away power. What changes is that information flows are greatly enhanced, and the potential is greater for joint action (figure 2.1). Collaboration is especially effective in the integrated design of urban areas because so many different parties may influence the results. Any particular system may be significantly affected by land use policies, private development projects, on-site systems, demand-side management programs, efficiency standards, the use of shared rights-of-way, and so on. Collaborative committees begin by agreeing to a simple set of rules or principles.

It also helps to have a common vision of the desired long-term outcomes.

One key rule is that, wherever agreement on strategies is reached within the collaboration, all the members must use their unique mandate and resource base in a more or less coordinated fashion to contribute to the agreed strategy.

Balancing the membership and structuring inputs of varying levels of authority

Ideally, a collaborative working group is composed of a balance of sectors: the government, the private sector, civil society, and academia (knowledge institutes). A balanced membership means that a collaborative working group needs to be carefully constructed to include a full range of perspectives: short-term and long-term, private and public. A convenient approach is to establish roughly proportional representation from various sectors: the government, the private sector, civil society, and academia. Each sector contributes different priorities and perspectives that help create balance. For

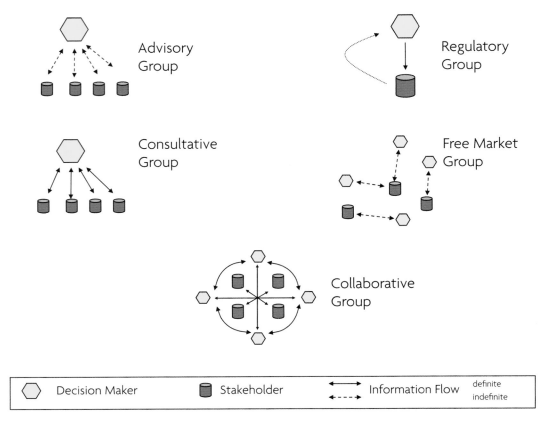

Figure 2.1 The Collaborative Model

Source: Author elaboration (Sebastian Moffatt).
Note: A collaborative model replaces hierarchical structures and increases the potential for exchange and cooperation.

example, if the government sector is often the most well informed, it is usually the least willing to take risks. Civil sector groups, if well represented, may help provide the motivation and vision to keep everyone from seeing only problems and barriers. Input from the academic and knowledge sector may be especially useful in expanding the scope of discussions and, at later stages, incorporating high-quality research and expertise into design exercises and planning proposals. The precise mix of stakeholders from each sector must be carefully considered because each city will be characterized by different political relationships and institutional structures. The composition should also vary to reflect the scope of the planning and the projects under consideration.

In the public sector, the stakeholders may include all the agencies and departments with the scope to affect cities. These may be national, state, municipal, or district. At each level, there may be additional regulatory, infrastructure development, and service delivery agencies responsible for land, water, energy, transportation, and waste management. Some of these agencies may participate in public-private partnerships, and this requires the involvement of networks. Neighboring jurisdictions are also potential stakeholders. Collaborating with adjacent cities and regions may result in strong synergies in such areas as integrated planning for the reuse of waste materials, the coordination of transportation and land development, and cooperative economic development.

The private sector and households are key players in energy and resource use, as well as in creating local pollution and global greenhouse gas emissions, and they need to be considered in the Eco² process. A recent report on sustain-

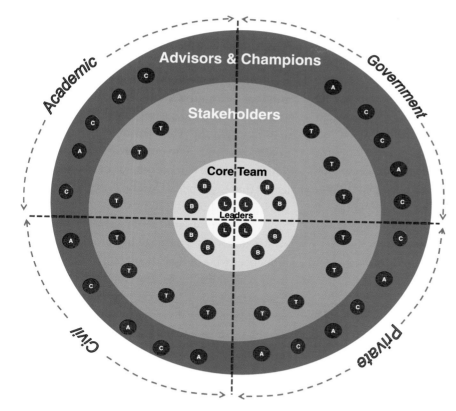

Figure 2.2 The Collaborative Working Group

Source: Author elaboration (Sebastian Moffatt).

Note: The makeup of a collaborative working group should seek to balance input from government with equal measures of leadership and expertise from the private, civil, and knowledge sectors.

able urban infrastructure in London strongly reflects this view:

> As this report makes clear, many different stakeholders are involved in making sustainability-related decisions. Success will require cooperation, rather than dictation from any one of these. Certain things can take place at the national and municipal government levels, but the most powerful actors in all of this are consumers [including households and businesses—Ed.], who may through their purchasing decisions bring about 70% of all possible CO2 abatement. Absolutely crucial to lowering emissions, therefore, will be removing the barriers to them doing so. (EIU 2008: 64–65)

The urban poor are also stakeholders in the city. Good urban planning creates more access for public and nonmotorized transportation and supports lower-cost services and the reduction, reuse, and proper treatment of harmful waste:

all outcomes that directly and tangibly improve conditions among the poor. At the same time, the fiscal gains achieved by the utilities or a city may be applied to benefit the poorer sections of society. For example, the congestion tax in London not only reduced traffic by 21 percent (70,000 fewer cars per day) and increased the use of buses and bicycles in the zone, but also raised £137 million during fiscal year 2007, of which a large portion is being reinvested to improve public transportation.

Collaborative committees function well if they have a strong champion, a hard-working secretariat, and a balanced membership. Figure 2.2 presents an example of how a citywide collaborative might be organized. A single all-policy working group provides a new institutional structure for collaboration. At the center are one or more leaders who direct the process and provide everyone with a sense of purpose and confidence. The secretariat is a small group that serves the collaboration by undertaking research on critical issues, facilitation at meetings, communications between meetings, and event planning. An effective secretariat may help build confidence and make the process feel productive, fun, and worthwhile to all members. The collaborative process may require a core team if the collaboration is undertaking to create a pact or strategic plan (figure 2.3). Alternatively, the collaboration may guide and direct the planning work accomplished by the staffs of the various stakeholders.

Figure 2.3 The Core Team and Sector Advisers

Source: Author elaboration (Sebastian Moffatt).

Note: The core team may be supported by a ring of sector champions, each of whom connects the working group to a larger network of experts and stakeholders. New urban infrastructure places special emphasis on sectors such as energy, transportation, water, the environment, and materials management. However, other sectors may also offer substantial contributions.

Developing a Shared Framework for Aligning Visions and Actions

A common framework may greatly enhance communications and coordination

Frameworks are the mental maps that we use to make sense of the work schedule: what comes first and second, how our particular contribution fits with the work of others, and so on. Typically, there are big differences in the frameworks that different individuals bring to a project. These differences may reside in how groups understand the goals of a project, or who influences who, or how their plans are expected to fit with other plans. Developing a common framework can help overcome these disconnects and make a diverse group of largely autonomous stakeholders operate more as a team. A framework for urban planning and design covers all steps from start to finish. An example of such a framework is the pyramid shown in figure 2.4. At the pinnacle is the scope of the framework, which clarifies the extent of the urban area to be included, identifies the types of urban systems to be considered, and diagnoses the strengths and weaknesses of the system as it currently operates.

After scoping and diagnostics, a framework typically expands to include a shared vision statement and a set of long-term goals. These broad statements are then unbundled into more specific and immediate targets, strategic plans, actions, and ongoing learning processes. The framework may include any principle, goal, or strategy that the users desire, and it may be easily molded to fit any current planning framework, method, and terminology. In this sense, it is a type of methodological pluralism: everything fits inside the framework.

Perhaps most important, a framework builds in accountability, thereby helping to avoid short-term political decisions that are inconsistent with goals and targets. It also creates the opportunity to monitor performance against specific goals and targets and to update plans and adapt to changes without losing sight of the original intentions. If the vision changes over time, then all subsequent layers of the framework may be adjusted accordingly. Or, if the implementation actions encounter surprise or produce inappropriate results, then the problem may be traced back to the choice of strategy, and corrective changes may then be made at all subsequent levels.

The first stage involves defining boundaries and understanding the current performance

Because a long-term planning framework supports collaborative decision making, the scope of the framework must match the platform for collaboration. If a city is leading a three-tier collaboration process, for example, then the planning framework will need to be extended to include visions and actions relevant for the entire urban area and for all participating stakeholders.

Whatever the platform, scoping and diagnostics help set the stage. Clear boundaries inform all participants about what is included or excluded from the planning framework. An extensive inventory or information collection process clarifies what is now known and not known. Some basic analysis of existing system performance may establish how well various systems are performing relative to systems in similar cities or in best practice case studies. This is sometimes referred to as a city profile. Often, the amount of work involved in scoping and profiling a city exceeds the work required for all other stages of the framework. Nonetheless, it is an extremely worthwhile investment because the results serve to direct all further activity.

Vision statements are elaborated into end-state goals

The vision may be a simple statement or even an artist's drawing; its purpose is to be inspirational and broad. If the scope is limited to infrastructure design and land use planning, then the vision should focus primarily on these areas.

Figure 2.4. A Long-Term Planning Framework

Source: Author elaboration (Sebastian Moffatt).

Note: The framework connects visions to actions and includes a process for learning and adaptation.

A set of end-state goals may elaborate on the vision by adding stand-alone goal statements. End-state goals define the ultimate condition that is desired by a city, even if this is something that may not be realized for many years. Typically, an end-state goal is expressed in a single definitive statement, followed by a commentary. Canada's capital region has a number of end-state goals. A few relate directly to infrastructure performance, such as achieving a sustainable urban metabolism or extending the use of green infrastructure. These goals are as follows:

1. The natural resource demand by each neighborhood is consistent with the long-term capacity of the city's infrastructure and the region's resource base.

2. Trees, gardens, ponds, wetlands, hedgerows, streams, greenways, green roofs, and engineered ecologies have become the elements of a cost-effective green infrastruc-

ture that cleans and constrains storm water flows, contributes to a quieter and more pleasant microclimate, shades buildings in summer, improves air quality, and generally contributes to the livability and biodiversity of neighborhoods.

Although such goals describe a long-term condition, they serve an immediate strategic purpose by providing a common reference point for all design and planning and a basis for collaborative decision making.

End-state goals for Eco² cities should address, at a minimum, the basic urban services (energy, water, and so on) and the ecological performance of the urban region. Cities may choose to use their own format and language, or they may adapt their framework goals from examples provided by the Eco² Cities Initiative. Either way, goals should reflect local conditions and cultural values and need to be discussed and endorsed by key stakeholders. Because the goals are long term, the process of building consensus around the goal statements tends to be a positive experience, creating a common purpose among stakeholders and residents.

Target setting may help translate goals into clear objectives

Sometimes, it helps to develop intermediate targets to support specific end-state goals. The targets are based on indicators that quantify the city's desired performance with respect to one or more goals. By setting targets for specific time periods, the city helps direct the pace of change and the priorities for investment. For example, Stockholm has set a target for all new construction to be carbon neutral by 2030; over 70 percent of New Zealand's cities and towns have adopted a zero waste landfill target, with a timeline for each milestone on the journey; San Diego and Irvine, California, have achieved their targets for the comprehensive coverage of reclaimed water for commercial properties.

As part of the adoption of end-state goals (and targets if desired), one may assess performance and set priorities. Using expert judg-

ment and local knowledge, one may subject each goal to a series of questions: How close is the city to achieving its goal today? What forces are likely to influence future success? What direction is the city now taking? Is the situation getting better or worse? How rapid is the pace of change? This type of rapid evaluation is helpful in setting priorities for Eco² projects.

Strategic planning requires that planners evaluate alternative scenarios

This exploratory stage in a planning framework offers the opportunity to develop a range of alternative scenarios or approaches and to assess their relative values in terms of how well they achieve guiding targets and goals. While city governments and departments may already have a strategic plan, the framework may help to extend and align the time horizons for such plans and to integrate strategies that address the long life cycles of such investments. At the scale of the urban region, strategic planning is especially useful, although many urban regions in developing countries are currently operating without a shared strategic framework.

In a growing urban region, the umbrella plan that sets the context for all other planning is sometimes referred to as a regional growth strategy (RGS). The RGS ensures that the various infrastructure plans—transportation, water, and energy—all share the same assumptions about land use, demand, and development priorities. The RGS takes regional population growth and employment projections into account and gives the region, including its component parts (towns, counties, and cities) long-term planning direction. It is the RGS that ensures the integration of the parts into a functional whole. In addition to providing the big picture on how a city fits into its surroundings, the RGS provides the broad-brush strategies for connecting neighborhoods and directing new growth and investment. The RGS should always address the critical issues that must be solved at the scale of the urban region; these issues might include restricted water supply, air quality, and transportation management. The RGS may also identify priorities for housing, regional services, parks, economic development, and climate change initiatives. The most effective regional growth strategies are developed through a consensus-building process that achieves agreement (sign-off) from the surrounding regions and from the mix of towns or stakeholders within the region.

To be effective over the long term, the RGS must provide a phased approach to accommodate the projected growth in population and jobs, including the identification of areas suitable for infill and densification and the timelines for the development of specific urban reserve areas. To ensure that the various elements of the city interact and support each other, the RGS typically adapts some of the best practices from successful regions, including the following:

- A hierarchy of regional growth centers connected to each other and to the growth concentration area via transportation corridors with efficient and convenient transit

- One or more growth concentration areas that provide the city with a destination center for shopping, business, and the arts

- Medium- or high-density developments located along the transportation corridors and at all transportation hubs

- Distinct, complete neighborhoods and districts that include a mix of land uses, a healthy ratio of jobs to housing, and well-defined open spaces

- Clearly defined containment boundaries, with permanent, functional edges that separate and protect urban areas, rural areas, and natural areas

- A fine-grained network of greenways and blueways that connect all residential areas to a network of parks and to a representative cross-section of the region's native ecologies

An RGS need not be a complicated undertaking. Some of the most well known RGSs have been produced quickly and have been initially introduced as a simple vision and a map. However, most often, the process takes a couple of years from start to finish, plus the added time to initiate and secure funding. The process requires major investments in capacity building, field research, mapping and analysis, collaboration, and public process. Consequently, the completion or renewal of an RGS may occur in parallel with other Eco² projects or at a later date.

Absent a long-term, complete, and up-to-date RGS, the shared planning framework may not function so well. Without an RGS, it might be more difficult, for example, to integrate Eco² projects into long-term land use and development, and some opportunities for design and policy integration may be lost. However, the Eco² pathway may incorporate interim solutions that offer a significant amount of guidance without a major investment in time or resources. One such solution is the organization of a regional design charrette, followed by the use of the outputs of the charrette as the first iteration or first cut of an RGS.

The implementation of the key strategies should begin with catalyst projects

The implementation of strategies may be achieved through project planning and investment. The first projects implemented in accordance with the Eco² pathway are referred to as catalyst projects. The function of a catalyst project is to accelerate learning and to promote the acceptance and understanding of the Eco² pathway. A catalyst project may be site specific or citywide. It should be designed to demonstrate the potential for the greater integration of designs and polices. Almost any type of infrastructure investment or land development may be adapted for this purpose. However, the best choices are catalyst projects that work through people or in locations that are already taking steps in the proper direction. It also makes sense to choose a catalyst project based on a city's priorities for change. If the end-state goal is to provide everyone with affordable housing and the actual price of housing is becoming less affordable every day, then some kind of intervention is clearly warranted. As long as the project details are not predetermined, the process of collaboration and integrated design, supported by new methods and tools, will lead to a more efficient multipurpose system design and a more coordinated set of enabling policies.

Because of the focus on learning and integration, catalyst projects are not strictly pilot or demonstration projects. The emphasis is on learning and on catalyzing change by influencing all subsequent projects. The catalyst projects help transform a city into a learning society.

A city might plan for one active catalyst project in every neighborhood as a way to begin implementing the Eco² pathway and as a contribution to local pride and place making. Figure 2.5 evokes a neighborhood catalyst project

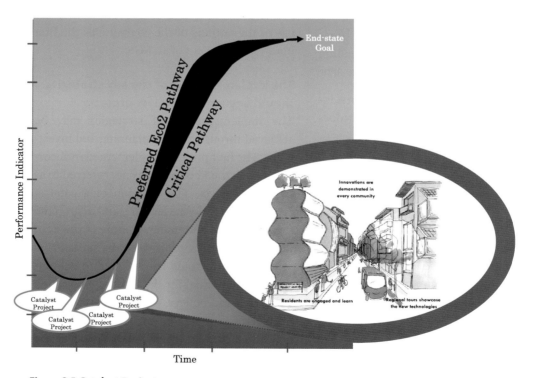

Figure 2.5 Catalyst Projects

Source: Author elaboration (Sebastian Moffatt).

Note: Catalyst projects are interventions in the short term designed to accelerate the changes needed to create an Eco² pathway that will reach the targets and end-state goals of the long-term planning framework.

and illustrates how the project serves to anticipate and redirect performance in an area where the trend is otherwise in the wrong direction.

Implementation policies should be integrated across each policy tool and stakeholder

To ensure that all stakeholders are engaged and that a full set of policy tools and instruments has been considered, one can create a matrix of stakeholders and policies by category. An example of such a matrix is presented in table 2.1. The various policy tools and instruments are listed across the top, and the stakeholders down the left side. Developing such a matrix is the output of a collaborative exercise based on a shared planning framework. The matrix is a tool for strategic planning and also a way for any collaborative working group to visualize the potential of teamwork. Each stakeholder tends to have different levers of control or influence, and these produce different, but complementary actions for implementation.

Conducting a Regional Systems Design Charrette

At every tier, a collaborative committee provides an important institutional structure for promoting and facilitating integrated design. Unlike the traditional planning and design process (which begins with a small team led by an architect, planner, or engineer who is later joined by experts as needed), integrated design engages a wide range of specialists, local stakeholders, and partners at early stages. The objective is to use the expertise to influence seminal design decisions before opportunities are constrained and to find the synergies and out-of-the-box solutions that lead to practical and affordable responses.

The prior existence of a formal collaboration process among senior decision makers means that the groups are likely to be more comfortable with participation in the integrated design process. Ideally, the collaborative committee will agree that design workshops are worthwhile and will contribute their best designers. The collaborative agreements may also ensure that the results of such workshops are properly assessed and integrated into final project plans.

Many kinds of design workshops may be used to facilitate the Eco^2 pathway. One of the most important kinds of workshops is the systems design charrette (figure 2.6). A charrette is an intensive workshop that may last four to seven days and that typically brings together a diverse group of specialists, designers, and residents. During the charrette, a number of small mixed teams work side by side, day after day, with occasional interaction with each other and with scheduled visits by the public and respected personalities.

Techniques for conducting design charrettes have evolved over the past few years. Initially, the charrette was a tool used primarily to stimulate creative design solutions in building form and the use of interior space. A new building or group of buildings would be drawn in various configurations with input from many experts. More recently, the techniques have been applied to entire neighborhoods, cities, and regions. The results have been excellent. Larger spatial areas may be treated as three-dimensional spaces with attention given to

Figure 2.6 Design Workshop: Systems Design Charette
Source: Photo by Sebastian Moffatt.

Catalyst Strategy 1: Protect & Connect Ribbons of Blue, Webs of Green

Rehabilitate and protect natural and built systems through reconnecting viable networks of blue ways green spaces and corridors

	Investment & Planning Initiatives	Research, Demonstration & Leadership	Education & Inspiration	Legislation, Regulation & Enforcement	Market Instruments
Federal Agencies	• Continue Fraser River Estuary Management Plan • Continue support for watershed and stream stewardship initiatives	• Start using federal lands to showcase sustainability strategies such as green roofs • Start using alternative development standards for developments on federal lands • Support establishing showcase sustainable communities	• Draft and pass a federal Environmental Bill of Rights • Continue regulating and enforcing instream works alternations to riparian zones; and discharge of deleterious substances into **fish habitat**	• Stop taxation of capital gains on lands donated for conservation purposes • Continue allowing gifts of ecologically sensitive lands and easements to qualify for charitable tax receipts	• Continue stewardship outreach • Continue making available the Lost Streams of the Lower Fraser Valley • Continue supporting education about marine and freshwater ecosystems
Provincial Agencies	• Continue development of strategic land use plans, and land and resource management plans • Continue watershed restoration programs	• Start or continue use of provincial lands and buildings to showcase sustainability strategies such as green roofs • Sponsor research at universities to **define characteristics** of well-functioning greenways	• Start process to develop and adopt provincial groundwater legislation • Draft and pass a provincial Environmental Bill of Rights	• Continue tax exemption status for privately owned **environmentally significant lands** protected by conservation covenants • Stop property transfer tax for donations of ecological lands	• Incorporate discussion of sustainability and ecosystems into school curriculums • Develop provincial recognition programs for organizations showing leadership in habitat and stewardship work
Regional Agencies	• Develop a regional stream stewardship centre • Coordinate stream stewardship activities in conjunction with expanding and maintaining the Green Zone • Develop a habitat and watershed atlas for the region	• Prepare model bylaws, such as stormwater bylaws for customization by municipalities • Improve understanding of aquifer protection and restoration, particularly in the eastern part of the region	• Strengthen commercial and industrial waste bylaws for discharges to the sanitary sewer system and direct discharges to streams	• Start increasing rates for sewage treatment services • Create an endowment fund for daylighting streams of **regional significance** • Leverage resources for habitat acquistion with NGOs	• Prepare & disseminate educational materials on the value of streams & habitat • Make habitat atlas available in print and on-line formats for schools, developers, community groups and NGOs
Municipal Agencies	• Include environmental goals and objectives in Official Community Plans and specify development permit areas • Develop strategies for daylighting streams • Develop watershed management plans	• Conduct pilot projects for daylighting streams through municipally-owned lands • Naturalize parts of municipal parks • Naturalize boulevards	• Use the full range of legislative tools to protect riparian and sensitive habitat areas • Adopt tree protection bylaws • Accelerate seperation of combined sewers to eliminate combined stormwater over-**flows**	• Use provincial enabling legislation to exempt landowners holding conservation covenants from property taxes for those portions of their land • Use density bonusing to facilitate habitat protection	• Continue or start sponsoring and supporting stream stewardship groups and education activities
Private Sector	• Work with all levels of government on planning initiatives	• Start voluntary conservation practices by landowners • Enter into voluntary stewardship agreements with NGOs • Sponsor greenspace acquisition and restoration of streams			
NGOs	• Work with all levels of government on planning initiatives • Provide volunteers for helping with implementation • Form habitat and stream stewardship groups	• Start land trusts to acquire habitat lands and hold conservation covenants • Work with all levels of government and the private sector on research and demonstration projects	• Continue to hold outreach activities and prepare publications to raise awareness and foster education on habitat and stream stewardship • Form habitat and stream stewardship groups		

ANYBODY (Implementation Measures for *Anyone* to Implement)

EVERYBODY (Implementation Measures for *Everybody* to Implement)

Stop, Start or Continue?

Stop - measures that need to be halted

Start - measures that need to be initiated

Continue - measures that should be continued or expanded

Governance Reform:

Mandates of each level of government need to change to enable more proactive initiatives and preventative measures.

A reformed governance system would include a shift towards a watershed based approach to planning and implementation with mechanisms that allow for the establishment and support of subwatershed stewardship groups.

Role of Citizens:

• Be environmental watchdogs during the development process

• Report infractions by development proponents if activities impact fish when sites are developed

• Advocate density levels that promote creation of greenspace

Table 2.1 A Policy Matrix

Source: Author elaboration (Sebastian Moffatt). Originally produced for Cities^PLUS (www.citiesplus.ca).

Note: A policy matrix indicates how each participant within a collaboration may use various policy instruments to support a specific catalyst strategy. NGO = nongovernmental organization.

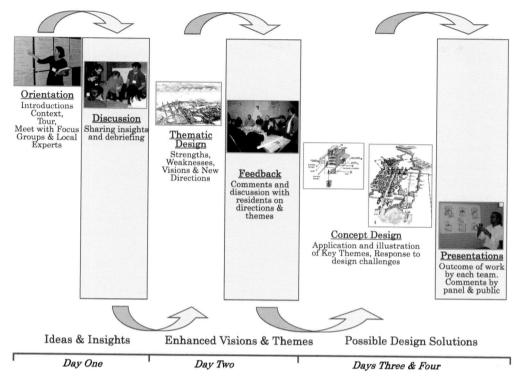

Figure 2.7 A Regional Design Charrette

Source: Adapted from Lennertz and Lutzenhiser (2006); photos by Sebastian Moffatt.

Note: A regional design charrette is an intense exercise that progresses over several days through orientation, thematic design, and concept design with lots of opportunities for discussion, feedback, and presentations.

scale, walkability, streetscapes, and public space. Specific locations may be used as case studies. Engineers and planners may address urban resource flows and include schematics and plans for alternative infrastructure. In this way, the design charrette expands to address all urban systems at the city scale.

In a charrette, the scale will vary to match the project. If the objective is to create a long-term regional plan, the scale will need to encompass the entire urban area and the rural fringe. One team might focus on boundaries and connections, another on the formation of complete neighborhoods, and another on infrastructure systems (urban systems). All teams are directed by the end-state goals. At first, the small teams talk and share information and ideas. Then the process progresses from simple drawings to complete plans, layered maps,

sketches, meta diagrams, and schematics. The pace of the workshop accelerates until it ends with a surprising amount of work accomplished. In the words of Patrick Condon (2008), a Canadian expert who has led many such workshops, a charrette is "the best way to get the most creative proposals to address the most difficult problems from the most accomplished designers in the most compressed period."

A charrette is a collaborative approach to design that may offer much more creativity and interdisciplinary thinking than is normal in city planning. At the beginning of the charrette, the teams review and discuss the city's or region's long-term planning framework (figure 2.7). During the workshop, the teams engage frequently with invited members of the public and specialists by means of many small presentations and intense discussions and drawing sessions. This broad and meaningful engagement contributes to a positive outcome, with less fear and resistance from stakeholders, and with significant potential for reaching consensus on contentious issues such as the way to apply best practices in the local context.

A regional design charrette concludes with a plenary presentation to stakeholders, important personalities, and the public and the preparation of a well-illustrated publication with recommendations on the RGS. (For manuals and case studies on charrettes, see Condon 2008; Lennertz and Lutzenhiser 2006; Swanepoel, Campbell, and Moffat 2003.)

References

Condon, Patrick M. 2008. *Design Charrettes for Sustainable Communities.* Washington, DC: Island Press.

EIU (Economist Intelligence Unit). 2008. "Sustainable Urban Infrastructure: London Edition; A View to 2025." Siemens AG, Munich. http://w1.siemens.com/entry/cc/en/sustainablecities.htm.

Lennertz, Bill, and Aarin Lutzenhiser. 2006. *The Charrette Handbook: The Essential Guide to Accelerated Collaborative Community Planning.* Chicago: APA Planners Press.

Meadows, Donnella. 1999. "Leverage Points: Places to Intervene in a System." Sustainability Institute, Hartland, VT.

Swanepoel, Lourette, Elisa Campbell, and Sebastian Moffat. 2003. "Tools for Planning for Long-Term Sustainability: The CitiesPLUS Design Charrettes." Research report, Canada Mortgage and Housing Corporation, Ottawa.

CHAPTER 9

Methods for Analyzing Flows and Forms

Meta Diagrams and Material Flow Analysis

The meta diagram is one of the most powerful tools available for systems thinking about infrastructure design and performance. It has two dimensions: it is a visualization tool that illustrates complex information in simple and standard ways; and it is a calculation method that tracks the flows of energy, water, and materials through cities. This section explores both dimensions of the meta diagram and the way that the meta diagram helps develop a systems perspective and may contribute in many ways to the process of integrated infrastructure design.

The visualization tool is a type of Sankey diagram. Like all Sankey diagrams, its function is to illustrate flow directions and quantities. Figure 2.8 provides an explanation of how a Sankey is constructed and interpreted. By illustrating quantity and the direction of flows, the Sankey displays more information on a single page than does any other graphic. As is often said, one Sankey is worth a thousand pie charts.

The calculation method is referred to as material flow analysis. The method tracks flows as a balanced set of input-output accounts. The inputs are either resources derived directly from nature (rainwater, for example, or local sunshine and biomass) or resources imported from other regions. Inputs are then tracked as they flow through the city's infrastructure and buildings. Typically, the input flows are first processed; for example, rain might be filtered, sunshine converted to electricity, or biomass burned to create heat. After processing, the flows are used to satisfy the demand for services, such as drinking, lighting, and cooking. After servicing demand, the flows may again be processed; for example, sewage might be treated, or biogas captured and recycled. Finally, the flows are returned to nature as waste and emissions to the air, water, and land, or they might be stored or exported to other regions. Whatever the resource or pathway, the inputs always equal the outputs.

If analyzed through the material flow analysis method, a city's infrastructure appears simi-

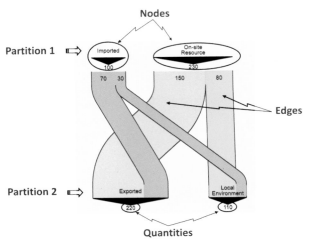

Figure 2.8 A Sankey Diagram

Source: Author elaboration (Sebastian Moffatt).

Note: A Sankey diagram is comprised of partitions, nodes, edges, and arrows. A partition represents the transitions or stages within the flow where transformations may occur. The nodes are the divisions within a partition; they represent processes or events that regulate or transform the quality of flows. Edges are the paths (or noodles) that emerge from nodes and that direct flows to nodes on the next partition. The width of the edges is proportional to the flow quantity. Arrows indicate flow direction.

lar to the metabolism of a living organism that consumes natural resources to stay alive. If Sankey diagrams are used to illustrate these nature-to-nature flows, they are referred to as meta diagrams. Flows of resources may be illustrated for individual developed sites or for whole cities. Flows are typically averaged over one year, although both the time period and the spatial scale may be selected to answer whatever questions are of the most interest.

Figure 2.9 provides an example of a meta diagram for water flows through a parcel (a house site) in New Delhi. The Sankey has five preestablished partitions: sources, converters, demands, reconverters, and sinks. Converters and reconverters are general terms for the on-site urban infrastructure or appliances that store, convert, regulate, separate, process, or recycle any flow. A converter is on-site and upstream of all service demands, while a reconverter is on-site and downstream of at least one service demand. In the example shown, the majority of water flowing through this parcel arrives as rain, about 60 percent of which passes directly through the site to be absorbed in the ground. The remaining rainwater is captured by the roof and stored in a cistern, from which it is mixed with a neighborhood groundwater system and used to supply many household needs. The greatest single use of freshwater is the cooling system. The diagram quickly reveals advanced looping systems: water from the kitchen and baths is reclaimed and used for toilet flushing, and water from the septic tank is reused for pipe irrigation.

Meta diagrams constructed at the parcel level, such as the one shown, may be summed to create a Sankey for a collection of parcels, the neighborhood, or a city. An example of a citywide meta diagram shows baseline water flows for Irvine, California, a community of 180,000 people south of Los Angeles (figure 2.10). The climate in Irvine is dry (13 inches of rain per year), and the city has developed one of the most complex and advanced water systems in the United States. The diagram provides all the key information on a single page.

Five Reasons for using meta diagrams in systems analysis and design

1. **Understanding the whole picture.** A meta diagram may be designed to convey quickly many aspects of a system to people of diverse

"The built environment as a self-organizing system functions as a 'dissipative structure' requiring a continuous supply of available energy, material, and information necessary to produce and maintain its adaptive capacity and rejecting a continuous stream of degraded energy and waste back into the ecosystem (entropy)."

Source: Rees (2002: 253).

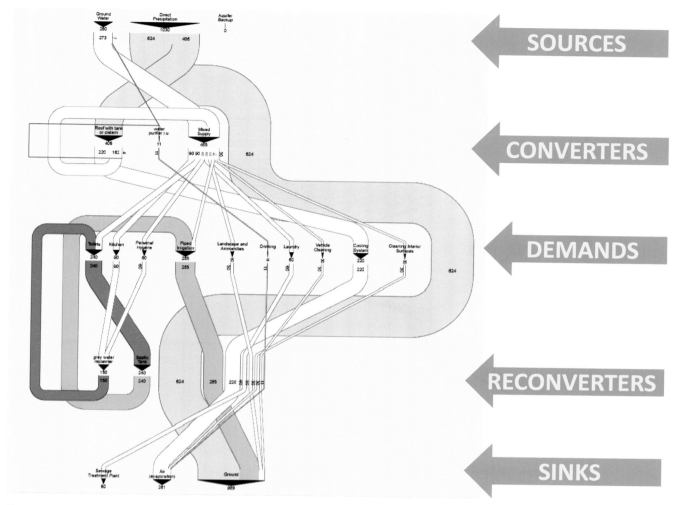

Figure 2.9 An Example of a Meta Diagram

Source: Author elaboration (Sebastian Moffatt).

Note: This meta diagram uses the five standard partitions to visualize water flow (liters per day) for a new, advanced detached home in New Delhi.

backgrounds. Few people understand the whole picture. In fact, in most cases, there is nobody in the entire city who is able to describe the energy system, including the diversity and relative weights of primary energy inputs, the relative importance of each energy demand, the amount of fossil fuels embodied in local electricity, and the share of energy that is cascaded to secondary uses. However, after a few minutes with a meta diagram, people will likely understand the basics (figure 2.11).

Typically, a meta diagram is constructed to reflect the substances, processes, and time periods relevant to specific decisions. Sometimes, a combination of meta diagrams is most effective. If energy flows are averaged over a year, for example, they provide a good benchmark for tracking overall efficiency and understanding ecological footprints. However, annual energy flows fail to reveal the seasonal and daily peaks that affect costs, which are often key determinants in system design. Thus, a meta diagram based on peak-hour energy flows for the peak month (or daily flows for water during the driest month) might be useful for understanding the big picture, especially in evaluating alternative system designs.

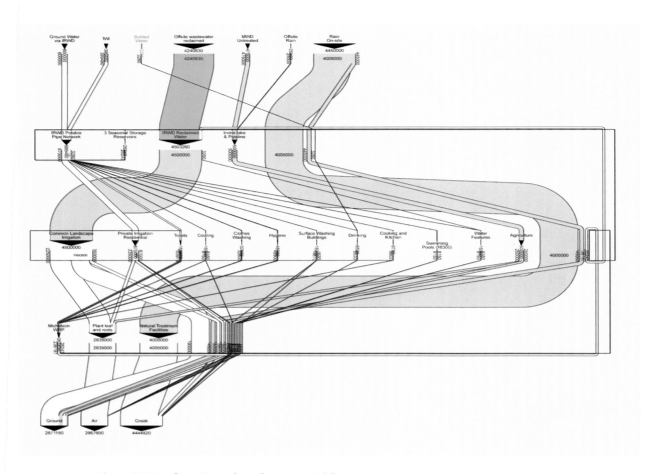

Figure 2.10 Baseline Water Flows for Irvine, California

Source: Author elaboration (Sebastian Moffatt), with approximate data provided by Mike Hoolihan and the Irvine Ranch Water District (2008).
Note: This diagram illustrates the effective use of reclaimed water for the irrigation of commercial and public land (average cubic feet per day). Irvine has the most advanced city water system in the United States. Note the diversity of water sources, including water harvested off-site and stored in an artificial lake and large amounts of piped water in the metropolitan water district imported from northern California. Freshwater is stored in ground aquifers and then harvested by the Irvine Ranch Water District, combined with expensive imported water, and used for hygiene, cooking, surface washing, and so on. Most of the water flowing through the city's land is rainfall, which is treated in constructed wetlands and released into a creek. A second large flow of water is reclaimed wastewater, which is used to irrigate landscapes on public and commercial properties during the driest periods. The most significant use of fresh water (imported and groundwater) is in the irrigation of the lawns around private homes. This diagram illustrates the importance of finding a way legally and safely to use reclaimed water to irrigate residential properties, a strategy the water district is now exploring.

2. **Creating a common language for interdisciplinary groups.** Meta diagrams help everyone understand infrastructure as a whole system and then focus on those parts of the system in which resource use is high and in which opportunities may exist for significant efficiency, reuse, or substitution. The diagrams provide a common language for exploring the key opportunities for integrated, holistic solutions.

By compiling and comparing a diversity of meta diagrams, one may identify a simple pattern language for physical flows at any scale (figure 2.12). The first pattern, *traditional,* is typical of the oldest and also the poorest houses in China and India. Total resource use is relatively small, but the mix of primary resources is complex. For example, among energy flows, each fuel is carefully matched to the requirements of the end use for optimum efficiency and lowest cost. Thus, coconut husks are used for heating water; liquefied petroleum gas is used for stove cooking; wood is used for open cooking; solar energy is used for clothes drying; kerosene is used for lighting; electricity is used for refrigeration; and petroleum is used for motor scooters. The traditional home may be poor and old, but the energy systems are relatively sophisticated.

The second pattern, *modern,* is based on the newer tract-built houses in the urban suburbs surrounding Shanghai, China, but the pattern is typical of suburban homes worldwide. Total resource use is almost an order of magnitude greater than the resource use associated with traditional homes, even though family size has typically dropped 60 percent or more. The primary mix of energies is simple because almost all energy demands involve gas- or coal-generated grid electricity, with the exception of the demands related to cooking and transportation.

The third pattern, *ecological,* is typical of more sustainable integrated systems that incorporate demand-side management and reuse. The resource load is midway between the traditional and modern. It combines the complexity of the traditional with the convenience of the modern. Some energy recycling (cascading) increases the service value of the flows, which raises the total flow at the demand partition relative to the other sources and sinks. The primary mix is even more complex than the traditional pattern because of the use of hybrid systems with intelligent controls and because of the greater diversity offered by networked local energy services. However, the greatest difference may be the increased flexibility and adaptability of the ecological home.

3. **Developing and communicating alternative development scenarios.** Scenarios for future development may be presented as meta diagrams and compared with the base case or other scenarios. Figures 2.13 and 2.14 portray scenarios for energy use in Jinze, a town in Shanghai Municipality; they convey a radical change in the electricity mix. Figure

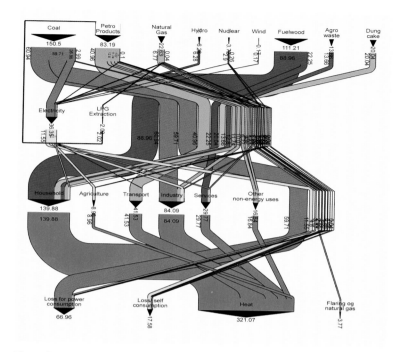

Figure 2.11 An Example of a Countrywide Meta Diagram

Sources: Data from TERI (1997); analysis by S. J. Prakash and Associates, Delhi; noncommercial biomass data from Society for Environmental Communications (2002).
Note: This meta diagram portrays the energy flows for India. Note the predominance of coal, which is used primarily for manufacturing, and oil, which is used for transport. In addition, note how much coal is wasted as heat and the secondary rank of informal biomass as fuel. Electricity use is relatively low, and per capita consumption is low, but emissions are high. LNG=liquefied natural gas.

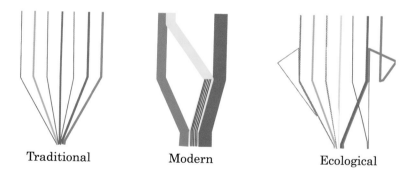

Traditional   Modern   Ecological

Figure 2.12 Meta Diagrams Patterns: Physical Flows

Source: Author elaboration (Sebastian Moffatt).
Note: A meta diagram's pattern language shows the possible evolution in technology for mass and energy flows at the parcel and regional levels.

2.15 is a schematic that elaborates on the scenario for Jinze, showing the system components for a typical downtown neighborhood on a canal. The schematic provides information on the spatial configuration of the technologies referenced in the meta diagram.

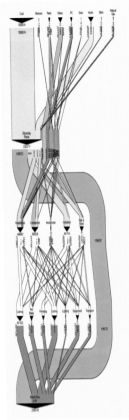

Figure 2.13 Meta Diagram for Jinze, Shanghai: The Current Energy System

Source: Author elaboration (Sebastian Moffatt) with approximate data provided by Professor Jinsheng Li, Tongji University, Shanghai. More details available at www.bridgingtothefuture.org.

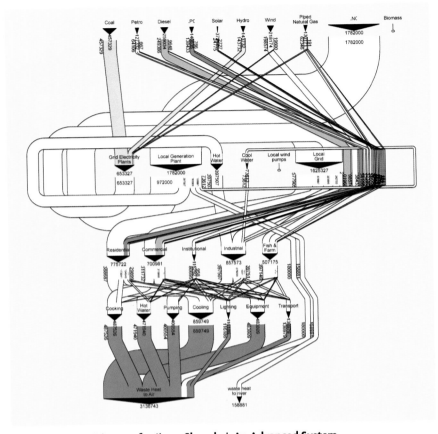

Figure 2.14 Meta Diagram for Jinze, Shanghai: An Advanced System

Source: Author elaboration (Sebastian Moffatt) with approximate data provided by Professor Jinsheng Li, Tongji University, Shanghai. More details available at www.bridgingtothefuture.org
Note: This meta diagram provides a scenario for an advanced system that helps reduce emissions and costs and increase local jobs and energy security. The advanced system represents a substantial change. For example, a local electricity generation facility is powered by liquefied natural gas and provides a majority of the electricity needs and hot and cool water for industry (cascading).

Generating scenarios with meta diagrams may be rather simple once base cases are completed. A new energy source or converter may be added and connected to parcels. The population of each category of parcel may be adjusted to reflect plans for upgrading buildings. For example, we might replace 1,000 older dwelling units with 1,000 retrofitted dwelling units and instantly see the impacts on water, energy, and material flows and on total economic costs and carbon emissions. Because every parcel uses the same database structure, meta diagrams may be joined. For example, it is easy to combine parcels to create a systems perspective on resource use in a specific neighborhood, development project, or category of housing. A parcel may be any discrete surface area (for example, a park, a house on a private lot, a shopping mall, or a sewage treatment plant, or a roadway). All parcels are connected. Each parcel demands resources from other parcels, and if infrastructure is distributed, the infrastructure may provide other parcels with resources.

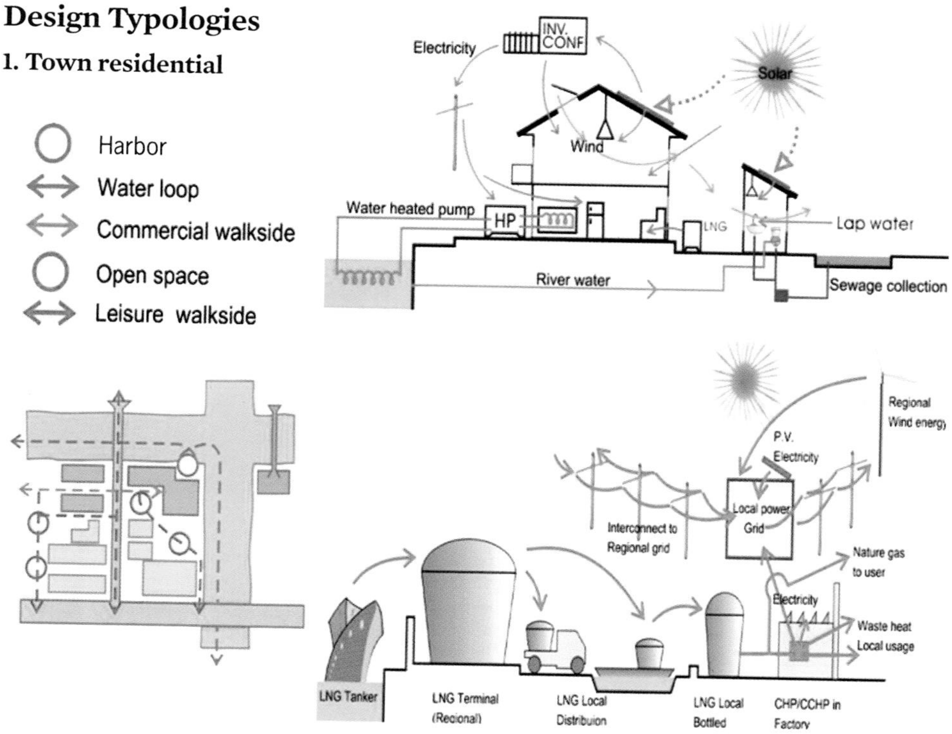

Design Typologies

1. Town residential

- ◯ Harbor
- ⟷ Water loop
- ⟷ Commercial walkside
- ◯ Open space
- ⟷ Leisure walkside

Figure 2.15 A Schematic for a Downtown Neighborhood

Source: Li (2006).

Note: This schematic elaborates on one of the design typologies that underlie the advanced energy systems of Jinze, Shanghai (see figure 2.14). Note the combination of a centralized electricity grid (liquefied natural gas) with distributed infrastructure, including solar photovoltaic installations with grid, solar domestic hot water, a river heat pump, and wind-driven ventilation.

4. **Setting priorities for research and design.** Understanding where waste is occurring and the relative importance of various resources and demands is essential for establishing research and design priorities. Each node presents opportunities for substitution, efficiency, looping, and cascading. Figure 2.16 portrays an energy analysis for a proposed town of 50,000 people in southern India. In this case, the combination of meta diagrams helps emphasize the importance of addressing transportation demand in future plans. Often, a combination of meta diagrams helps zero in on particular issues. A meta diagram for the driest month helps assess the potential for self-reliance. A meta diagram that shows only residential demands in precise detail is helpful in preparing policy for a residential neighborhood.

5. **Calculating performance indicators in transparent and comparable ways.** The meta diagram may be used not only for systems analysis, but also for generation of specific indicators of performance. In fact, every flow portrayed on a meta diagram is a potential indicator that may be monitored over time or compared with other locations or other scenarios. The balanced nature-to-nature flows on a meta diagram may be converted to money or emissions where relevant and, thus, provide an average life-cycle inventory of all costs. A material flow analysis offers a consistent method for tracking all consumption, emissions, and expenses at each stage in the life cycle and is thus the preferred method for assessing internal and external costs. The meta diagram helps clarify exactly what is included and excluded in calculations. For example, the total water consumption for any

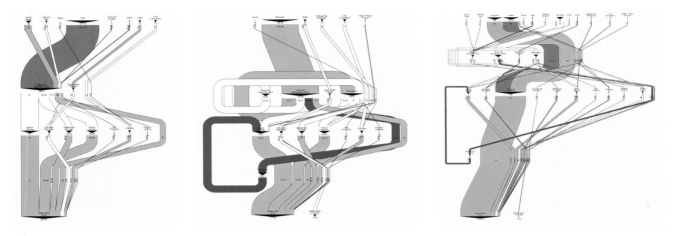

Figure 2.16 Meta Diagrams on Energy for a Proposed New Town

Source: Author elaboration (Sebastian Moffatt).

Note: This series of energy meta diagrams was used to guide development plans for a proposed new town near Poona, India. The first diagram represents a business-as-usual scenario. It shows how current development practice in southern India encourages greater use of coal-generated electricity. The second meta diagram portrays an advanced system with biomass brought by train and used in a local district energy plant, with the cascading of energy. The third meta diagram includes the transportation energy that was ignored by the designers and is missing from the other meta diagrams. Note that, because residents are expected to commute, transportation-related energy exceeds all other energy uses combined. The third meta diagram suggests that a reduction in the need for commuting and the provision of incentives for the creation of quality transit systems must be a priority in urban design in affluent new towns.

particular use may be clearly subdivided into off-site potable water, on-site water (the roof catchment), and reclaimed water. Without this type of separation, it is impossible to understand a water consumption indicator. By standardizing the meta diagram format, one may directly compare results from different locations or time periods and create comparable benchmarks for assessing system performance and trend lines. Comparable benchmarks also help in the important process of establishing long-term targets for resource use. For example, authorities in the resort municipality of Whistler, which represents one of Canada's leading examples of sustainable planning, were unable to agree on long-term performance targets for a set of indicators until they benchmarked their current performance relative to other leading resorts in North America (figure 2.17).

Creation of meta diagrams if data are lacking

Creating meta diagrams is easy once the data are properly stored in a database or spreadsheet. In fact, simple software applications may be used to draw the diagrams automatically. The difficulty arises in collecting baseline data to portray existing conditions or to construct a business-as-usual scenario. Two kinds of baseline information may be used (figure 2.18).

1. **Top-down data** establish how much of any given resource (energy, water, and material) was actually sold, delivered, or imported during the most recent period. If one is dealing with a greenfield development, then top-down data may be used from a neighboring site as a proxy for business as usual. Once the inputs are known, the rest of the database may be constructed using population data and default values for demand by end-use category. For example, we might imagine a situation in which the population is 10,000 and the average person uses 200 liters per day of municipal water, divided into toilets (40 percent), showers (5 percent), surface washing (8 percent), and so on.

2. **Bottom-up data** aggregate the flows of any given resource by beginning with detailed flows generated at the scale of various types

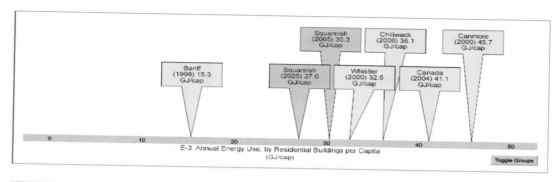

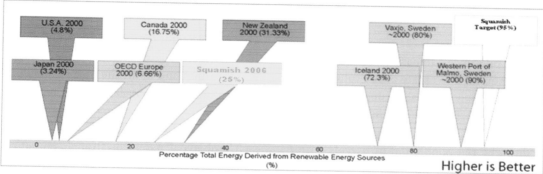

Figure 2.17 Annual Energy Use as an Indicator in Squamish, Canada

Source: Author compilation (Sebastian Moffatt), adapted from Sheltair Group (2007).

Note: On both of these benchmarking scales, it is possible to compare the performance of Squamish today with the performance in other locations. In the top chart, the annual energy use in residential buildings is compared with the corresponding energy use in other mountain resort communities. In the bottom chart, the percentage of total energy derived from renewable energy sources is compared with the percentages in countries around the world. Note that Squamish has set a target for renewables of 95 percent by 2025. OECD = Organisation for Economic Co-operation and Develpment.

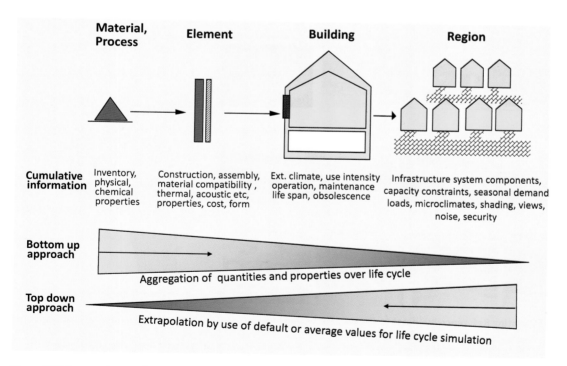

Figure 2.18 Approaches to the Development of Meta Diagrams

Source: Author elaboration (Sebastian Moffatt), with assistance from Niklaus Kohler.

of parcels or discrete pieces of land with their attendant buildings and end uses. This approach provides much greater precision and is preferred in dealing with existing stocks of buildings. Parcels are grouped into categories based on land use and demand profile, (for example, prewar low-rise multiunit residential or recent strip mall commercial). Aggregating parcel information requires that experts visit and audit several typical parcels within each category and use these parcels to create a solid reference database. Reference parcels are then used to create proxy values for all the parcels within each category. The total (aggregate) flow for the meta diagram is calculated simply by multiplying the proxy flows by the population of parcels within each category. Using such shortcuts, it is possible to determine an accurate baseline flow quickly (+/– 10 percent). Figure 2.19 shows an example from Squamish, Canada, where a diverse collection of reference

parcels was audited and used to create an energy meta diagram for the whole region.

Tools for aggregation

In the development of a reference database on existing stock, all data collection occurs at the level of an individual parcel. Flows are recorded in a predefined matrix that corresponds with the Sankey diagram structure. Thus, the flows for each node on the parcel are connected with nodes upstream or downstream to account for all sources and destinations. By cross-referencing flows in and out of each partition and node, the matrix functions as the numerical equivalent of the meta diagram. A matrix may be automatically generated either from empirical (field) data collected on each archetypal parcel or from hypothetical data deduced from a theoretical parcel design.

Field data and hypothetical data for reference parcels need to be converted into flows of resources. The conversion is accomplished using standard models for predicting thermal

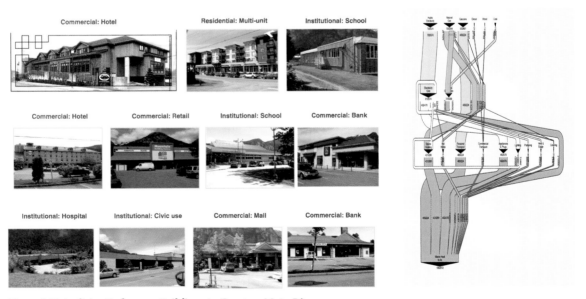

Figure 2.19 Auditing Reference Buildings to Create a Meta Diagram

Source: Author compilation (Sebastian Moffatt), adapted from Sheltair Group (2007).
Note: Carefully selected reference buildings in Squamish, Canada, were visited, audited, and used as proxies for the various categories of building stock. With the use of these reference buildings, a complete energy meta diagram was generated for the region. The result reveals a simple energy mix, with almost no cascading or on-site generation. This condition is typical of an area, such as Squamish, in which energy prices are low. The large share of energy used for personal transportation is typical of a dormitory community; in Squamish, two-thirds of the working population is employed elsewhere.

loads, water demand, and so on. For example, a data collection form may keep track of primary data such as type of appliance and number of occupants, and these data may then be used to calculate the probable flows of water, energy, materials, and people for each purpose. Data collection forms need to accommodate a wide variety of lifestyles and building types. Table 2.2 shows excerpts from data collection forms developed for water flows. Similar forms may be used for energy and organic materials. The forms are fairly simple, but they require that connectivity be recorded. For example, the forms shown in the table record exactly where the roof drainage water is output: to the ground, cistern, street, garden, sewerage, storm drain, or some combination of such destinations.

The data collected on each parcel may be used to generate the inputs for a universal flow matrix automatically (figure 2.20). The matrix may then be used to produce files to generate meta diagrams with the aid of various diagramming tools.

A parcel may be any discrete surface area (for example, a park, a house on a private lot, a shopping mall, a sewage treatment plant, or a roadway). The single format for data structure allows for each parcel to demand flows and to serve other parcels as a supply (or service) node. Thus, the data structure allows for transformer parcels that evolve over the long term to become part of integrated and distributed infrastructure systems. For example, a single-family home may begin as a water or energy demand node in the regional system, but if the roof is retrofitted to catch rainwater or solar energy, the database easily accommodates the changes. The use of this standardized data structure also helps visualize the process of stock aggregation. Designers may move from a Sankey at the parcel or building scale to a Sankey at other scales simply by stacking the database for each parcel within the larger area and adding the cells. The systems perspective is always maintained.

Effective Overlay Mapping

When a picture is worth a thousand words

The best way to communicate complex information to planners and designers is through pictures, including maps, photos, schematics, or a combination of these. The history of using maps quickly to convey complex relationships between the built and the natural environments begins with *Design with Nature* (McHarg 1969). Although Ian McHarg's simple overlays of transparencies are still good tools, the options have evolved considerably with geographic information systems (GISs) on computers and the Web. GIS is now a mature, affordable, and widely used technology for mapping and spatial analysis that will soon be part of the standard practice in all cities in all countries. All large metropolitan areas now have GIS departments and routinely use GIS to assist in design and management.

In the context of building capacity for Eco^2 projects, cities require GIS and related visualization technologies to support the interdisciplinary planning process. Initially, GIS applications need not be demanding or time-consuming. All that is required is (1) the capacity to produce simple overlay maps that consolidate spatially referenced information and help planners recognize relationships and patterns on the landscape and (2) the capacity to calculate a few simple spatial indicators, such as density, diversity, and proximity (figure 2.21). Such capacities are absolutely essential in supporting charrettes, foresight workshops, and other integrated design exercises.

Unlike many GIS applications, the generation of overlay maps and the calculation of spatial indicators provide exceptional value in exchange for a small investment in time and human resources. Moreover, new technology is now allowing for visualization in a wider variety of formats that also contribute to decision making. For example, simple contour maps (also referred to as digital elevation models)

WATER DEMAND	Units	Values	List of Options					
Laundry								
Clothes washing system		None	None	Full size stnd top loading	Full size, side load or short cycle	Compact (<45 liters) top loading	(<45 liters) side load	Advanced water efficient
Number of full loads created	per person week	0	0	0.5	1	1.5	2	2.5
Washing appliance	liters per load	0						
Personal Hygiene								
Shower use	per person week	0	0	1	2	3	4	5
Bath use	per person week	0	0	1	2	3	4	5
Shower system and length		None	None	Standard long (8 minutes)	Standard Short (5 minutes)	Low Flow Long	Low Flow Short	Bucket
Bath		None	None	Full	Normal	Bucket		
Hand & face, shaving, brushing		None	None	Tap on constantly & long	Tap on constantly short	Tap on constantly	Tap off except when essential	
Showering system	liters per shower	0						
Bath	liters per bath	0						
Hand & face, shaving, brushing	liters per person	0						
Kitchen								
Cooking Frequency	meals per person day	0	0	1	2	3	4	5
Dishwashing System		None	None	Basin or Sink	Stnd Machine	Water Eff. Machine		
Number of full loads created	per person week	0	0	0.5	1	1.5	2	2.5
Dishwashing system	liters per load	0						
Toilets								
Primary toilet water system		Standard	None	Standard flush	Low flush	Low volume dual flush	Extra low w' dual	Compost Toilet
Primary toilet use	flushes per person day	4	0	1	2	3	4	5
Secondary toilet water system		None	None	Standard flush	Low flush	Low volume dual flush	Extra low w' dual	Compost Toilet
Secondary toilet use	flushes per person day	0	0	1	2	3	4	5
Primary toilet category	liters per flush	22						
Secondary toilet category	litres per flush	0						
Drinking								
Irrigation								
Cummulative operating time for all irrigation pipes & outdoor watering taps (excluding reuse of	hours per month	0	0	0.5	1	1.5	2	2.5
Potted Plants and Pools								
Typical quantity of water per	liters per month	0	0	2	4	6	8	10
Interior Surface Cleaning								
Frequency of interior surface	times per week	7	0	1	2	3	4	5
Quantity of water used (excluding reuse of wash water)	liters per event	4	0	1	2	3	4	5
Exterior Surface Cleaning								
Days per month exterior	number of days	0	0	1	2	3	4	5
Duration of watering	minutes per cleaning	0	0	5	10	15	20	30
Vehicle Cleaning								
Number of 4 wheel vehicles cleaned on-site	number of vehicles	0	0	1	2	3	4	5
Number of 2 wheel vehicles cleaned on-site	number of vehicles	0	0	1	2	3	4	5
Frequency of cleaning	per vehicle each week	0	0	1	2	3	4	5
Evaporative Cooling								
Typical frequency of use during	hours per month	0	0	50	100	150	200	250
Category of cooling system		None	None	Small (residential), no bleed	Multi-unit, no bleed	Large	Small with bleed	Multi-unit with bleed
Consumption of water by cooler	liters per hour	0						
Humidification								
Typical monthly water consumption	liters of water per month	0	0	5	10	15	20	25
Client Demand								

Table 2.2 Sample Forms for the Collection of Standardized Data on Water Flows

Source: Author compilation (Sebastian Moffatt).

Note: The table shows a complilation of sample computerized forms used to collect standardized data on water demand and water flow connections at the level of a land parcel.

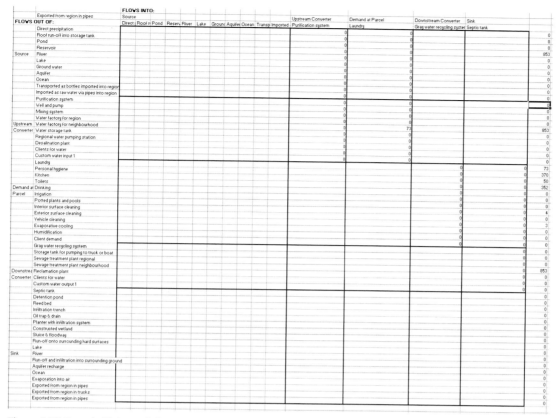

Figure 2.20 Sample Universal Flow Matrix for Water

Source: Author elaboration (Sebastian Moffatt).

Note: The figure shows an example of a universal matrix that identifies all water flows at the parcel level by quantity and direction from source to sink.

may be combined with air photos (for example, Google Earth) to produce three-dimensional imagery. Such techniques provide planners and others with the ability to fly through a digital landscape that has acquired the look and feel of a proposed development. With additional training, specific objects in the GIS database may be given attributes related to resource consumption, and GIS may evolve into a scenario development tool (for example, CommunityViz; see elsewhere below).

Getting real value from mapping

One of the challenges in overlay mapping is avoiding the common problem of GIS for the sake of GIS. Traditionally, GIS work has been far removed from decision making and has been surrounded by a mystique that obscures the simple nature of the tool and its role in planning. Without lots of advance notice and direction, GIS departments produce maps that are complex and colorful, but provide little added value. To realize the great potential of overlay mapping, one may find it helprul to consider the following suggestions.

Provide clarity on the key questions decision makers are asking

For example: Where are the ecological assets? Where are the threats to urban systems?
Integrated design workshops depend on the use of maps to inform interdisciplinary groups about many factors that influence the performance of infrastructure systems. For instance, maps may help create an integrated understanding of the potential for taking advantage of existing eco-

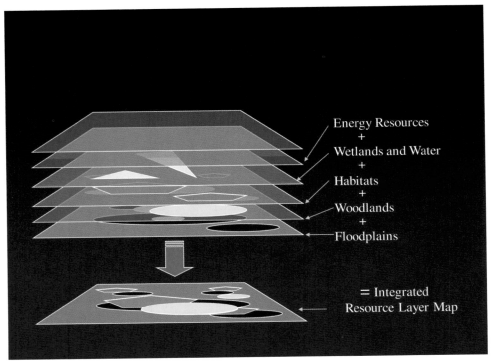

Figure 2.21 Layering Data

Source: Author elaboration (Sebastian Moffatt), with assistance from Lex Ivy, Terra Cognitio GIS Services.
Note: This example illustrates how a number of layers of information on the natural capital of a region have been visually integrated on a single map. This information may be useful for strategic land use decisions and in identification of infrastructure system design options.

logical assets inside or near a city. A map might show the locations with potential for generating renewable energy from wind, microhydro, biomass, geothermal, tide, industrial process, and other sources. Overlay mapping also provides a useful way of evaluating existing infrastructure systems and comparing their capacity in specific locations relative to the projected demand arising from growth in populations and economic activities. With such information consolidated on a single map, groups may integrate local energy assets relatively easily as they plan new energy infrastructure systems and urban settlements (figure 2.22). The same overlay process works in all sectors.

An example of hazard mapping at a larger scale is the work recently completed by Gujarat State Disaster Management Authority in India on a composite risk atlas aimed at assisting the various departments involved in disaster mitigation planning in areas that are

the most vulnerable to natural or constructed hazards. The GIS consulting team has compiled one of the largest and most detailed digital GIS databases ever prepared in India and then generated an atlas that reveals the relative risks to life and capital from earthquakes, cyclones, storm surges, floods, chemical accidents, and droughts.

An example of overlay mapping on risk is shown in figure 2.23. In this example, regional growth planning for energy and transportation in the City of Squamish is informed by the distribution and intensity of risks on the landscape, including risks associated with landslides, earthquakes, floods, and unstable soils. All risks have been consolidated as layers on a single multihazard risk assessment map. This community is located in a flood-prone and geologically active area, and the map reveals few locations for the safe development of land for residential and commercial uses. Moreover, the existing

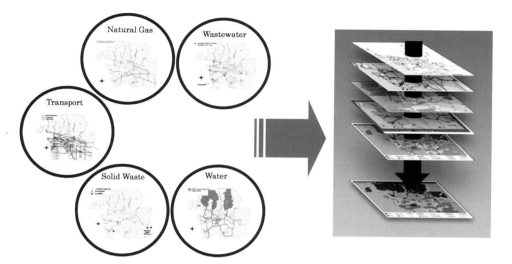

Figure 2.22 Overlay Mapping

Source: Author elaboration (Sebastian Moffatt), with assistance from Metro Vancouver GIS Department.

Note: In this example of overlay mapping, multiple infrastructure systems have been layered onto a single map to depict the location of manufactured capital assets. This information may also be useful in land use planning and in optimizing the use of existing infrastructure systems.

natural gas and electricity infrastructure systems and the major transportation routes for rail and road are already exposed to significant risk because they are located in the fan of projected debris from landslides off the east slope of the mountains. Even the electrical substation has been incorrectly located in this hazardous area because of the lack of prior overlay mapping.

Another example of innovation in overlay mapping is the collection of maps in figure 2.24, which displays a series of maps for the same Squamish region. Each map focuses on a different renewable energy asset. By overlaying the energy asset maps, one produces a tool that makes it easier to plan development based on local renewable energy resources. For example, a new residential development may be located in areas in which the sunshine permits year-long use of solar water heaters and in which new buildings stay clear of the windy ridge north of town, where average wind speeds are sufficient to support a wind farm.

It is not unusual for land use and infrastructure plans to proceed without any reference to the risk of natural disasters or to other critical landscape relationships such as local resource

assets, ecological functions (such as rainwater catchment, food production, and wind protection), and unique or ecologically sensitive areas that contribute to local biodiversity and ecological health. Design teams and policy experts are able to adapt their policies and designs to this information only if it is available in a timely and easy-to-understand format.

A more common and especially useful application of overlay mapping involves mapping the capacity of existing infrastructure and comparing this with the projected demand for services. Many urban regions are now using this type of overlay mapping to assist in growth management. Areas with surplus capacity within their infrastructure systems are the most appropriate locations for new development or infill, all other factors being equal. Areas with especially high demand may be appropriate for localized infrastructure systems; for example, high energy demand makes district energy systems cost-effective. Because maps are available on such areas, one may more easily implement the policies that make buildings suitable for hooking up to a local network. This kind of forward-looking policy helps create a municipal ecology in which many

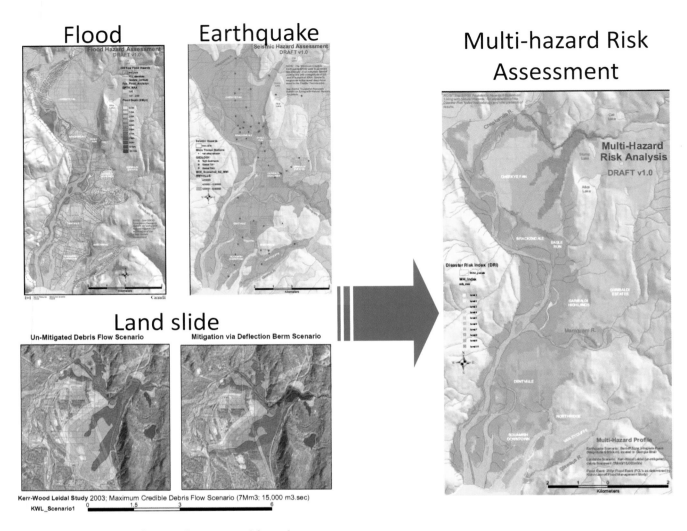

Figure 2.23 An Example of an Overlay Map Used for Risk Assessment

Source: Overlay maps completed by Pathways Group at Natural Resources Canada and presented as a contribution to the Canadian team in the "Bridging to the Future Project" (Sheltair Group 2007).

Note: Combinations of landscape risks may be overlaid to create a map for multihazard risk assessment that quickly and easily communicates which landscapes are suitable for specific uses.

locations function as both supply and demand nodes for flows of resources.

Focus on quality inputs

Similar to the case of meta diagrams, the difficulty in overlay mapping is the scarcity of reliable data. Maps may be beautiful, but their utility depends on the accuracy and scope of the data that have been supplied. In fact, the mapping of ecological resources within and around a city, for example, is a relatively simple task that may be performed by any recent college graduate.

However, inventorying resources is not so simple. A substantial investment is sometimes required to survey resources and document conditions throughout a region. The process involves experts in many disciplines. Because every location is unique, there are few shortcuts. One technique for speedy data collection is the use of photography from the air, in combination with a global positioning system, to create data rapidly on the length and area of the natural and built elements of a region, including building footprints, the length of key streets and shorelines, and the characteristics of open

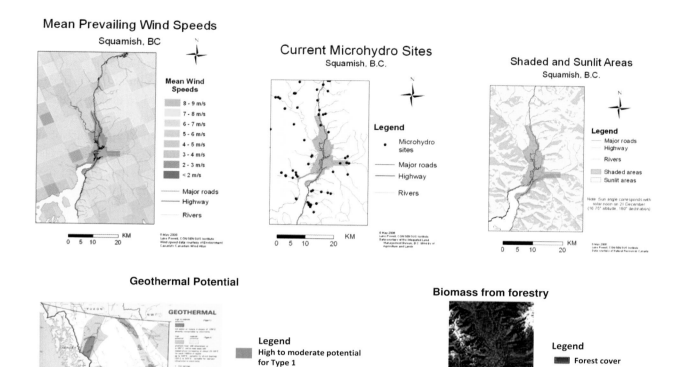

Figure 2.24 An Example of an Overlay Map of Renewable Energy Sources

Source: Overlay maps completed by Pathways Group at Natural Resources Canada and presented as a contribution to the Canadian team in the "Bridging to the Future Project" (Sheltair Group 2007).

Note: Individual energy asset maps may be viewed together or overlaid to produce a picture of all areas in a region that have easy access to renewable energy resources. With appropriate growth management and the aid of the maps, a city may become well positioned to achieve energy independence and carbon neutrality. KM=kilometer.

spaces. The information collection and storage plan are the most important elements of the method and need to be addressed as part of an Eco² pathway. A good example of how a broad information strategy leads to more effective mapping tools is found in the World Bank's primer for cities on climate change, which includes a step-by-step approach to identifying hot spots that are especially vulnerable and then exploring mitigation options (see Prasad and others 2009).

Integrate local knowledge

Another key part of an information strategy is engagement with knowledgeable local resi-

dents in the mapping work. Workshops may be organized for tapping into this information and creating maps that are far more informative through a process that is more inclusive. Recently, this sort of community mapping has been successfully applied in many locations. Communities often have a truly surprising amount of information to contribute, much of which cannot be obtained in any other fashion.

Take advantage of technologies for sharing results

The use of Web-based GIS applications is an emerging technology that leverages the benefits of mapping. Colorful maps and images on the

Internet offer greater potential for public and stakeholder participation and help decision makers collect diverse viewpoints on how to improve the quality and the acceptability of plans. Also, the maps are open to scrutiny by a wider audience that may provide ideas for additional spatial inquiry, have access to more information or updated information, and point out inadequacies in the maps.

Work toward scenario-based GIS

As capacity builds within a city, overlay mapping methods may evolve to include powerful scenario-based GIS. Such applications can rapidly alter maps to reflect changes in design and automatically generate precise calculations of spatial indicators and resource flows. An exam-

ple is CommunityViz, a software package developed for cities and made available at a minimal cost through the Orton Family Foundation, a charitable foundation. CommunityViz greatly reduces the time required to create plausible scenarios for urban system design and for establishing a standard protocol for the use of indicators and benchmarks (figure 2.25). It also provides a convenient basis for sharing data and integrating results across departments or institutions. In the Squamish application described above, CommunityViz has been used as a common platform by three separate design teams: *smart growth* (urban form and transportation), *pathways* (risk management and natural hazards), and *bridging to the future* (30-year pathways for sustainability).

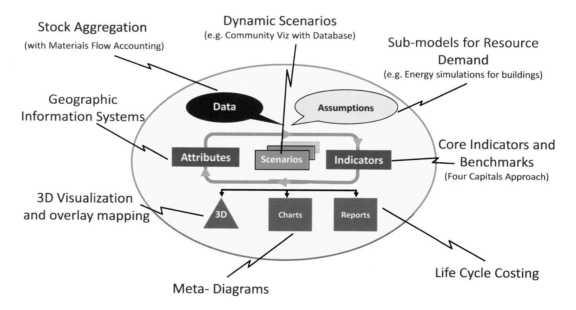

Figure 2.25 CommunityViz

Source: Author elaboration (Sebastian Moffatt) with a core schematic adapted from Orton Family Foundation (2009).

Note: CommunityViz is a GIS application based on scenarios and indicators that may be used in concert with other methods to produce much of the information required for assessing development options.

References

Li Jingsheng, ed. 2006. "Bridge of Jinze, Bridging to the Web." Bridging to the Future Project. http://www.bridgingtothefuture.org/sites/default/files/China%20Bridging%20to%20the%20Future%20presentation.pdf.

McHarg, Ian L. 1969. *Design with Nature*. Wiley Series in Sustainable Design. Garden City, NY: Natural History Press.

Orton Family Foundation. 2009. "CommunityViz User's Guide." Orton Family Foundation, Middlebury, VT.

Prasad, Neeraj, Federica Ranghieri, Fatima Shah, Zoe Trohanis, Earl Kessler, and Ravi Sinha. 2009. *Climate Resilient Cities: A Primer on Reducing Vulnerabilities to Disasters*. Washington, DC: World Bank.

Rees, William E. 2002. "Globalisation and Sustainability: Conflict or Convergence?" *Bulletin of Science, Technology and Society* 22 (4): 249–268.

Sheltair Group. 2007. "Bridging to the Future in Squamish, BC: Summary Report—New Directions for Energy System Design." Prepared for District of Squamish, BC, March 2007. Available at http://www.squamish.ca/downloads/community-energy-action-plan.

Society for Environmental Communications. 2002. *Down To Earth: Science and Environment Online*. December 15, 2002. Center for Science and Environment, New Delhi. http://www.downtoearth.org.in/default.asp?foldername=20021215.

TERI (Tata Energy Research Institute). 1997. *TERI Energy Data Directory and Yearbook 1997/98*. New Delhi: Teri Press.

CHAPTER 10

Methods for Investment Planning

Investing wisely in urban development is a complex process. A large number of professionals must be engaged (architects, designers, suppliers, engineers, economists, and financial planners), each of whom brings a different concept of what is important and how it may be measured (figure 2.26). Construction occurs in many phases and over many years (programming and planning, design and engineering, construction, operation, and dismantling). The final product is composed of many levels of subsidiary products (materials, components, technologies, whole buildings, infrastructure systems, and open space). At each phase and on every scale, different actors are involved in the decision-making process. The complexity of interactions among these actors and among the many elements is one of the most significant challenges faced by someone trying to assess the real costs and benefits of alternative plans for development. Investing wisely in urban development revolves around coping with complexity.

A number of assessment methods are available to help cities cope with complexity. The following methods are worth considering. Each of them may be applied using simple and scalable tools that adjust to the needs and capacities of the user.

First and foremost is life-cycle costing (LCC), which is used to understand many of the indirect and contingent costs associated with any project design over the expected lifetime of the facilities. Some LCC tools work with complete urban environments, including spatial elements and infrastructure. Other LCC tools work exclusively with specific types of infrastructure facilities, such as treatment plants and power plants. We look at both types.

The second method to consider is environmental accounting, which attempts to add up the lifetime environmental impacts of a project. Environmental accounting includes material flow analysis, but also expands the scope to encompass the broader impacts of specific projects on the environment, such as resource use and depletion and the costs of emissions.

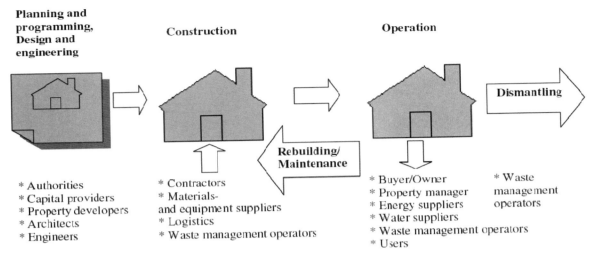

Figure 2.26 The Life Cycle of a Building

Source: Brick (2008).

Note: The life cycle of a building is long and complex and influenced by many different professionals.

This method may be applied in demonstration projects because it helps reveal the whole picture; life-cycle environmental accounting helps identify areas in which the problems are the greatest. For example, ecological priorities among rapidly industrializing countries include cutting back on the excessive amounts of heavy masonry used in construction, converting the inefficient and polluting energy systems used to produce materials, and enhancing the durability of concrete and other key materials that affect the lifetimes of buildings and infrastructure. The challenges involved in undertaking environmental accounting on projects are centered on the time and effort needed to consider all the inputs and outputs, including those that are embodied in the materials and services procured for the project.

A third method is risk assessment, which is especially important during times of rapid change and, yet, is largely ignored by urban professionals. A full assessment of risk requires that cities consider a variety of future possibilities and research the likely impact of trends in many areas, from climate change to technology. Scenario planning and adaptation are difficult, but they are also rewarding exercises and are a key part of investing in resilience and sustainability.

A fourth method is the overall valuation of a city's performance on an Eco^2 pathway and on a long-term and project-by-project basis; this involves using a holistic set of indicators to assess costs and benefits. It is important for cities to review ways to select performance indicators that are useful to decision makers at varying levels and that cover the full range of assets within a city: manufactured capital, natural capital, human capital, and social capital. It is important that these indicators include qualitative attributes such as historic, traditional, or cultural dimensions that cannot be represented monetarily. The challenge is to find ways to measure these assets that are balanced, affordable, dependable, and comparable and to present the results in a format that may be easily understood by decision makers.

Life-Cycle Costing

One of the most significant factors influencing decision making on urban area development is the long-term impact on city finances and the costs for residents and businesses.

Unfortunately, in complex integrated designs that include new land uses and new infrastructure, costs and revenues are often difficult to estimate. Good commonsense arguments might exist for alternative Eco² designs, but, without accompanying financial analysis, decision makers may be understandably reluctant. This is especially true because of the prevailing misconceptions and widespread misinformation. Moreover, it is not always easy to recognize the win-win potential, and the assumption may be that change is costly, at least until a financial analysis is complete. The debate over costs has two dimensions: What are the true costs over the long term? And how equitably are costs and benefits distributed? An LCC method is the only way to address both these questions.

LCC for urban infrastructure applies to all components, from buildings to roads; pathways; rights-of-way; parking lots; wires; pipes; ditches; bridges; and the associated treatment plants, substations, open spaces, and facilities. Most of these infrastructure elements have exceptionally long lifetimes and may account for large quantities of life-cycle material and energy flows. By adopting LCC, one may adjust the design and purchasing choices to optimize the system over its entire life. The composition of sewerage pipes, for example, may be changed from concrete to welded steel if due consideration is given to durability, cleaning, maintenance, and other recurring costs over the full life of the system and to the potential for adaptability and recycling. Road surfaces may be changed from asphalt to concrete if due consideration is given to the improved efficiency and fuel savings achieved by truck tires on concrete. Linear in-ground infrastructure grids may reflect the adoption of combined trenching and easily accessed utilidors if these are shown to offer greater adaptability and lower operating costs over the long life cycle of such systems. LCC can produce even greater changes in practice in terms of spatial planning, as described in part 1.

LCC applications in integrated land use and community infrastructure planning

LCC may assist with the evaluation of alternative spatial planning activities by providing credible estimates of the full long-term costs for infrastructure and the impacts on taxes, fiscal health, affordable housing, and commercial space. The benefits are most effectively understood through a specific example of the application of LCC. The example presented here is based on a tool developed by the Canada Mortgage and Housing Corporation, a Canadian public agency, for use in integrated design exercises. The tool, the life-cycle costing tool for community infrastructure planning, allows users to compare alternative development scenarios and to estimate the major costs of community development, particularly the costs that change depending on the form of urban development (for example, linear infrastructure). The tool is geared toward estimating planning costs and the revenues associated with the residential component of a development. The financial impacts of commercial development and other types of development may be incorporated, provided infrastructure requirements have been specified correctly. The tool is suited to assessing development projects ranging in size from a collection of houses to a block-by-block infill development to an entire subdivision or neighborhood. A good measure of the applicability of the tool to a given project is whether or not alternatives may be conceived that would result in significantly different densities or infrastructure requirements or make use of different green infrastructure alternatives.

The tool is a spreadsheet-based application (Microsoft Excel) that makes the estimation of life-cycle costs quick and easy for almost any type of land use and infrastructure alternative. It includes costing variables with default values that may be adjusted to match local or national costs according to a city's location. Outputs in-

Life-cycle costing software is available from Canada Mortgage and Housing Corporation. www.cmhc-schl.gc.ca.

clude integrated financial assessments with monetary values for the following major categories:

- Hard Infrastructure, including roads, sewerage, storm water facilities, schools, and recreation centers
- Municipal services, including transit services, school transit, fire services, police services, and waste management services
- Private user costs, including driving costs and home heating costs
- External costs, including air pollution, climate change, and motor vehicle collision
- Green infrastructure alternatives

Two scenarios for Fort St. John

In 2008, Fort St. John, Canada, conducted a design charrette to create a sustainable neighborhood concept plan for a 37-hectare greenfield site on the edge of the urban area. The city had three goals:

1. Adopt a more proactive and engaging planning process for managing growth.
2. Create a new demonstration neighborhood that embodies the community's long-term goals and objectives.
3. Field-test new approaches for guiding future mixed use development throughout the region.

As part of the design charrette, the development costs and value of the sustainable neighborhood concept plan were compared against a baseline scenario that was grounded on the typical low-density neighborhoods currently existing in Fort St. John. The sustainable neighborhood scenario is an alternative incorporating the principles and recommendations that arose from the charrette process. The analysis using the LCC tool for community infrastructure planning allowed the scenarios to be run for an entire neighborhood. The calculations are comprehensive, including typical capital and operating costs for utilities and services such as roads, water, sewerage, garbage, schools, recreation facilities, public

transit, private vehicle use, fire protection, and police. Interest rates for lending, tax rates, and service revenues are also calculated. Life-cycle costs are annualized (converted into an annual cost) over a 75-year period and allow for the operation, maintenance, and replacement of all utilities. All costs may be allocated on a per household basis.

All cost and service demand assumptions were the same for the two calculations. The only differences were that the sustainable neighborhood scenario has a smaller neighborhood street width, green storm water infrastructure, green roofs on public buildings (reducing the size of the storm water infrastructure), and higher energy efficiency building standards. The sustainable neighborhood scenario also has a higher density and greater mix of housing types and land uses. Table 2.3 compares the two scenarios.

Baseline scenario: low-density, primarily single-family residential and apartment

A simple mask method was used to assess the road and lot layout and the number of units that would typically be developed on this site. A scaled mask was created and placed over an existing neighborhood near the site (figure 2.27). The neighborhood has parks, mostly single-family homes, and so on. The approximate number of lots and the length and type of streets captured inside the mask were counted; allowance was made for the inclusion of a school and community center. Three elements were then added: a small strip of commercial use buildings, some duplex lots, and several small three-story apartment buildings. Public open space was limited to parks, school grounds, and street rights-of-way.

Sustainable neighborhood scenario: medium-density, varied housing forms, and mixed use

The sustainable neighborhood plan developed in the 2008 charrette was the basis of the second scenario. A three-dimensional model was

Table 2.3 City of Fort St. John, Canada: Comparative Statistics for Two Scenarios

FACTOR	BASELINE SCENARIO: LOW DENSITY	SUSTAINABLE NEIGHBORHOOD SCENARIO: MEDIUM DENSITY
Site area	37 hectare (93 acre)	37 hectare (93 acre)
Residential area ([a] %)	94	90
Commercial and community (%)	6	10
Service area of parks	approximately 2.7 hectare	not estimated; multiuse open space
Total residential units	368	932
Single-family units	188	56
Duplex units (large lot)	72	0
Minilot duplex units	0	84
Townhouses (two stories)	0	108
Townhouses (three stories, stacked)	0	138
Apartments (three or four stories)	144	516
Apartments above commercial	0	30
Commercial units	8	15
Gross unit density, units/hectare (units/acre)	11 (4)	28 (12)
Total population	888	1,922
Adult population	682	1,542
Child population	206	380
Total Population	888	1,922
Total roads (linear meters)	4,120	5,260
Neighborhood roads, compact (meters)	0	2,410
Collector roads[b] (linear meters)	3,200	1,930
Arterial roads (linear meters)	920	920

Source: Fort St. John (2009).

a. Residential area includes roads, parks, schools, and so on.

b. Collector roads includes two types: one with a 17-meter right-of-way and one with a 15-meter right-of-way.

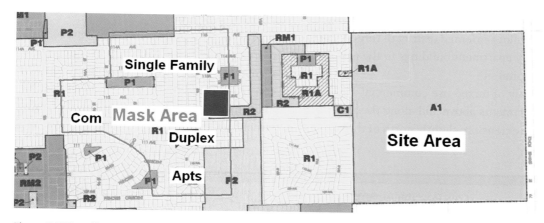

Figure 2.27 Baseline Low-Density Scenario Developed Using a Mask

Source: Fort St. John (2009).

Note: Apts = apartments; Com = commercial use.

developed from the workshop sketches using a variety of the housing forms listed in the residential section (see above). Though a precise, detailed neighborhood design was not established, zoning areas, building forms, street types, and public buildings were scaled accurately so that the outcome was a plausible result. Generally, this is a much more compact plan than the baseline scenario. Buildings are more tightly packed, and there are more open, intermediary public spaces. There are several more uses and building types than the case in the baseline scenario. Neighborhood streets and roads are also narrower, in line with Canadian Alternative Development Standards.

The sustainable neighborhood plan includes the following:

- A small area of large-lot single-family residences
- A number of small areas of minilot duplexes
- A few small areas of two-story townhouses
- Several areas of three-story stacked townhouses
- Three- or four-story apartment buildings along the east part of the 112th Avenue extension
- A seniors-oriented district of three- or four-story apartment buildings to the east of the hospital site
- A row of mixed use commercial units with apartments above and along the 112th Avenue extension, slightly north of the hospital site

- A school and a community center
- A public lookout at the water tower
- Energy-efficient homes

Open public space is integrated and multiuse, including greenways, community gardens, bicycle paths, cross-country ski trails, and a large commons around the school and community center.

Analysis of Costs and Value in the Scenarios

Baseline scenario: Costs and value

Using typical costs for roads, sewerage, water supply, schools, and other services, one sees that the baseline scenario results in about US$36,000 in initial capital costs for each residential unit (figure 2.28). The estimated cost of a storm water pond is included under green infrastructure. The estimated cost of a water pumping station is included under user-defined costs. Note that roads dominate in the capitals costs. Using these capital costs, the baseline scenario results in about US$6,500 in operating costs for each residential unit (figures 2.29 and 2.30).

All infrastructure assets depreciate; so, a true representation of the costs must include replacement costs over time, including inflation in construction costs. Figure 2.31 illustrates the annual operating costs for all types of infrastructure if these are spread over a 75-year life-

INITIAL CAPITAL COSTS

	Total Development $	Residential Portion $	Household Costs $/hh	Percent Breakdown %
HARD INFRASTRUCTURE Local Roads	7,560,890	7,107,236	19,313	54%
Regional Roads	0	0	0	0%
Water Distribution and Water Treatment	672,000	672,000	1,826	5%
Sanitary Sewers and Wastewater Treat.	766,000	766,000	2,082	6%
Storm Sewers and Water Management	3,178,600	2,987,884	8,119	23%
Schools	1,347,446	1,347,446	3,662	10%
Recreational Facilities	6,771	6,771	18	0%
MUNICIPAL SERVICES Transit Services	112,530	105,778	287	1%
Fire Services	20,976	20,976	57	0%
Police Services	11,040	11,040	30	0%
GREEN INFR. Green Infrastructure Items	80,000	75,200	204	1%
USER DEFINED User Defined Costs	160,000	150,400	409	1%
Total Costs	**$13,916,253**	**$13,250,732**	**$36,007**	**100%**

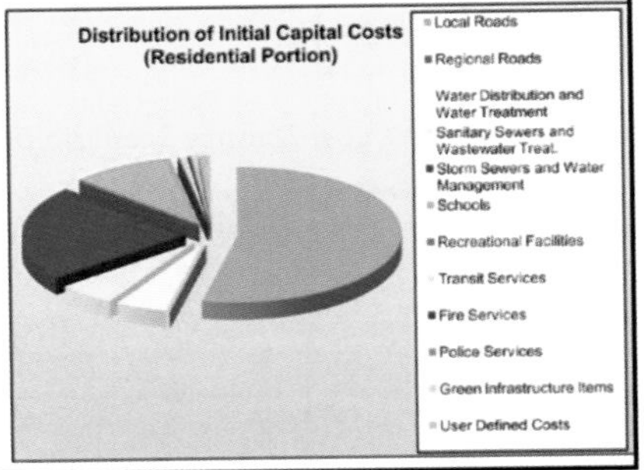

Distribution of Initial Capital Costs (Residential Portion)

Legend: Local Roads, Regional Roads, Water Distribution and Water Treatment, Sanitary Sewers and Wastewater Treat., Storm Sewers and Water Management, Schools, Recreational Facilities, Transit Services, Fire Services, Police Services, Green Infrastructure Items, User Defined Costs

Figure 2.28 Baseline Scenario: Initial Capital Costs

Source: Fort St. John (2009).

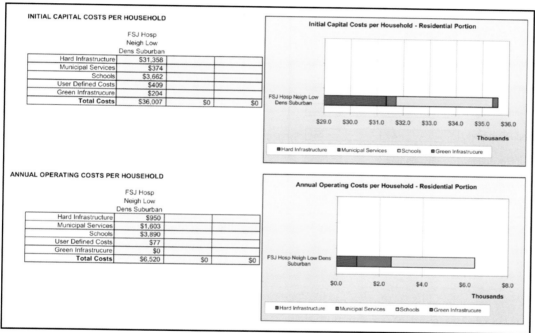

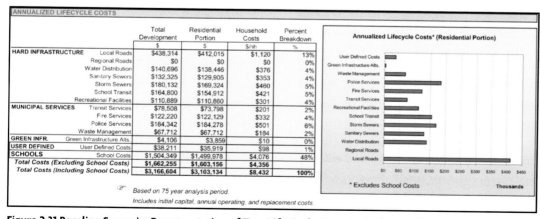

ANNUAL OPERATING COSTS

		Total Development $	Residential Portion $	Household Costs $/hh	Percent Breakdown %
HARD INFRASTRUCTURE	Local Roads	$34,000	$31,960	$87	3%
	Regional Roads	$0	$0	$0	0%
	Water Distribution and Water Treatment	$103,200	$103,200	$280	11%
	Sanitary Sewers and Wastewater Treat.	$92,000	$92,000	$250	10%
	Storm Sewers and SWM	$12,800	$12,032	$33	1%
	School Transit	$164,800	$154,912	$421	16%
	Recreational Facilities	$110,400	$110,400	$300	11%
MUNICIPAL SERVICES	Transit Services	$67,518	$63,467	$172	7%
	Fire Services	$120,704	$120,704	$328	12%
	Police Services	$183,264	$183,264	$498	19%
	Waste Management	$67,712	$67,712	$184	7%
GREEN INFR.	Green Infrastructure Items		N/A		0%
USER DEFINED	User Defined Costs	$30,000	$28,200	$77	3%
SCHOOLS	School Operating Costs	$1,431,494	$1,431,494	$3,890	-
Total Costs (Excluding School Costs)		$986,398	$967,851	$2,630	100%
Total Costs (Including School Costs)		$2,417,892	$2,399,345	$6,520	

Distribution of Annual Operating Costs* (Residential Portion)

Local Roads; Regional Roads; Water Distribution and Water Treatment; Sanitary Sewers and Wastewater Treat.; Storm Sewers and SWM; School Transit; Recreational Facilities; Transit Services; Fire Services; Police Services; Waste Management; Green Infrastructure Items; User Defined Costs; School Operating Costs

* Excludes School Costs

Figure 2.29 Baseline Scenario: Annual Operating Costs per Unit

Source: Fort St. John (2009).

INITIAL CAPITAL COSTS PER HOUSEHOLD

	FSJ Hosp Neigh Low Dens Suburban		
Hard Infrastructure	$31,358		
Municipal Services	$374		
Schools	$3,662		
User Defined Costs	$409		
Green Infrastrucure	$204		
Total Costs	$36,007	$0	$0

Initial Capital Costs per Household - Residential Portion

ANNUAL OPERATING COSTS PER HOUSEHOLD

	FSJ Hosp Neigh Low Dens Suburban		
Hard Infrastructure	$950		
Municipal Services	$1,603		
Schools	$3,890		
User Defined Costs	$77		
Green Infrastrucure	$0		
Total Costs	$6,520	$0	$0

Annual Operating Costs per Household - Residential Portion

Figure 2.30 Baseline Scenario: Graphic Representation of Initial Capital Costs and Annual Operating Costs per Unit

Source: Fort St. John (2009).

ANNUALIZED LIFECYCLE COSTS

		Total Development $	Residential Portion $	Household Costs $/hh	Percent Breakdown %
HARD INFRASTRUCTURE	Local Roads	$438,314	$412,015	$1,120	13%
	Regional Roads	$0	$0	$0	0%
	Water Distribution	$140,696	$138,446	$376	4%
	Sanitary Sewers	$132,325	$129,905	$353	4%
	Storm Sewers	$180,132	$169,324	$460	5%
	School Transit	$164,800	$154,912	$421	5%
	Recreational Facilities	$110,889	$110,860	$301	4%
MUNICIPAL SERVICES	Transit Services	$78,508	$73,798	$201	2%
	Fire Services	$122,220	$122,129	$332	4%
	Police Services	$184,342	$184,278	$501	6%
	Waste Management	$67,712	$67,712	$184	2%
GREEN INFR.	Green Infrastructure Alts.	$4,106	$3,859	$10	0%
USER DEFINED	User Defined Costs	$38,211	$35,919	$98	1%
SCHOOLS	School Costs	$1,504,349	$1,499,978	$4,076	48%
Total Costs (Excluding School Costs)		$1,662,255	$1,603,156	$4,356	
Total Costs (Including School Costs)		$3,166,604	$3,103,134	$8,432	100%

Annualized Lifecycle Costs* (Residential Portion)

* Excludes School Costs

☞ Based on 75 year analysis period.
Includes initial capital, annual operating, and replacement costs.

Figure 2.31 Baseline Scenario: Representation of True Life-Cycle Costs, Including Replacement

Source: Fort St. John (2009).

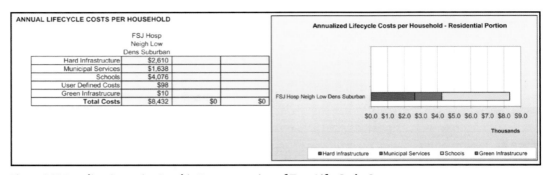

ANNUAL LIFECYCLE COSTS PER HOUSEHOLD			
	FSJ Hosp Neigh Low Dens Suburban		
Hard Infrastructure	$2,610		
Municipal Services	$1,638		
Schools	$4,076		
User Defined Costs	$98		
Green Infrastrucure	$10		
Total Costs	$8,432	$0	$0

Figure 2.32 Baseline Scenario: Graphic Representation of True Life-Cycle Costs

Source: Fort St. John (2009).

REVENUES	Total Residential ($)	Revenues per household or unit ($)
Annual Property Taxes	$1,800,368	$4,892
Annual User Charges	$144,159	$392
Total Initial Development Charges	$2,736,000	$7,435
Annual User Defined Revenues	$0	$0
Annualized Value of Revenues*	$2,084,943	$5,666

* Based on 75 year analysis period. Includes annual property taxes, initial development charges, user charges and user defined revenues

Figure 2.33 Baseline Scenario: Estimate of Taxes, User Fees, and Initial Development Cost Charges

Source: Fort St. John (2009).

time. The true annual life-cycle cost per household (US$8,432) is about 30 percent more than the initial operating cost calculated for each residential unit (US$6,520) (figure 2.32).

The taxes, user charges (such as the charges for garbage collection), and initial development cost charges were roughly estimated for the baseline scenario (figure 2.33). It had not been decided how Fort St. John would share or recover the development costs from the private sector; so, these results are hypothetical.

Sustainable neighborhood: Costs and value

Using typical costs for roads, sewerage, water supply, schools, and other services, one sees

that the sustainable neighborhood scenario results in about US$16,500 in initial capital costs for each residential unit; this is less than half the costs of the baseline scenario (figure 2.34). The estimated cost of a green roof for the school and community center, as well as the (now smaller) storm water pond, is included in green infrastructure. The estimated cost of a water pumping station is included in user-defined costs. Note that, as in the baseline scenario, roads still dominate in the capitals costs.

Using these capital cost estimates, the sustainable neighborhood scenario results in about US$5,200 in operating costs for each residential unit, which is about 25 percent less than the baseline scenario (figures 2.35 and 2.36). The estimated operating cost of the water pumping station is included in user-defined costs.

Annualized life-cycle costs for the sustainable alternative are estimated at US$6,053 per unit, or about 17 percent more than the initial operating costs (US$5,185) (figures 2.37 and 2.38). This difference is about half the difference for the baseline scenario, mainly because of the larger number of households sharing the infrastructure more efficiently.

These initial results indicate that per unit taxes in the sustainable neighborhood may be slightly lower (mainly because of the smaller homes) and that the initial development charges would also be lower than those in the baseline scenario (figure 2.39).

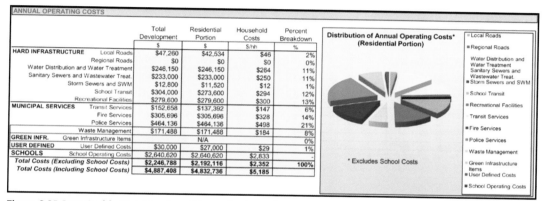

INITIAL CAPITAL COSTS		Total Development	Residential Portion	Household Costs	Percent Breakdown
		$	$	$/hh	%
HARD INFRASTRUCTURE	Local Roads	8,233,564	7,410,208	7,951	48%
	Regional Roads	0	0	0	0%
Water Distribution and Water Treatment		843,000	843,000	905	5%
Sanitary Sewers and Wastewater Treat.		965,500	965,500	1,036	6%
Storm Sewers and Water Management		3,198,200	2,878,380	3,088	19%
Schools		2,485,580	2,485,580	2,667	16%
Recreational Facilities		17,149	17,149	18	0%
MUNICIPAL SERVICES	Transit Services	254,430	228,987	246	1%
	Fire Services	53,124	53,124	57	0%
	Police Services	27,960	27,960	30	0%
GREEN INFR.	Green Infrastructure Items	326,500	293,850	315	2%
USER DEFINED	User Defined Costs	160,000	144,000	155	1%
	Total Costs	**$16,565,007**	**$15,347,737**	**$16,468**	**100%**

Figure 2.34 Sustainable Neighborhood Scenario: Initial Capital Costs per Unit

Source: Fort St. John (2009).

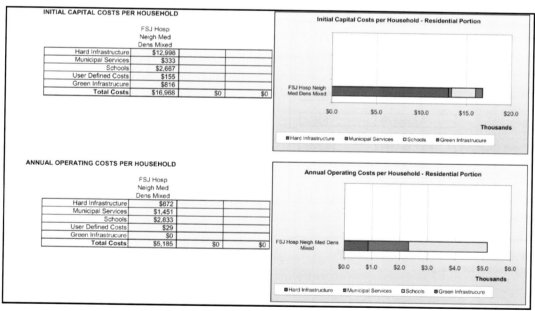

ANNUAL OPERATING COSTS		Total Development	Residential Portion	Household Costs	Percent Breakdown
		$	$	$/hh	%
HARD INFRASTRUCTURE	Local Roads	$47,260	$42,534	$46	2%
	Regional Roads	$0	$0	$0	0%
Water Distribution and Water Treatment		$246,150	$246,150	$264	11%
Sanitary Sewers and Wastewater Treat.		$233,000	$233,000	$250	11%
Storm Sewers and SWM		$12,800	$11,520	$12	1%
School Transit		$304,000	$273,600	$294	12%
Recreational Facilities		$279,600	$279,600	$300	13%
MUNICIPAL SERVICES	Transit Services	$152,658	$137,392	$147	6%
	Fire Services	$305,696	$305,696	$328	14%
	Police Services	$464,136	$464,136	$498	21%
Waste Management		$171,488	$171,488	$184	8%
GREEN INFR.	Green Infrastructure Items		N/A		0%
USER DEFINED	User Defined Costs	$30,000	$27,000	$29	1%
SCHOOLS	School Operating Costs	$2,640,620	$2,640,620	$2,833	-
	Total Costs (Excluding School Costs)	**$2,246,788**	**$2,192,116**	**$2,352**	**100%**
	Total Costs (Including School Costs)	**$4,887,408**	**$4,832,736**	**$5,185**	

Figure 2.35 Sustainable Neighborhood Scenario: Annual Operating Costs per Unit

Source: Fort St. John (2009).

INITIAL CAPITAL COSTS PER HOUSEHOLD	FSJ Hosp Neigh Med Dens Mixed		
Hard Infrastructure	$12,998		
Municipal Services	$333		
Schools	$2,667		
User Defined Costs	$155		
Green Infrastrucure	$816		
Total Costs	**$16,968**	**$0**	**$0**

ANNUAL OPERATING COSTS PER HOUSEHOLD	FSJ Hosp Neigh Med Dens Mixed		
Hard Infrastructure	$872		
Municipal Services	$1,451		
Schools	$2,833		
User Defined Costs	$29		
Green Infrastrucure	$0		
Total Costs	**$5,185**	**$0**	**$0**

Figure 2.36 Sustainable Neighborhood Scenario: Graphic Representation of Initial Capital Costs and Annual Operating Costs per Unit

Source: Fort St. John (2009).

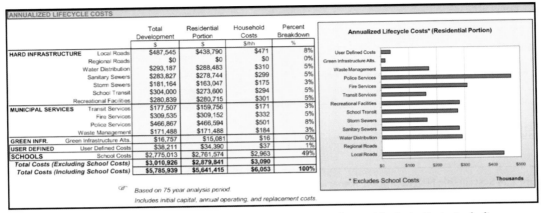

Figure 2.37 Sustainable Neighborhood Scenario: Representation of True Life-Cycle Costs, Including Replacement

Source: Fort St. John (2009).

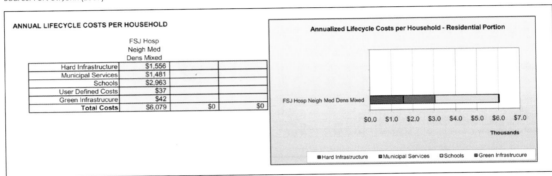

Figure 2.38 Sustainable Neighborhood Scenario: Graphic Representation of True Life-Cycle Costs

Source: Fort St. John (2009).

REVENUES

	Total Residential ($)	Revenues per household or unit ($)
Annual Property Taxes	$3,951,140	$4,239
Annual User Charges	$355,929	$382
Total Initial Development Charges	$5,712,000	$6,129
Annual User Defined Revenues	$182,196	$195
Annualized Value of Revenues*	$4,782,414	$5,131

* Based on 75 year analysis period. Includes annual property taxes, initial development charges, user charges and user defined revenues

Figure 2.39 Sustainable Neighborhood Scenario: Estimate of Taxes, User Fees, and Initial Development Cost Charges

Source: Fort St. John (2009).

Comparative analysis of costs and value

Figure 2.40 demonstrates the huge savings in initial capital costs in the sustainable neighborhood scenario relative to the baseline scenario. Decision makers may emphasize this substantial savings in their effort to overcome reluctance and pursue innovative solutions.

Figure 2.41 illustrates the modest reductions in operating costs per household for the sustainable neighborhood scenario mainly because more households share the infrastructure. In addition, the costs of schools are distributed among more homes with fewer children.

Figure 2.42 illustrates the significant reductions in estimated annual municipal costs and necessary revenues for the sustainable neighborhood over a 75-year period.

Figure 2.43 summarizes the estimated annual life-cycle costs per household for the two neighborhoods over a 75-year period.

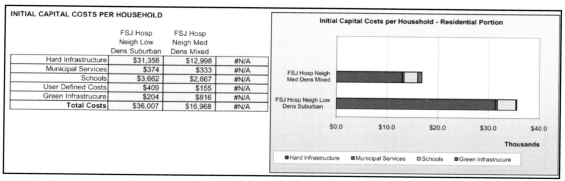

Figure 2.40 Comparison of Baseline and Sustainable Neighborhood Scenarios: Initial Capital Costs

Source: Fort St. John (2009).

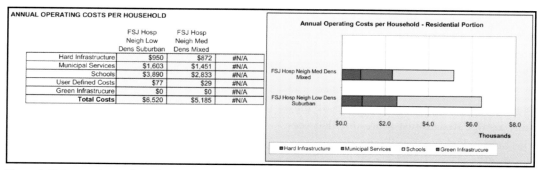

Figure 2.41 Comparison of Baseline and Sustainable Neighborhood Scenarios: Annual Operating Costs

Source: Fort St. John (2009).

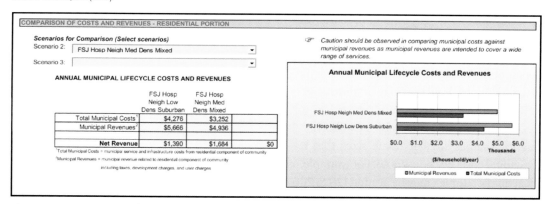

Figure 2.42 Comparison of Baseline and Sustainable Neighborhood Scenarios: Annual Municipal Costs and Necessary Revenues over 75 Years

Source: Fort St. John (2009).

Insights from the LCC Case Study

Specific development cost inputs must be updated as design continues

There are many hard and soft cost assumptions built into the model that are fairly reliable because they are taken from national databases and localized to Fort St. John. However, the specifics of development are not sufficiently certain at this conceptual level to obtain highly accurate results. Furthermore, the way in which Fort St. John, as the landowner, will handle development costs and municipal service costs by recovering them through sales and development cost charges has not been determined. For the moment, it has been assumed

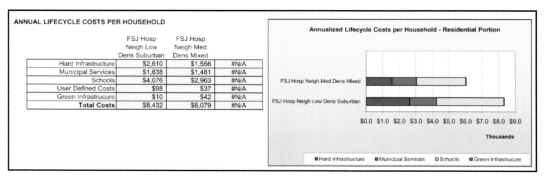

ANNUAL LIFECYCLE COSTS PER HOUSEHOLD

	FSJ Hosp Neigh Low Dens Suburban	FSJ Hosp Neigh Med Dens Mixed	
Hard Infrastructure	$2,610	$1,556	#N/A
Municipal Services	$1,638	$1,481	#N/A
Schools	$4,076	$2,963	#N/A
User Defined Costs	$98	$37	#N/A
Green Infrastrucure	$10	$42	#N/A
Total Costs	$8,432	$6,079	#N/A

Figure 2.43 Comparison of Baseline and Sustainable Neighborhood Scenarios: Annual Life-Cycle Costs per Household

Source: Fort St. John (2009).

that the city will carry approximately 20 percent of the service costs and pass 80 percent on to the developer. Municipal costs and revenues are thus preliminary. However, once the spreadsheet model has been set up, it may readily be updated to test more detailed scenarios before decisions are finalized.

LCC helps clarify the greater affordability and value of sustainable options

In the sustainable neighborhood scenario, the effects on home prices and operating costs are not explored in detail. Clearly, the compact development form results in savings on municipal services per household that may be passed on to residents through lower purchase prices or rents. In addition, the sustainable neighborhood scenario assumes smaller home sizes that reduce capital costs per household. For example, a typical minilot duplex may be around 120 to 200 square meters (1,300 to 2,200 square feet); this compares with 220–300 square meters (2,400–3,200 square feet) for a typical detached single-family home (median 60 percent less floor area). Furthermore, green, energy-efficient building standards are proposed for the sustainable neighborhood that will result in lower operating costs and lower repair and replacement costs (because these are more durable homes).

It has been estimated that the median price of a minilot duplex is lower than the price of a standard single-family lot in the following order of magnitude:

1. Land price: this is approximately 25 percent below a single-family lot because of the smaller land area and lower servicing costs.

2. Home price: this is approximately 35 percent below a single-family lot because of the smaller home size and the economies of duplex construction. This also accounts for the slightly higher cost of higher-quality energy-efficient construction.

3. Operating cost: for a minilot duplex home may, this be about 50 percent less than the corresponding cost of a single-family home because of energy efficiency, water savings, more durable construction, and the reduced yard area to maintain.

In today's world of economic uncertainties and unstable costs for energy and services, one may confidently say that the more compact, more energy-efficient, and more durable home is likely to retain its value much better than the large, inefficient homes of the past.

LCC is especially effective in helping the municipality cope with future costs

Every established city today is facing trouble managing the replacement costs of declining infrastructure. Some are in a more serious situation than others because of low revenues in a declining economy and deferred replacements that are long overdue. At the same time, there have been major increases in capital costs over the past few years because of the demand for global construction materials, energy prices, and other factors. In short, times are difficult for financing infrastructure, but opportune for innovative solutions.

As cities look to the future, it will be increasingly important to adopt development solutions that reduce future municipal costs and increase resiliency. The sustainable neighborhood example offers much lower capital costs per unit, reduced municipal costs, lower costs to residents, and better value in the long term. It is also a more adaptable model, offering more options for a wider demographic with high environmental quality and social amenities. The simple LCC tool has helped the city clarify these benefits prior to making any decisions about how to proceed. A number of other tools are available that perform similar functions; adopting an appropriate choice of tool is a key part of capacity building for the Eco² pathway.[1]

LCC for a single infrastructure facility

While Fort St. John has used LCC for community-wide planning, the LCC tool may also be applied to infrastructure facilities on a case-by-case basis. One of the challenges in integrated design is the ability to assess quickly a range of engineering options in infrastructure. How does one obtain informed estimates of the performance of different technologies without commissioning an expensive series of feasibility studies? Where might one find engineers and economists with sufficient experience to compare the alternatives fairly over the life cycle? The solution usually involves the application of scalable spreadsheet tools that allow users to plug in default values established in previous projects in other locations and to alter assumptions rapidly as the design concept evolves and new information becomes available.

An example of an LCC tool for infrastructure facilities is the RETScreen Clean Energy Project Analysis Software, a decision support tool that builds the capacity of planners, decision makers, and industry to implement renewable energy, cogeneration, and energy efficiency projects (figure 2.44). The software, provided free of charge, may be used worldwide to evaluate the energy production, energy savings, costs, emission reductions, financial viability, and risks associated with various types of renewable energy and energy-efficient technologies. Available in multiple languages, the software includes product, project, hydrology, and climate databases, a detailed user manual, a case study–based college- or university-level training course, and an engineering e-textbook. By using the software to explore options at the outset, cites may greatly reduce the cost of prefeasibility studies. The rigorous structure of the software model also helps ensure that decision makers are fully informed and that analysts are trained in assessing the technical and financial viability of projects of all kinds. The software is sponsored by the government of Canada and has received contributions from many universities. It has been used in almost all countries.

Figure 2.44 RETScreen Software

Source: Natural Resources Canada.

As part of the RETScreen Clean Energy Project Analysis Software, an emission analysis worksheet is provided to help the user estimate the greenhouse gas emission reduction (mitigation) potential of the proposed project. A cost analysis worksheet is used to help the user estimate the costs (and credits) associated with the proposed case. These costs are addressed from the initial or investment cost standpoint and from the annual or recurring cost standpoint. The user may refer to the RETScreen Product Database for supplier contact information to obtain prices or other required information.

A financial analysis worksheet is provided for each project evaluated (figures 2.45 and 2.46). This financial analysis worksheet contains six sections: financial parameters, annual income, project costs and savings–income summary, financial viability, yearly cash flows, and cumulative cash flow graphs. One of the primary benefits is the fact that the RETScreen software facilitates the project evaluation process for decision makers. The financial analysis worksheet covers financial parameter input items (for example, the discount rate and the debt ratio) and calculated financial viability output items (the internal rate of return, simple payback, net present value, and so on) and allows project decision makers to consider various financial parameters with relative ease.

A sensitivity and risk analysis worksheet is provided to help the user estimate the sensitivity of important financial indicators in relation to key technical and financial parameters. This standard sensitivity and risk analysis worksheet contains a settings section and two main sections: sensitivity analysis and risk analysis. Each section provides information on the relationship between the key parameters and important financial indicators, showing the parameters that have the greatest impact on

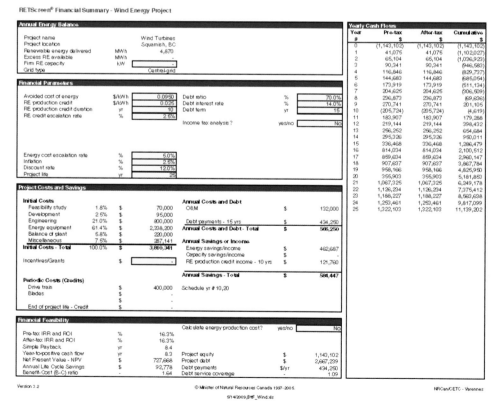

Figure 2.45 An Example of a RETScreen Financial Summary

Source: Author compilation (Sebastian Moffatt).

Note: The summary refers to the financial viability of a wind energy system for Squamish, Canada.

the financial indicators. The sensitivity analysis section is intended for general use, while the risk analysis section, which performs a Monte Carlo simulation, is intended for users with knowledge of statistics.

Environmental Accounting

In Stockholm, the municipal government, together with the Royal Institute of Technology and an engineering consultant firm, has created a tool to help plan and assess the development of Hammarby Sjöstad, a southern city district. The tool, called the environmental load profile (ELP), has proven successful in evaluating and providing critical feedback to Stockholm's cutting-edge initiative in sustainable development. Stockholm developed this tool to demonstrate how decisions and strategies in urban development might prove significantly more beneficial in enhancing long-term environmental and urban sustainability.

The environmental load profile

The ELP is a life-cycle assessment-based tool built on defining relevant activities from an environmental perspective and on quantifying the environmental loads originating from these activities, such as emissions to air, soil, and water, as well as the use of nonrenewable energy resources. It takes into account all activities related to a project, such as materials, transport (transport of materials, supplies, and persons), machinery, electricity, heating, and materials recycling.

RETScreen® Financial Summary - Wind Energy Project

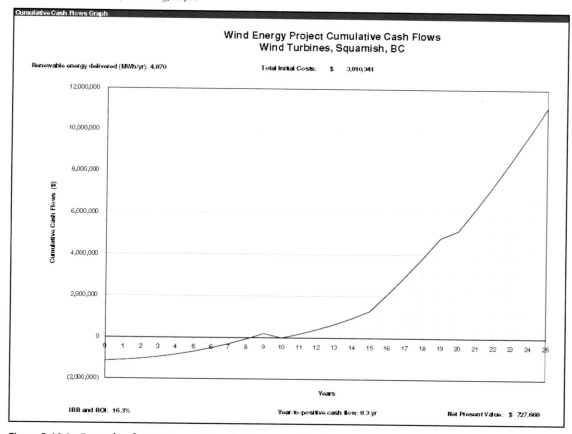

Figure 2.46 An Example of a RETScreen Financial Summary Visual
Source: Author compilation (Sebastian Moffatt).
Note: The financial summary refers to a wind energy system for Squamish, Canada.

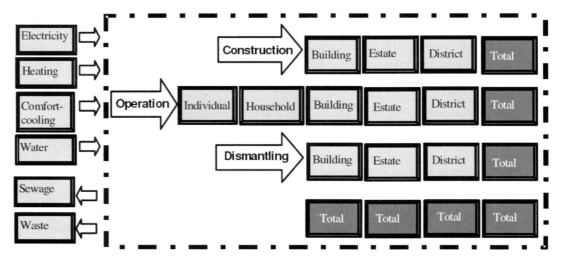

Figure 2.47 The Environmental Load Profile

Source: Brick (2008).

Note: Environmental loads are quantified for different services and levels, and the results may be totaled to reflect any combination of concerns.

The main strength of the ELP is that the tool is flexible and dynamic, which makes it suitable for application both as a planning tool and as an evaluation tool. By factoring in variables, one may use the ELP to calculate the environmental loads that various planning decisions will generate at various project phases, including construction, use, demolition, and redevelopment (figure 2.47). It is also possible to test scenarios. For instance, one may compare the environmental performance of different construction methods prior to making a decision on which method to use. Hence, it is possible for decision makers to include environmental concerns early in the process.

The ELP may also be used to evaluate the environmental performance of an existing city district or building based on the consumption of resources such as water and energy during the usage phase. The ELP enables analyses of environmental performance at multiple levels. The tool takes into account activities associated with individuals (for example, cooking and laundry); buildings (such as building materials, district heating, and electricity); the unbuilt real estate area (materials, working machines, and so on); and the common area (for instance,

materials, personal transportation, and the transport of goods). By aggregating all factors, the environmental load of the whole city district may be analyzed. If each factor is analyzed separately, the various activities of the city may provide useful information for city planning.

The ELP enables comparisons among alternative designs, construction, and infrastructure. The ELP encompasses two life-cycle calculations:

- Effects from each of the life-cycle stages: construction, operation, and dismantling

- Effects from the life cycle of the building materials and electricity flowing in and out of buildings and the city district

The ELP in the follow-up to Hammarby Sjöstad

In Stockholm's Hammarby Sjöstad project, the city imposed tough environmental requirements on infrastructure solutions and technical installations in buildings. Since 2002, when the first area, Sickla Udde, was completed, environmental goals and performance have been monitored using the ELP's different indicators. Figure 2.48 illustrates the results from four of the

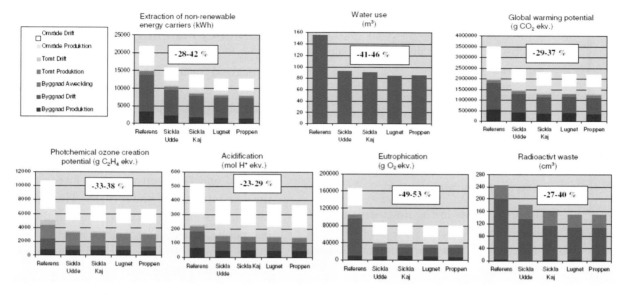

Figure 2.48 ELP-Related Achievements in Hammarby Sjöstad

Source: Grontmij AB.

Note: The figure shows environmental impacts per apartment and year compared with the reference. The ELPs illustrate the effect of the property developer's measures and the effects of improved energy production and wastewater management (Levin and Rönnkvist-Mickelsson 2004).

areas. Relative to a reference scenario, the results were a 28–42 percent reduction in nonrenewable energy use, a 41–46 percent reduction in water use, a 29–37 percent reduction in global warming potential, a 33–38 percent reduction in photochemical ozone creation, a 23–29 percent reduction in acidification potential, a 49–53 percent reduction in eutrophication potential, and a 27–40 percent reduction in radioactive waste.

The overall environmental goal set for Hammarby Sjöstad is to reduce environmental loads by half relative to urban development loads in the early 1990s. Even though this goal of twice as good has not yet been reached, the reductions in the environmental loads in the area are significant. The main contributor to the improvements has been effective planning in the area, such as planning for district heating, urban transportation, waste, and wastewater management.

Monitoring in the Hammarby Sjöstad environmental program has contributed to the technical and economic understanding of ap-propriate societal and financial environmental measures in the continued development of the district. The results of the monitoring may also be useful in planning or undertaking similar projects.

The ELP in the planning process

Environmental planning has not been the norm in the past, and it is apparent that there is still much room to improve the environmental impacts of urban development initiatives. The availability of a proactive approach and the ability to analyze the potential for enhancement in the planning process make it more likely that one will adopt cost-effective measures that contribute significantly to increased sustainability. Improvements may be possible in three areas: (1) the upstream system (streams of materials and services flowing into the area), (2) the core system (the project), and (3) the downstream system (the management of waste flows and the reuse of materials), as follows:

1. Improvements in the upstream system may be realized through enhanced energy production (electricity and heating) and raw materials production.

2. Improvements in the core system may be realized through advances in construction and maintenance; the installation of solar cells or heat recovery systems; and human behavioral change, particularly by promotion of energy conservation.

3. Improvements in the downstream system may be realized through improved waste and wastewater management, including recycling and reuse.

By using the ELP in the planning process, one may analyze various options and vet various interventions from an environmental point of view (figure 2.49). By adding the cost of environmental impacts to the analysis of alternatives, one may visualize the life-cycle perspective. The ELP allows for comprehensive assessments and more precise target setting. By following up on outcomes and providing feedback to stakeholders and actors, the ELP also contributes to building knowledge and fostering improvements. It is becoming more important to have good decision-making tools such as the ELP.

Foresight Workshops and Resiliency Planning

A method for mitigation and adaptation is needed

Credible forecasting is essential for all development planning. All cities require the capacity to forecast. A city's land use plan is typically driven by population and economic demand and relies on plausible forecasts of the demand for land and services. Thus, credible forecasting is a necessary part of directing public investments and is essential for gaining support from potential financial partners and other stakeholders.

Forecasting is always a challenge. The demand for services of any type may vary greatly, depending on the assumptions made regarding population growth, occupant lifestyles, new technology, and the pace of development. Transformative forces may also influence demand. For example, population migration, climate change, and globalization may lead to large-scale changes in the local demand for land and services. A rise in sea level might alter the location of shorelines and dislocate neighborhoods and infrastructure. The increased frequency of wind storms might require more space dedicated to tree breaks, pedestrian shelters, and underground services. A global economic crisis, fuel price increases, or changes in

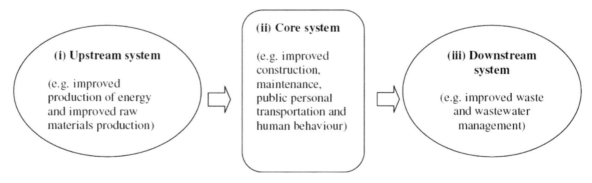

Figure 2.49 Opportunities to Reduce Environmental Impacts

Source: Brick (2008).

Note: The figure illustrates opportunities to reduce environmental impacts in the building sector in the upstream system, the core system, and the downstream system.

rainfall patterns might boost the need for food security and require more space and water for urban gardens and local agriculture. The range of threats and mitigation plans for cities today covers a broad spectrum and is subject to constant change.

In reality, the forecasting challenge far exceeds the capacity of any city. The best place to begin is to generate the best possible demand and supply forecasts, using whatever data and practical models are available. Over time, it becomes possible to augment such forecasts with insights on how changes in climate, technology, and other external factors might influence key assumptions.

The first step is developing the capacity for forecasting land use demand

At a minimum, standardized methods should be used to estimate the demand for housing, commercial space, and industry. The demand is driven by population growth and economic indicators. Generally, the process begins by assuming reasonable population growth rates under margins of uncertainty over a 30-year period. This creates both high- and low-growth scenarios.

Population growth is translated into housing demand by dividing the population into subsets based on age and socioeconomic status. One then associates a propensity for different types of housing with each subset: low-rise apartments, large detached houses, high-rise dwellings, and so on. Predictions may thus be generated on the demand for different types of housing.

The second challenge is to forecast supply. Ideally, this is accomplished using a GIS tool that allows for easy contingent analysis. Existing zoning or a set of zoning options is used as a basis for equating land areas with potential numbers of housing by type. Based on these assumptions, each area of land has a build-out capacity, and this limits the supply of units in the city. It is then a simple matter to com-

pare the demand forecast with the supply forecast and identify gaps for specific types of housing. The same forecast provides a basis for estimating gaps in infrastructure capacity in terms of roads, transport, water, and energy.

A similar forecasting process may be used for commercial and industrial demand and supply. Ideally, because of interdependencies, residential forecasts and commercial and industrial forecasts should be considered together.

Foresight workshops are helpful in understanding the impacts of external forces

A number of techniques have been developed specifically to engage large groups of experts in envisioning the long-term future and developing appropriate design strategies. Some traditional foresight tools have proven quite difficult to apply; for example, the Delphi technique developed by RAND researchers in the 1950s has not been particularly successful as a predictive method. However, a host of other visioning and exploring techniques may now be used to bring groups of experts together as part of groups of research and reflection or in workshops on future urban issues. Such techniques may be referred to as creativity tools; they include trial and error, brainstorming, morphological analysis, the method of focal objects, and lateral thinking. In urban planning and design, communicative planning has been promoted as a field-tested method for engaging stakeholders and experts in a more dynamic, open-ended enquiry. An example is the European Awareness Scenario of the Sustainable Cities Program (for example, see Bilderbeek and Andersen 1994). Extensive engagement exercises of this type have, at times, been intellectually and physically trying for participants. They are thus accompanied by some risk that stakeholders may lose interest. In this context, the success of collaborative planning may depend on tools that promote a simpler systems approach and involve stakeholders in more intense, time-limited exercises such as design charrettes and foresight workshops.

A foresight workshop consists of a progressive series of presentations and exercises intended to introduce designers and planners to the potential for proactive risk management through resilient land use and infrastructure design. Typically, it begins with an exploration of the possible impacts of external forces on urban and rural systems within the region. A summary presentation or a set of papers may be provided that covers the local context in terms of the five major forces: demographic, climate change, technological change, globalization, and sudden shocks. Foresight papers review the patterns and trends in each force and examine how this force might affect the urban region.

As part of the workshop, subteams may explore the possible influences of the forces on various urban systems: mobility, housing, buildings, land use, energy, materials and waste, water, health, information and communications, security, agrifood, and the economy. The subgroups may use graphical tools to assist in forecasting.

Decision trees, influence diagrams, and belief nets are examples of tools that support the front end of a decision analysis. A particularly effective technique is to use influence diagrams to structure and facilitate dialogue. For many people, an influence diagram is the easiest way to understand a series of chains of cause and effect, although, strictly speaking, the causality is not always direct or restricted to the elements shown. Thus, terms such as influence or relevance are used. These diagrams are easy to draw, and they are intuitive; they allow straightforward numerical assessments. Most important, they visually communicate independencies among variables. By visually displaying changing assumptions, they allow groups to focus on internal dependencies as a whole, rather than in disjointed sections. Aspects of inference, prediction, and decision may be drawn using simple nodes and arrows

and then discussed at a nontechnical level with a view to reinforcing a systems perspective. These diagrams may also provide a foundation for more complex modeling among groups.

An example template for an influence diagram is shown in figure 2.50. Each unique chain of cause and effect leads to potential impacts on the economic, social, and environmental features of the region. With help from specialists, subteams may use such diagrams to map chains of cause and effect and the impacts on the four capitals in the case of each major force and urban system.

Through influence diagrams, the interdisciplinary subteams may then explore specific interventions or alternative designs for mitigating any significant negative impacts. In this way, the influence and intervention diagrams become a framework or mind-map, helping interdisciplinary groups to explore the longer-term vulnerabilities of the region and then to develop mitigation strategies. A foresight workshop may orient design teams to unfamiliar topics, such as security and resiliency. Such a workshop also initiates capacity building in the larger field of resiliency planning. Most designers and planners have little understanding of future studies, including on such topics as technology scans, S curves and innovation cycles, risk management, and the accelerating pace of change in many urban systems. While many of these concepts are difficult to grasp and integrate into daily practice, the foresight workshop exercises allow pragmatic issues to be discussed and complex concepts to be presented in visual formats that are easy to understand and reference.

A foresight workshop also creates the possibility for generating initial design solutions that promote resiliency. The workshops represent opportunities to explore adaptable designs. Designs that are versatile and durable favor simplicity, factor in redundancy, allow upgrades, opt for independence, and minimize

Climate Change and Our Urban Region

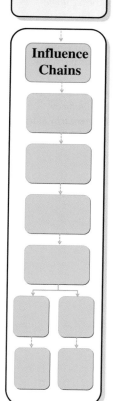

Forecast

➢Forecasting climate change is difficult at all times, not only because climate systems are complex by nature, but also because difficult assumptions are required about economic growth, new technology and pollution control efforts. It is best to prepare for surprises, since climate change has never been forecast with much success. However it is also prudent to plan for the most likely changes, based upon historical trends, and well researched climate change models for Southern Ontario and the National Capital Region.

➢Models project a 2-3°C rise in summer temperatures by mid century and a 4-5' rise by 2071. Summers will be longer. The growing season in southern Ontario could be 4-7 weeks longer.

➢Winters will be shorter, with less snow; skating season cut in half.

➢By 2020 the region will see 62-70 days of below zero temperatures versus the current normal of 80-89 days.

➢An overall drier climate is expected because rainfall cannot compensate for the increase in evaporation resulting from greater temperatures. Thus the region may see drier soils and more droughts.

➢Extreme heat will be more common, and the frequency of heavy rainstorms and windstorms will increase. Days when the temperature is greater than 30°C will increase from 11 to 28. Precipitation will be harder and more intense, but also more sporadic.

Mega-Trends

Warmer Summers	Longer, Drier Summers	More Extreme Weather Events	Shorter Winters less snow and ice	

Influence Chains

Water shortages

Intervention
Program to renovate existing stock with water efficient fixtures

Narrative

Graphic

Empty reservoirs

Health Crisis

Industrial work stoppages

Impacts

Manufactured Capital ±

Natural Capital ±

Social Capital ±

Cultural Capital ±

Figure 2.51 Template for an Influence Diagram

Source: Author elaboration (Sebastian Moffatt).

Note: The figure shows a typical template for the development of influence diagrams visually to represent chains of cause and effect in forces and urban subsystems.

destructive change. The workshops are also chances to demonstrate the benefits of ecological design solutions, such as compartmentalization and modularization, which help reduce the vulnerability of systems to the failure of any single part.

Note

1. InfraCycle is another example of a commercial spreadsheet application that helps cities calculate the capital, maintenance, replacement, and operating costs of all municipal infrastructure and estimate future revenues. See http://www.infracycle.com/.

References

Bilderbeek, Rob H., and Ida-Elisabeth Andersen. 1994. "European Awareness Scenario Workshops: Organizational Manual and Self-Training Manual." Report STB/94/045, Sustainable Cities Program, Center for Technology and Policy Studies, Apeldoorn, the Netherlands.

Brick, Karolina. 2008. "Barriers for Implementation of the Environmental Load Profile and Other LCA-Based Tools." Licentiate thesis, Royal Institute of Technology, Stockholm.

Fort St. John. 2009. "Sustainable Neighbourhood Concept Plan." Prepared by Sheltair Group, LCC analysis by David Rosseau. http://www.fortstjohn.ca.

Levin, Per, and Therése Rönnkvist-Mickelson. 2004. *Rapportsammanfattning—Uppföljning av miljöbelastning och ekonomi i Hammarby Sjöstad, Sickla Udde.* Stockholm: Carl Bro AB.

PART 3

The Field Reference Guide

A City-by-City and Sector-by-Sector Lens on Urban Infrastructure

The Eco² Field Reference Guide is a technical resource especially tailored to build ground-level and technical knowledge. It contains background literature designed to support cities in the development of in-depth insight and fluency on relevant issues at two levels. This section provides a city-by-city and sector-by-sector lens on urban infrastructure. It begins with a series of case studies on good-practice cities around the world. Each city offers a separate example of ways in which the various elements of the Eco² approach may be applied. The next section comprises a series of sector notes, each of which explores sector-specific issues that pertain to urban development. The sectors include energy, water, transportation, and solid waste. The section also includes a note on managing the spatial structure of cities. Together, these sector notes provide insights on how each sector functions and how they interrelate. As we view these issues through a city-by-city and sector-by-sector lens, the bigger picture starts to emerge. Finally, part 3 concludes with information on the specific financial instruments of the World Bank Group and some multidonor funds.

Eco^2 Case Studies
Good-Practice Cities

CASE 1
Curitiba, Brazil

Cost Is No Barrier to Ecological and Economic Urban Planning, Development, and Management

The case of Curitiba, Brazil, shows that cost is no barrier to ecological and economic urban planning, development, and management. Curitiba has developed a sustainable urban environment through integrated urban planning (figure 3.1). To avoid unplanned sprawl, Curitiba directed urban growth linearly along strategic axes, along which the city encouraged high-density commercial and residential development linked to the city's integrated master plan and land use zoning. Curitiba adopted an affordable but innovative bus system rather than expensive railways that require significant time to implement. Curitiba's efficient and well-designed bus system serves most of the urban area, and public transportation (bus) ridership has reached 45 percent.[1] The city now has less traffic congestion, which has reduced fuel consumption and enhanced air quality. The green area has been increased, mainly in parks that have been created to improve flood prevention and through regulations that have enabled the transfer of development rights to preserve green areas and cultural heritage zones. As part of efforts to concentrate shops and facilities in the city center and along dense axes, Curitiba's car-free central city zone (including its main streets and recreational facilities such as parks) has become more walkable, lively,

Figure 3.1 Curitiba Cityscape
Source: Institute for Research and Urban Planning of Curitiba (IPPUC).

and attractive to citizens. Crime has also decreased. In addition, citizens, particularly the poor, are provided with opportunities to participate in environmental activities and educational programs.

The social, economic, and environmental elements of sustainable development in Curitiba have been facilitated by integrated land use, public transportation, and street network plans (figure 3.2). Much of the success may be attributed to the Institute for Research and Urban Planning of Curitiba (IPPUC), an independent public authority that handles not only research and planning, but also the implementation and supervision of urban plans. IPPUC has coordinated the various aspects of urban development and ensured continuity and consistency in planning processes amid turnover in city administrations. This is an illustration of successful path dependency in urban development in terms of the spatial, institutional, and cultural aspects.

Profile of Curitiba and the Curitiba Metropolitan Region

Curitiba

- The capital of the State of Paraná, in the south of Brazil
- Land area: 432 km²
- Population (2008): 1.83 million
- Annual population growth rate: 1.86 percent
- The city is bordered by the Iguaçu River to the east and Passaúna Park to the west.
- The city is located at the center of Brazil's largest economic corridor, which includes Brasília, Porto Alegre, Rio de Janeiro, and São Paolo, and near major cities, such as Buenos Aires and Montevideo, in other South American countries.

Curitiba Metropolitan Region

- Consists of 26 municipalities, including Curitiba
- Land area: 15,622 km²
- Population (2008): 3.26 million
- Population growth rate: 2.01 percent

Map 3.1 Location of Curitiba
Source: Map Design Unit, General Services Department, World Bank.

Population Growth in Curitiba

YEAR	1960	1970	1980	1991	2000	2007	2008
Population (1,000s)	361	609	1,025	1,315	1,587	1,797	1,828
Population density (persons per km²)	836	1,410	2,373	3,044	3,674	4,161	4,232
Green area (km² per person)	—	<1	—	—	—	—	51.5

Source: IPPUC, http://ippucnet.ippuc.org.br (accessed January 15, 2009); data for 2008 from IPPUC (2009a).
Note: — = not available; km² = square kilometers

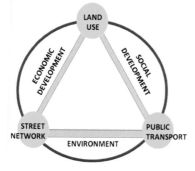

Figure 3.2 Policy Integration in Curitiba
Source: IPPUC

Approaches and Ecological and Economic Benefits

Curitiba took various innovative approaches to ecological and economic urban planning. The following are the seven major approaches.

Innovative land use planning integrated with transportation planning

Urban sprawl and concentrated traffic in Curitiba's downtown area were anticipated because of rapid population growth. The city formulated a master plan in 1966 that integrated land use and transportation plans. Curitiba decided to direct urban growth linearly by designating structural axes radiating from the city center (figure 3.3). Major economic activities are concentrated along these corridors, and the city appears to have a linearly formed downtown. At the same time, the city center was reinforced with high-density development (figure 3.4). The structured corridors became major public transportation routes under a bus rapid transit (BRT) system that includes dedicated lanes and bus stops nearly every 500 meters.

To realize this plan and guide linear urban growth, Curitiba implemented detailed zoning plans that reflect the master plan's strategic vision, geographical and geological constraints, water and wind directions, Curitiba's industrial profile, and urban cultural and social factors. In 2000, Curitiba had 50 types of specific zoning categories (figure 3.5). Each zoning category defines requirements related to land use, building-to-land ratios, floor area ratios, and maximum building heights. For example, in the city

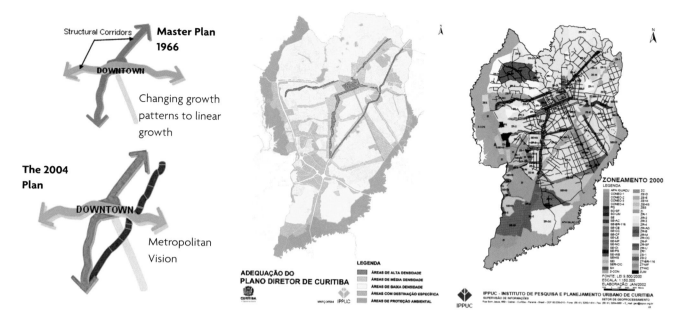

Figure 3.3 Urban Growth Axes in Curitiba

Source: IPPUC (2009a).

Figure 3.4 Density of Curitiba, 2004

Source: IPPUC, http://ippucnet.ippuc.org.br (accessed January 15, 2009).

Figure 3.5 Zoning in Curitiba, 2000

Source: IPPUC, http://ippucnet.ippuc.org.br (accessed January 15, 2009).

center area, the zone ZC category allows the development of residential apartments and commercial and service facilities (except supermarkets) subject to specific parameters: floor area ratios up to 5, first-floor building-to-land ratios up to 100 percent, and no limit on building heights in most areas. (However, to ensure aesthetics, buildings are normally limited to 20 floors, and some areas are subject to building height limitations to secure flight routes.) In addition, many zones facing structural axes (that is, zone SE) allow the development of residential apartments and commercial and service facilities with floor area ratios up to 4, first-floor building-to-land ratios up to 100 percent, and no limit on building heights in most areas. (As in zone ZC, buildings are normally limited to 20 floors to ensure aesthetics, and some areas are subject to building height limitations to secure flight routes; see Hattori 2004, Prefeitura Municipal de Curitiba 2000.)

To shift the land use and growth pattern into linear forms and to provide good access to transportation services, new development was permitted only in areas reachable by public transportation. Because Curitiba was designed for people, not cars, public transportation coverage and service frequency are critical. Bus service reaches almost 90 percent of the city area, and all users may access public transportation services by walking less than 500 meters (figure 3.6) (IPPUC 2009a). Bus routes are serviced nearly every five minutes. Curitiba initially acquired land and reserved rights-of-way along the strategic axes, which enabled the city to build social housing in these areas. Subsequently, major economic activities and urban functions, including residential neighborhoods and schools, were reorganized densely along these axes.

To accommodate BRT routes and fulfill transportation needs along the axes, the city designated functions to existing roads under its trinary road system. The five major axes now accommodate both dedicated BRT lanes and roads to access buildings. Cars that do not need to access services along the axes may bypass these areas by using roads parallel to the axes

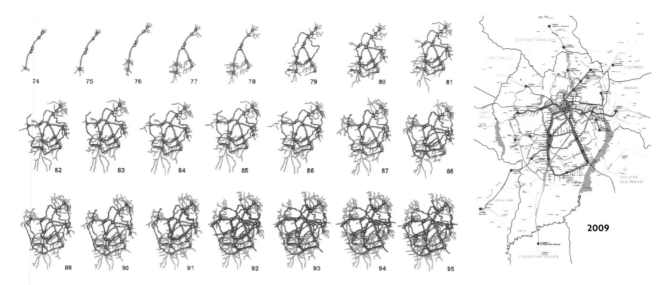

Figure 3.6 Evolution of the Integrated Bus Network in Curitiba, 1974–95 and 2009

Source: IPPUC (2009a).

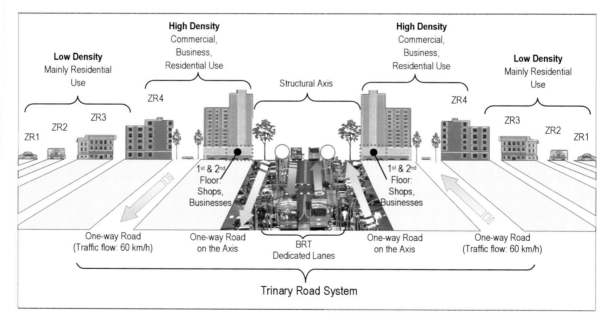

Figure 3.7 The Trinary Road System in Curitiba

Source: Author compilation (Hinako Maruyama) based on IPPUC (2009a), Hattori (2004), and pictures supplied by IPPUC.
Note: km/h = kilometers per hour.

(figure 3.7). In addition, to avoid concentrated traffic in the city center, a previous mayoral administration transformed selected streets in the city center into pedestrian walkways on which cars are prohibited.

Through these measures, Curitiba's spatial growth and urban land use patterns have been efficiently controlled and defined. Traffic is diverted from the city center or the axes thanks to an effective mixture of land use planning and a well-conceived public transportation network. Because housing, service facilities, and job centers have been incrementally developed along the axes and linked to the BRT system, the distances between homes, jobs, and schools have shortened, and many people travel by bus. Bus ridership as a share of all commuting trips reaches 45 percent, and 70 percent of these bus trips bypass the downtown area (IPPUC 2009b). As a result, the city has reduced car emissions and traffic congestion, thereby saving time and enhancing economic activity. A calculation based on 2002 data has estimated that Curitiba loses R$2.55 million annually (US$1.20 million) because of time lost to severe traffic congestion (table 3.1). Curitiba's per capita loss from severe congestion is about 6.7 and 11 times less than the corresponding losses in Rio de Janeiro and São Paulo, respectively. In 2002, Curitiba's annual fuel losses from severe traffic congestion equaled R$1.98 million (US$0.93 million). On per capita terms, this is about 4.3 and 13 times less than the losses in Rio de Janeiro and São Paulo, respectively (CNT 2002; Vassoler 2007). In contrast, in 2000, congestion in 75 metropolitan areas in the United States caused fuel and time losses valued at US$67.50 billion (Downs 2004). Curitiba's fuel usage is also 30 percent lower than the usage in Brazil's other major cities (Friberg 2000). Reduced car emissions have decreased air pollution, which threatens public health. Curitiba now has one of the lowest rates of ambient air pollution in Brazil (Leitmann 1999). In addition, emissions of greenhouse gases that affect climate change have declined.

Traffic flow has been diversified by assigning a logical and efficient road hierarchy, which has obviated the need to undertake substantial remedial works, such as widening street space (which may entail destroying buildings and disrupting neighborhoods). By making the most of infrastructure and adding new functions and traffic rules, the city has saved on construction costs. By avoiding extensive unplanned urban sprawl, its investment in infrastructure has been minimized and concentrated along the axes, and the installation of water pipes or cables into new areas has been avoided. More people now come to the city center because they are able to walk pedestrian streets, increasing the economic opportunities for local shops relative to streets with a predominance of car traffic.

The integrated public transportation system

The construction cost of Curitiba's BRT system was US$3 million per kilometer, which was more affordable than a tram system (a cost of US$8 million to US$12 million per kilometer) or a subway (US$50 million to US$100 million per kilometer) (Friberg 2000). Along the main axes, the BRT system functions much as a surface subway system would. Moreover, com-

Table 3.1 The Time and Fuel Losses Caused by Congestion

LOSS	CURITIBA, BRAZIL 2002	SÃO PAULO, BRAZIL 2002	RIO DE JANEIRO, BRAZIL 2002	UNITED STATES 2000	TOKYO, JAPAN 1994
Time loss					
Total, US$ millions/year	1.20	79.94	27.48	—	—
Per capita, US$/year	0.67	7.34	4.51	—	—
Fuel loss					
Total, US$ millions/year	0.93	73.23	13.47	—	—
Per capita, US$/year	0.52	6.72	2.21	—	—
Time and fuel losses					
Total, US$ millions/year	2.13	153.17	40.94	900[a]	49,000[b]
Per capita, US$/year	1.19	14.07	6.72	—	4,100[b]

Sources: For Brazil: CNT (2002), Vassoler (2007); for the United States: Downs (2004); for Tokyo: TMG (2000).
Note: Data are for reference only. Calculation methods for Brazil, United States, and Tokyo may be different and are not necessarily comparable. — = not available.
a. Average across 75 metropolitan areas. The total for these areas was US$67.5 billion.
b. Calculation based on the loss of travel speed from 30 kilometers per hour to 18 kilometers per hour.

pared with a normal bus system, BRT run times are two-thirds less, while the costs are 18 percent less owing to several factors, including a 72-kilometer dedicated BRT lane, a fare system requiring payment before boarding, bi-articulated buses (articulated buses with three cabins instead of two), and a tube-shaped bus station that eases bus entry and exit (figure 3.8) (Hattori 2004).

The bus system is color coded and designed for various scales and levels of service (interdistrict, feeder, intermunicipal, and so on) to reach more areas of the city (figure 3.9). The bus system has adopted a flat-rate "social" fare. No matter how far a passenger rides or how many times the passenger transfers, the fare is the same. The poor tend to live in the urban periphery and need to travel long distances to commute, while the wealthy tend to live in the cen-

ter and need to travel shorter distances. About 80 percent of all residents are estimated to benefit from the flat-rate fare (Hattori 2004). Frequent high-quality services and inexpensive fares encourage people to use buses. Of all trips, 45 percent are made in buses, 5 percent by bicycle, 27 percent on foot, and 22 percent by private car, which is surprisingly low given that Curitiba has the second-highest rate of car ownership in Brazil (IPPUC 2009a).

Buses running on the BRT dedicated lanes are bi-articulated, and the fleet is kept relatively young. The average age is a little more than 5 years, and no bus is more than 10 years old. The buses are well maintained and are less polluting. The greater carrying capacity of Curitiba's bi-articulated buses (270 people) and the reduced travel times associated with the use of these buses have resulted in 50 percent less energy consumption relative to nonarticulated conventional bus services (Hattori 2004, IPPUC 2009c,).

The BRT system pays for itself. Bus fares finance the system, generating profit for the bus companies and covering the costs of human resources and the maintenance and depreciation of buses without government subsidies. According to a law established in 1990, transportation revenue is exclusively dedicated to paying for the BRT system (Friberg 2000). In comparison, in some German cities with light rail, fare revenue covers only 30 percent of the operating costs; federal government subsidies are thus required. In the United States, subsidies for light rail are often generated through consumption taxes (Hattori 2004). The operation of the BRT system in Curitiba is managed by Urbanização de Curitiba (URBS), a city agency, but is served by private bus companies. The bus companies are paid on the basis of distance traveled, not by number of travelers; so, they are encouraged to operate even in areas with relatively few riders. Moreover, people are more motivated to use the buses because the service is frequent, affordable, and convenient.

Figure 3.8 Bi-articulated BRT Bus and Bus Station in Curitiba

Source: IPPUC.

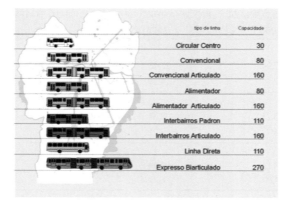

Figure 3.9 Color-Coded Buses in Curitiba

Source: IPPUC.

Green area enhancement and flood control

To improve the quality of the lives of its citizens, Curitiba decided to enhance the green areas and recreational facilities within the city, including parks and bicycle paths. Because Curitiba is surrounded by rivers such as the Iguaçu, flooding has been a big problem. However, instead of controlling water flow using concrete structures, Curitiba has created natural drainage systems. Riverbanks have been converted into parks where overflow water may be absorbed in the soil, and lakes have been constructed to contain floodwaters. River and rainwater flooding may be held naturally in the lakes and parks surrounding the lakes (figure 3.10). The ecosystem is thus preserved naturally. Because the park areas only gradually release flooded water that has been absorbed into the ground, rather than draining water rapidly through concrete drainage conduits into rivers, downstream flooding can be avoided. In addition, people are less exposed to flood-linked environmental hazards and diseases. Enormous expenditures are also avoided because there is less need for drainage canals, flood control measures, and flood damage repairs, including disease control measures. The cost of building parks and relocating favela (slum) dwellers has been estimated at five times less than the cost of building concrete canals (Vaz Del Bello and Vaz 2007).

Flood control areas are normally used as parks and recreational areas. Green areas have been enhanced from less than 1 square meter per person in the 1970s to 51.5 square meters per person today (ICLEI 2002; IPPUC 2009a). There are 34 parks in the city, and green areas cover about 18 percent of the urban land (Curitiba S. A. 2007). Bicycle paths are provided along the streets and inside the parks. The total length of the bicycle network is about 120 kilometers. Though park area has been expanded, the city has lacked the budgetary resources to maintain park grass. Instead of hiring mowers, sheep are kept in the parks to eat grass and provide natural fertilizer, which has reduced park

Figure 3.10 Barigüi Park, Curitiba

Source: IPPUC.

Note: This area was once flood prone and occupied by slum dwellers. It is now a converted 140-hectare park with a 40-hectare lake.

Figure 3.11 Former Slums in Flood-Prone Areas in Curitiba

Source: IPPUC.

maintenance costs by 80 percent, while enhancing the ecological image of the city.

Flood-prone land used to be occupied by slum dwellers (figure 3.11). Curitiba acquired the land, relocated the slum dwellers to better land, and provided compensation. After the park was established, the zone facing the park became neighborhoods with high-end housing. Houses with good views of the park and lake have high real-estate values; thus, property tax revenue has increased. Property taxes collected from these high-end houses have been estimated at the equivalent of the cost of park construction, including slum relocation and compensation.

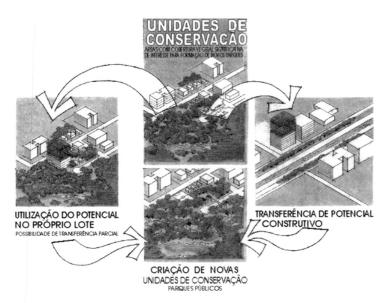

UNIDADES DE
CONSERVAÇÃO
ÁREAS COM COBERTURA VEGETAL SIGNIFICATIVA
DE INTERESSE PARA FORMAÇÃO DE NOVOS PARQUES

UTILIZAÇÃO DO POTENCIAL
NO PRÓPRIO LOTE
POSSIBILIDADE DE TRANSFERÊNCIA PARCIAL

TRANSFERÊNCIA DE POTENCIAL
CONSTRUTIVO

CRIAÇÃO DE NOVAS
UNIDADES DE CONSERVAÇÃO
PARQUES PÚBLICOS

Figure 3.12 The Transfer of Development Rights for Environmental Preservation in Curitiba

Source: IPPUC.

Many trees are found in Curitiba. There are 300,000 trees along public streets that create shade and prevent heating (IPPUC 2009b). Trees absorb pollutants and carbon dioxide Curitiba's reserved forest areas capture an estimated 140 tons of carbon dioxide per hectare, which helps reduce negative impacts on climate change (IPPUC 2009b). In addition, the shade from trees cools buildings and the environment, which saves energy.[2] City regulations restrict the area of private land for development depending on the ratio of land to forest or trees. To encourage urban trees, the city offers landowners compensation for planting, such as the relaxation of floor area ratios and tax reductions. For example, the city tax is discounted 10 percent if a private landowner has one Paraná pine tree on his land. Also, rights to develop forest areas may be exchanged for rights to develop other city areas (figure 3.12). Guided by market principles, IPPUC regulates and monitors the implementation, negotiation, and transfer of development rights among interested parties (that is, private developers and landowners). As such, the city does not need to undertake relocation or assume the land acquisition costs of creating green areas or preserving historical areas.

Solid waste management

Curitiba has several innovative programs in solid waste management. Curitiba's landfill was strained, and the city did not have sufficient revenue to build an incinerator. To slow the growth of waste, Curitiba initiated unique waste management programs that depend on citizens rather than constructing new and expensive waste treatment facilities. What is innovative is that these programs have not only reduced the growth of waste, but also offered opportunities for poor people, which is one of the critical aims of the city.

Curitiba's Garbage That Is Not Garbage Program encourages people to separate discards into recyclable and nonrecyclable waste (figure 3.13). To raise awareness of this program, Curitiba educates children to understand the importance of waste separation and environmental protection. Campaign mascots have been created, and school activities are regularly organized. One to three times a week, trucks collect paper, cardboard, metal, plastic, and glass that have been sorted at homes. This recycling saves the equivalent of 1,200 trees a day, and information displays in local parks show the numbers of trees that have been saved (Rabinovitch and Leitmann 1993). Money raised through the sale of recyclables supports social programs, and the city employs the homeless and people in alcohol rehabilitation programs in the garbage separation plant. Recycling also leads to other benefits. For instance, recycled fiber is used to produce asphalt for roads. Recycling has also eliminated piles of discarded tires, which attract mosquitoes that transmit dengue disease. Proper tire collection has decreased dengue disease by 99.7 percent (Vaz Del Bello and Vaz 2007). Nearly 70 percent of city residents participate in Curitiba's recycling program. Around 13 percent of Curitiba's waste is recycled, which greatly exceeds the 5 and 1 percent recycling rates in Porto Alegre and São Paulo, respectively, where education on waste dissemination has not translated into significant impacts (Hattori 2004).

The Green Exchange Program was also started in Curitiba's slum areas that are inaccessible to waste collection vehicles (figure 3.13). To encourage the poor and slum dwellers to clean areas, and thereby improve public health, the city began offering bus tickets and vegetables to people who brought garbage to neighborhood centers. In addition, children have been allowed to exchange recyclables for school supplies, chocolate, toys, and show tickets. The city purchases vegetables at discounted prices from farmers who have trouble selling abundant products. Through this program, the city saves the costs of arranging waste collection in slum areas, which often have inadequate roads, and helps farmers unload surplus produce. The program also helps improve nutrition, accessibility to transportation, and entertainment opportunities among the poor. Most important, slums are cleaner and have less disease incidence, and less garbage is being dumped in sensitive areas such as rivers.

The Industrial City of Curitiba

In the 1970s, Curitiba's economy was based mainly on the service sector. To attract investment, boost employment, and reduce poverty, IPPUC decided to introduce manufacturing industries. To further this goal, the local government established the Industrial City of Curitiba on the city's west side, taking into account wind direction to avoid polluting the central city. The industrial park has extensive green areas encompassing 4,300 hectares and is well connected to the bus network. Many employees at the industrial park live nearby and commute by bicycle. The industrial park has strict environmental regulations. Polluting industries are not allowed.

After three decades, the Industrial City of Curitiba now hosts more than 700 companies, including global firms such as information technology companies and an automaker producing BRT buses. It has created about 50,000 jobs directly and 150,000 jobs among secondary industries. About 20 percent of the exports of the State of Paraná originate in the industrial park, and the industrial park accounts for 25 percent of the industrial tax revenues (state value added taxes on sales and services) of the State of Paraná (Hattori 2004; Prefeitura Municipal de Curitiba 2009).

Social considerations

Although Curitiba's economy is relatively well developed compared with the economies of other Brazilian cities, many poor people still live in slums. To encourage the poor to obtain jobs and to promote an inclusive community, Curitiba has adopted various innovative social approaches.

The city converted the undeveloped land under a high-voltage line in a southern area of the city into a "job line" that helps people start businesses and encourages the growth of the local

Figure 3.13 Curitiba's Waste Program

Source: Photos courtesy of IPPUC.
Note: The Garbage That Is Not Garbage Program (left) and The Green Exchange Program (right).

economy. Two social incubators provide training and facilities for the establishment of local business, and 12 entrepreneur sheds were created (Guimarães 2009). In addition, these facilities offer entrepreneurial capacity building. Underutilized occupied land was cleared; people were relocated; and public transportation services were commenced, which represented steps toward land recovery (Hattori 2004).

One of the largest problems in Curitiba has been slums. Those who do not have their own land occupy and settle on private land. Often, these areas become derelict, causing river pollution and fomenting crime (figure 3.14). Rather than spending time and money on relocating squatters and restoring the areas that had been occupied, the city, at low cost, purchased privately held lands that could be occupied. It then provided this land for unofficial occupancy. A formal land use zoning category was developed for such land. Thus, these areas were integrated into city plans, and residents could feel included. Simple land arrangements and water and electricity are offered because such services

risk being illegally procured if not provided, which may lead to fatal accidents. Occupants feel some sense of ownership over the land and are able to arrange roads and create quality living environments. Under city agency coordination, the value of the occupied land may be reimbursed through long-term loans. In addition, legal mailing addresses may be provided for occupiers, which helps people find jobs (Hattori 2004; Nakamura 2007).

Curitiba provides social housing in the suburbs, where land prices are relatively cheap, and in the city, especially between the city center and industrial areas (figure 3.15). Rather than encouraging groups with similar incomes to settle in neighborhoods, Curitiba encourages a mix of income groups so that the neighborhoods become inclusive. Apartments and small detached homes are provided as social housing. Poor people who can afford to purchase small detached houses are given incentives to improve the properties and their overall living environment by building additions and extensions onto their houses. In Curitiba, development

Figure 3.14 Illegal Occupancy in Curitiba
Source: IPPUC.

Figure 3.15 Social Housing in Curitiba
Source: IPPUC.

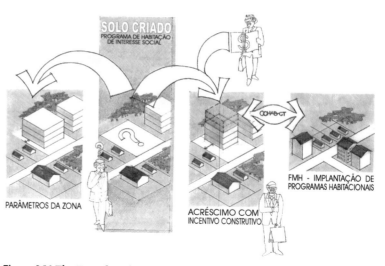

Figure 3.16 The Transfer of Development Rights for Social Housing in Curitiba
Source: IPPUC.

rights may be purchased. The money paid by developers to purchase the rights to develop sites may then be used to build social housing in other areas (figure 3.16).

City services are decentralized and are provided in major bus transit terminals. People are not necessarily obliged to travel to the city center for such services. Allowing people who live far from the city center to procure services close to home promotes equal opportunity. A flat bus fare also helps people reach bus terminals where city offices are located. In addition, city services such as educational, health, cultural, and social service facilities are distributed equally throughout the city. This system provides equal, high-quality, and accessible services to all citizens regardless of income.

Culture and heritage preservation

Curitiba maintains an attractive and lively cityscape. This is a result of well-planned urban design and successful cultural heritage preservation. Vehicular streets in the city center have been converted into pedestrian streets to allow people to enjoy the urban cultural atmosphere (figure 3.17). Under Curitiba's 1977 Metropolitan Area Heritage Plan, 363 buildings were identified for preservation. However, because most of these buildings were on private land, managing their preservation was difficult. The city thus adopted a policy under which development or

Figure 3.17 Pedestrian Streets in the Center of Curitiba
Source: IPPUC.

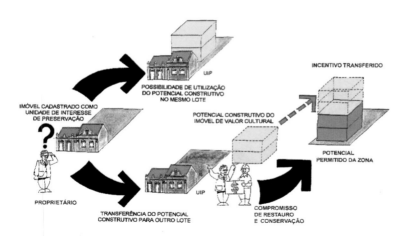

Figure 3.18 The Transfer of Development Rights for Heritage Preservation in Curitiba

Source: IPPUC.

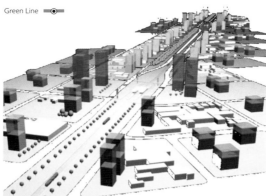

Figure 3.19 The Green Line in Curitiba

Source: IPPUC.

building rights may be transferred to other areas in the city. In 1993, the city identified special preservation units. Money earned from selling development rights over these structures must be used only to preserve buildings (figure 3.18). Through these measures, the money required for preservation is mainly market generated, and the city does not need to fund preservation. In addition, Curitiba's Coresda Cidade project has revitalized 44 historic buildings in the city center, and the buildings have been repainted in the original colors. The area targeted by this project used to be crime prone and run down. However, after revitalization, people came to the area; building owners took better care of the buildings; and the crime rate fell by an estimated 30 percent. Moreover, Curitiba provides a good case of a city revitalized by heritage preservation and effective urban design. In addition, cultural facilities, which were previously lacking in the city, have been established in innovative ways. A historical gunpowder house has been converted into a theater. An opera house constructed of metal tubes and glass has been established in the middle of a deactivated quarry crater, surrounded by a beautiful landscape. A botanical garden, one of the main tourist attractions, has also been created in once neglected open space (Hattori 2004).

Future Challenges for Curitiba

Green Line: Federal highway 116 used to cut through the city and impose its dangerously heavy traffic—mainly transport trucks traveling South America's economic corridor—on residents. This divided the city into two sections in an inefficient way. In response, a beltway was created to divert traffic outside the city's boundary, and a former federal highway was converted into Curitiba's sixth axis, called the Green Line. This line is expected to reduce traffic on the existing five axes. A new BRT route will be introduced, and mixed use high-density development is planned along the Green Line to make the area more attractive (figure 3.19). Land use is being carefully planned so as not to interfere with wind circulation by creating barriers of buildings. A linearly shaped biodiversity park will also be created along the Green Line, and only indigenous plant species will be cultivated.

Regional integration: Because the Curitiba Metropolitan Region is growing, Curitiba now faces a new challenge: how to integrate city and regional planning. Migration from surrounding areas has resulted in housing shortages, which might lead to more slums. In addition, even if Curitiba has a good primary BRT system and integrated land use, development in surrounding areas that is unconnected to the public transportation system (such as large shopping malls) may favor car use and increased traffic. In this context, Curitiba is taking steps to strengthen its regional planning capacity and is creating intermunicipal partnerships.

Lessons Learned in the Curitiba Case

Leadership and continuity: The mayors of Curitiba have focused on urban planning. Many mayors have had technical backgrounds in engineering or architecture, for example. Since the 1960s, when the Curitiba master plan was formulated, the direction of urban planning has been largely consistent across administrations. Curitiba places a premium on implementation and rapid action to address urban issues; if there is a 70 percent chance of success, the city undertakes plans quickly.

Institutionalized planning and expertise: Curitiba's practices affect the city in several positive ways. Curitiba's success is linked to strong mayoral leadership and people's active participation in city programs. It is also attributable to IPPUC. The integrated planning institute has been playing an important role as a municipally independent public authority that researches, formulates, implements, and supervises urban plans. IPPUC has provided integrated, cross-sector urban planning and oversight for implementation and monitoring, while ensuring consistency amid changes in political leadership. These holistic approaches to urban planning have been brought about by the creativity of planners and their imagination and good understanding of local culture. For more than 50 years, engineers and architects have undertaken urban planning to address key urban issues in integrated ways. The work of IPPUC ensures continuity and consistency in planning processes that have extended beyond the mayoral cycles since 1966, the year IPPUC was established. Curitiba has made the most of its existing infrastructure and local characteristics without spending much money on new construction. Although Curitiba's activities have been accomplished with few budgetary resources, tremendous benefits have been generated.

Citizen ownership and eco-consciousness: Citizens are encouraged and are provided opportunities to comment during urban planning processes. Public hearings with the mayor are held frequently, and proposed plans are evaluated and discussed with the community. People may speak directly to the mayor and city officials. More than 250 public hearings have been held since 2005. Citizens are actively involved in planning because people have made the link between good urban planning and a better quality of life. The city provides opportunities for people to participate in other urban activities—such as collecting garbage, constructing neighborhood roads, and maintaining green areas—which strengthen citizen ownership and the maintenance of urban facilities. Children are also enrolled in environmental education activities, such as the urban waste program. Moreover, behaving in environment-friendly ways is now the norm for Curitibanos.

Local character: Curitiba considers its local situation, including its budget, capacity, and social conditions, in devising urban strategies.

Taking into account municipal capacity, local officials develop innovative solutions to solve urban problems. For example, rather than waiting for adequate revenues to construct a subway, Curitiba implemented the BRT system, which proved affordable and could be implemented quickly without time-consuming construction work.

Notes

1. The modal shares are public transport (bus), 45 percent; bicycle, 5 percent; walking, 27 percent; and private automobile, 22 percent. The data are from IPPUC (2009a).

2. For example, in Houston, Texas the process of evapotranspiration from trees has been found to cool peak temperatures by 1.1 to 5 degrees Celsius. Tree shadings provide Houston with annual energy savings of US$26 million (HARC 2004).

References

CNT (Confederação Nacional do Transporte). 2002. "Pesquisa da Seção de Passageiros CNT, 2002; Relatório Analítico: Avaliação da Operação dos Corredores de Transporte Urbano por Ônibus no Brasil." Report, CNT, Brasília.

Curitiba S. A. 2007. *Bulletin 2007 of Socioeconomic Information.* Curitiba, Brazil: Curitiba S. A.

Downs, Anthony. 2004. *Still Stuck in Traffic: Coping with Peak-Hour Traffic Congestion,* rev. ed. Washington, DC: Brookings Institution Press.

Friberg, Lars. 2000. "Innovative Solutions for Public Transport: Curitiba, Brazil." *Sustainable Development International,* 4th ed., ed. Anna Pink, 153–56. Brighton, U.K.: ICG Publishing. http://www. brtchina.org/old/ReportE/Sustainable%20 Development.pdf.

Guimarães, Eduardo. 2009. "Curitiba: Liveable City; Transit and Sustainable Development." Presentation at the "Transportation Forum 2009," World Bank Group, Washington, DC, March 31.

HARC (Houston Advanced Research Center). 2004. "Cool Houston! A Plan for Cooling the Region." HARC, Woodlands, TX. http://files.harc.edu/ Projects/CoolHouston/CoolHoustonPlan.pdf.

Hattori, Keiro. 2004. "Ningen toshi Curitiba: Kankyou, koutsuu, fukushi, tochiriyou wo tougou shita machizukuri" 人間都市クリチバ―環境・交通・福祉・土地利用を統合したまちづくり [Human City Curitiba: urban planning integrating environment, transportation, social aspects, and land use]. Gakugei Shuppan Sha, Kyoto.

ICLEI (ICLEI–Local Governments for Sustainability). 2002. "Curitiba: Orienting Urban Planning to Sustainability." Case Study 77, ICLEI, Toronto.

IPPUC (Institute for Research and Urban Planning of Curitiba). 2009a. "The City of Curitiba: Planning for Sustainability; An Approach All Cities Can Afford." Presentation at "World Bank Energy Week 2009," World Bank, Washington, DC, March 31.

——. 2009b. "Energy Efficiency in Cities: Curitiba's Green Line." Presentation at "World Bank Energy Week 2009," World Bank, Washington, DC, April 1.

——. 2009c. "Public Transportation: Evolution of the Integrated Net of Transport." http://www.ippuc. org.br/pensando_a_cidade/index_transpcoletivo_ ingles.htm.

Leitmann, Josef. 1999. *Sustaining Cities: Environmental Planning and Management in Urban Design.* New York: McGraw-Hill.

Nakamura, Hitoshi. 2007. "Curitiba, Brazil ni okeru hito ni yasashii kankyou toshi zukuri no jissen" クリチバ（ブラジル）における人に優しい環境都市づくりの実践 [People-Friendly and Sustainable Urban Planning Practice in Curitiba, Brazil]. Presentation, July 13. http://www.sumai-machi-net.com// files/file/hitoshi(1).pdf.

Prefeitura Municipal de Curitiba. 2000. "Lei No 9.800 de 03 de janeiro de 2000, Annexos." Prefeitura Municipal de Curitiba, Curitiba, Brazil.

——. 2007. "Socioeconomic Information Bulletin 2007." Prefeitura Municipal de Curitiba, Curitiba, Brazil.

——. 2009. "Curitiba: Economic Changes." Prefeitura Municipal de Curitiba, Curitiba, Brazil. http://www.curitiba.pr.gov.br/siteidioma/ mudancaeconomica.aspx?idiomacultura=2.

Rabinovitch, Jonas, and Josef Leitmann. 1993. "Environmental Innovation and Management in Curitiba, Brazil." Working Paper 1, Urban Management Programme, United Nations Human Settlements Programme, Nairobi.

TMG (Tokyo Metropolitan Government). 2000. "TDM koutsuu jyuyou management Tokyo koudou plan" (TDM 交通需要マネジメント)東京行動プラン [Transportation demand management, Tokyo: Action plan]. Report, February, TMG, Tokyo.

Vassoler, Ivani. 2007. *Urban Brazil: Visions, Afflictions, and Governance Lessons.* New York: Cambria Press.

Vaz Del Bello, Giovanni, and Maria Terezinha Vaz. 2007. *A Convenient Truth: Urban Solutions from Curitiba, Brazil.* DVD. Directed by Giovanni Vaz Del Bello. Felton, CA: Maria Vaz Photography, in association with Del Bello Pictures.

CASE 2
Stockholm, Sweden

Integrated Planning and Management through Systematic Stakeholder Collaboration Can Lead to Greater Life-Cycle Benefits

The City of Stockholm, the capital of Sweden, has pursued integrated city planning and management to become a sustainable city (figure 3.20). The city has a comprehensive urban vision, environmental programs, and concrete action plans to reduce greenhouse gas emissions and tackle climate change. It implements integrated urban planning approaches that consider ecological benefits and efficient resource use.

The ongoing redevelopment in the city's southern district, Hammarby Sjöstad, is a good model for understanding integrated approaches to sustainable urban planning and redevelopment. The area aims to be twice as sustainable as Swedish best practice in 1995. The area implements integrated resource management (waste, energy, water, and sewage) through systematic stakeholder collaboration and has transformed the linear urban metabolism into a cyclical one known as the Hammarby Model. According to Grontmij AB, a private consultancy firm in Stockholm, primary assessments of the initially developed districts of Hammarby Sjöstad show that the area has achieved, for example, 28 to 42 percent reductions in nonrenewable energy use and 29 to 37 percent reductions in global warming potential.

Figure 3.20 Stockholm Cityscape

Source: Photo by Lennart Johansson, Stockholm City Planning Administration.

Stockholm provides great leadership in planning and implementing sustainable urban development strategies. The city's one-system approach to resource use has been successful. In addition, Hammarby Sjöstad has applied the environmental load profile (ELP) tool to assess and monitor environmental performance in the development project.

Profile of Stockholm

Stockholm

- The capital of Sweden, located in the northern part of Europe
- Total area: 209 km²
  (land: 188 km²; water: 21 km²)
- Population (2008): 795,000
- Expected increase in population by 2030: 150,000.

Source: USK (2008).

Map 3.2 Location of Stockholm

Source: Map Design Unit, General Services Department, World Bank.

Stockholm's Approaches to Sustainable Development

Stockholm pursues comprehensive sustainable development policies. In 2007, the city adopted a strategic project, Vision 2030, that charts the way forward to strengthen sustainable urban development (City of Stockholm 2007). This project indicates that, by 2030, the population of Stockholm will have grown to more than 1.0 million people, while the greater Stockholm region will have grown to nearly 3.5 million. The city is expected to face new demands from globalization, trade shifts, migration, an increase in the number of the elderly, and environmental challenges. Based on the Vision 2030 project and other strategies, Stockholm has adopted an approach to urban development that recognizes the strategic level and local levels (City of Stockholm 2007).

Aligned with Vision 2030, the Stockholm Environment Programme established six environmental goals or principles for 2008–11: (1) environmentally efficient transport, (2) safe goods and buildings free of dangerous substances, (3) sustainable energy use, (4) sustain-able land and water use, (5) waste treatment with minimal environmental impacts, and (6) a healthy indoor environment (City of Stockholm 2008).

In addition, Stockholm has implemented action programs on greenhouse gas emissions and climate change. The plans invite wide cooperation from public and private organizations and individuals who live and work in the city. Various measures have already been taken, including the adoption of biofuels, the expansion of districtwide heating and cooling management, and the promotion of vehicle driving behavior that is environment friendlier (City of Stockholm 2003). As a result, emissions of greenhouse gases fell from 5.3 tons to 4.0 tons of equivalent carbon dioxide (CO_2e) per person between 1990 and 2005 (City of Stockholm 2009). The city recognizes the importance of energy efficiency to reduce emissions and tackle climate change, but also prioritizes cost-effectiveness through resource conservation. Devising ways to engage stakeholders in actions that are environmentally and economically sustainable remains a challenge. Stockholm's long-term target is to be free of the use of fossil fuels by 2050 (City of Stockholm 2009).

Approaches to Sustainable Urban Development

Sustainable urban development is clearly a key aim. Stockholm can more easily implement integrated and sustainable land use and transportation plans because the city has traditionally exerted substantial authority over land use planning and ownership. In 1904, Stockholm started purchasing land for future development. As a result, around 70 percent of all urban land belongs to the city (Cervero 1998). This large share of city-owned land has prevented speculative land investments by developers and investors and empowered the city in planning and implementing development. The city thus has a solid platform for development. Developers construct buildings and housing on public land corresponding to city plans. Because rights-of-way are easily secured, transportation development has been straightforward, and other development has been promoted around transportation stations. Development benefits are now being returned to the public through planning in new town areas. In addition, parks and green areas cover 40 percent of Stockholm's land, and citizens enjoy an ecologically rich environment (USK 2008).

To promote sustainable development, Stockholm's planning strategy targets densification through the development of brownfield (already used) land inside the city before any unused greenfield land on the outskirts is developed (box 3.1, map 3.3). This is the overall objective of the comprehensive land use plan adopted by the city council in 1999.

Old and partly abandoned industrial and harbor areas (brownfields) adjacent to the inner city are being reused and redeveloped as part of the city development strategy. Several of these strategic development areas are directly linked to a new rapid tram system and also have direct access to other public transportation systems, such as the metro line. The areas have unique qualities because they are often located

close to water and natural areas. Some areas have been under construction for several years and will provide housing as part of the city's housing programs. Other areas are at the planning stage. The areas are being targeted for mixed use development, with attractive housing and business facilities; these dense structures create a more urban atmosphere in formerly suburban areas.

Hammarby Sjöstad, one of the original and ongoing redevelopment areas, is a full-scale demonstration site. It represents an example of an integrated urban development approach illustrating system solutions, innovative technology, environmental awareness, and active cross-sector collaboration.

Hammarby Sjöstad

The ongoing redevelopment project for Hammarby Sjöstad—the name of the district means "city on Hammarby Lake" in Swedish—is set on a former industrial and harbor brownfield area on the south side of Hammarby Lake and to the south of the city center. The aim of the project is to expand the inner city into an at-

---

BOX 3.1

The Development Strategies of Stockholm

- Reusing developed land (brownfields)
- Locating new development in areas with good access to public transportation
- Respecting and enhancing the character of the city, for example, the cityscape, the built environment, and the green structure
- Redeveloping semicentral areas and transforming industrial areas into urban areas of mixed uses characterized by variation
- Establishing focal points in the suburbs
- Meeting local demand
- Developing public spaces

Source: City of Stockholm.

Profile of Hammarby Sjöstad

Hammarby Sjöstad

- A city district in southern Stockholm
- Total area: 200 hectares, including 50 hectares of water
- Planned population: 25,000
- 11,000 apartments projected
- 200,000 km² of retail and office area projected

- One of three ecocycle districts in Stockholm: Hammarby Sjöstad, Östberga, and Skärholmen
- About 35,000 people are expected to live and work in the area.
- Today, more than half of the development has been completed, and it is anticipated that the district will be fully developed by 2017.

Residential Area in Hammarby Sjöstad

Source: Photo by Lennart Johansson, Stockholm City Planning Administration.

Hammarby Sjöstad Cityscape

Source: Photo by Lennart Johansson, Stockholm City Planning Administration.

Map 3.3 The Inner City of Stockholm and Adjacent Development Areas

Source: Stockholm City Planning Administration.

tractive water setting, while converting a rundown industrial area into a modern, sustainable, mixed used neighborhood. Soil will be decontaminated by removing tons of oil, grease, and heavy metals (Fryxell 2008). The ecosystem will be revitalized, and existing eco-assets, including trees and parks, will be preserved. The redevelopment will unlock land and property values by revitalizing brownfield land. Moreover, a once-shattered area will be reinvigorated, and about 11,000 new residential units and 200,000 square kilometers of new office and service area will be created.

The urban vision and concept for this new district was born in the early 1990s. The area's natural continuation of Stockholm's inner city toward the waterfront has influenced planned infrastructure and building designs. Hammarby Sjöstad adds a new layer to Stockholm's development: a modern, semiopen zone comprising a mix of traditional inner-city perimeter blocks and open and contemporary urban zones. Inner-city street dimensions, block lengths, building heights, and densities are well harmonized and offer openness, sunlight, parks, and water views (map 3.4).

The area is also well connected to public tramlines. According to a 2005 survey, two-thirds of all resident trips were made via public transportation, bicycles, and walking, and one-third by car (CABE 2009). Significant public transportation ridership and bicycling and walking have helped to reduce car emissions and the associated greenhouse gases. Mixed land uses are promoted, and land policy requires that ground floors along main streets be used for commercial applications. This encourages people to walk or cycle to visit streets with lively shop fronts. To attract shops and services to the new development area, the city has provided financial subsidies. Moreover, the area's economic activities were established in the development's early phases. Urban and building designs make the most of the waterfront. Myriad designs were created by different architects to provide diverse, lively, and high-quality urban environment.

Stockholm desired that Hammarby Sjöstad be twice as more sustainable than Swedish best practice in 1995 on a range of indicators (the environmental program was adopted in 1995), most notably energy efficiency per square meter. In Sweden, the average annual rate of energy use in some regular new developments is 200 kilowatt-hours per square meter. Cutting-edge Swedish developments and practices produce an efficiency of 120 kilowatt-hours per square meter. The Hammarby Sjöstad project aims for 100 kilowatt-hours per square meter. The project also sets other targets: water conservation, waste reduction and reuse, emissions reduction, the reduced use of hazardous construction materials, the application of renewable energy sources, and the implementation of integrated transportation solutions. Stockholm

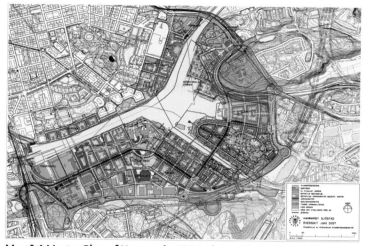

Map 3.4 Master Plan of Hammarby Sjöstad, Stockholm

Source: Stockholm City Planning Administration.
Note: For details of the Master Plan, see http://www.hammarbysjostad.se.

is already a sustainable city, but the city council expects this project to demonstrate additional innovations in sustainable urban development.

The Hammarby Model

The environmental goals for Hammarby Sjöstad, which was originally intended to be the Olympic Village in Stockholm's bid for the 2004 Summer Olympics, are audacious. The area's integrated environmental solutions may be understood as an ecocycle known as the Hammarby Model (figure 3.21). The ecocycle addresses energy, waste, water, and sewerage for housing, offices, and other commercial structures. Core environmental and infrastructure plans for this area have been developed jointly by three city agencies: the Stockholm Water Company, the energy company Fortum, and the Stockholm Waste Management Administration. Project management was spearheaded by a project team comprising representatives from city departments overseeing planning, roads and real estate, water and sewerage, and waste and energy. The project team is housed in the Department of Roads and Real Estate (now called the Development Administration).

The model is an attempt to turn a linear urban metabolism, which consumes inflowing resources and discards outflowing wastes, into a cyclical system that optimizes the use of resources and minimizes waste. The model streamlines infrastructure and urban service systems and provides a blueprint for achieving sustainability objectives. For instance, it shows the interaction between sewage processing and energy provision, the way refuse should be handled, and the added value to society of modern sewage and waste processing systems. Highlights are as follows:

- *Building materials:* Environmental considerations apply to all materials, whether used visibly in facades, underground, or internally. This includes structural shells and installed equipment. Only sustainable and tested eco-friendly products are used. Potentially hazardous materials, such as copper and zinc, are avoided to prevent leakages of unwanted substances into the environment.

- *Water and sewerage:* Storm water is unconnected to sewerage systems to improve the quality of wastewater and sludge. Rainwater from streets or nondomestic storm water is collected, purified through a sand filter, and released into the lake. This reduces pressure on the wastewater treatment plant. Rainwater from surrounding houses and gardens flows through open drains to the channel. This water runs through a series of basins, known as an equalizer, and then to the lake. Hammarby Sjöstad has its own wastewater treatment plant built to test new technology. Four new and different processes for purifying water are currently being tested.

- *Biogas:* Biogas is produced in the wastewater plant from the digestion of organic waste and sludge. The wastewater from a single household produces sufficient biogas for the household's gas cooker. Most biogas is used as fuel in eco-friendly cars and buses.

- *Green spaces:* Roofs covered in stonecrop or sedum plants are attractive. In addition, the plants absorb rainwater that would otherwise drain into the sewerage system, adding pressure on the wastewater treatment plant.

Moreover, the region's carefully preserved oak forests, green areas, and other planted trees help collect rainwater instead of draining it into the sewerage system. This vegetation also ensures cleaner air and balances the dense urban landscape.

- *Waste:* Combustible waste, food waste, newspapers, paper, and other discards are separated and deposited in different refuse chutes in or adjacent to buildings. The refuse chutes are linked to underground vacuum-powered pipes that lead to a central collection station. An advanced control system sends the waste to large containers, one for each waste category. Refuse collection vehicles thus collect the containers without driving into the area, and refuse collection workers avoid heavy lifting.

- *District heating and cooling:* Treated wastewater and domestic waste become sources for heating, cooling, and power. A combined heat and power plant uses domestic waste as fuel to produce district heating and electricity. Wastewater from the treatment plant fuels the production of district heating in the Hammarby heat plant. Cooled by heat pumps, the treated and cooled wastewater may also be used in the district cooling network.

- *Electricity (solar energy):* Solar energy is transformed into electrical energy in solar cells. The energy from a single solar cell module covering one square meter provides around 100 kilowatt-hours per year, which is equivalent to the energy used by three square meters of housing space. There are solar panels on many roofs used to heat water. Solar panels on residential buildings often provide sufficient energy to meet half of the annual hot water requirements of the buildings.

Hammarby Sjöstad has its own Environmental Information Center, GlashusEtt. This center facilitates communications on environmental considerations to area inhabitants and showcases Hammarby to international visitors.

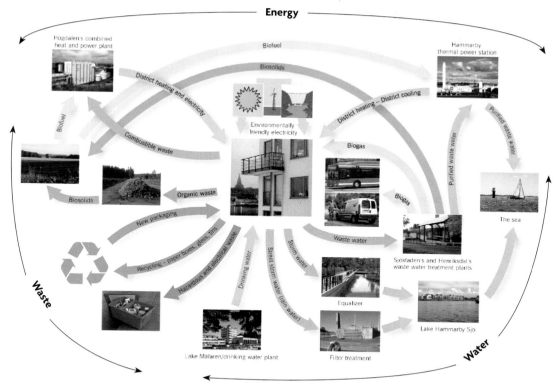

Figure 3.21 The Hammarby Model

Source: Fortum, Stockholm Water Company, City of Stockholm.

The environmental load profile

An environmental assessment tool, the ELP, has been developed through a cooperative effort by the City of Stockholm, the Royal Institute of Technology, and the consultancy firm Grontmij AB. The ELP assesses environmental performance and follows up on the targets set in the project's environmental program. It is a life-cycle assessment tool that defines relevant activities from an environmental perspective and quantifies the environmental loads originating from these activities, such as emissions, soil pollutants, waste, and the use of water and nonrenewable energy resources. It accounts for all project development and implementation activities, including material acquisition, the transport of inputs and people, construction methods, electricity, heating, and materials recycling.

The main strengths of the ELP are that the tool is flexible and dynamic, which makes it suitable for application under any conditions in planning, simulation, and evaluation. By factoring in well-constructed variables, one may use the ELP to calculate the environmental loads of various planning decisions during a project's construction, operation, demolition, or redevelopment. The tool thus facilitates a life-cycle approach. Testing scenarios are facilitated. For instance, different construction methods may be compared prior to taking decisions. Hence, decision makers understand environmental issues early in project planning. The ELP may also be used to evaluate the environmental performance of existing city districts or buildings based on the consumption of resources such as water and energy. The ELP enables analyses of environmental performance at multiple levels. The tool takes into account the activities and impacts of individuals (for example, cooking and laundry), buildings (building materials, district heating, electricity, and so on), unbuilt areas (such as materials and working machines), and common areas (including materials and the transport of people and goods). By aggregating these factors, the environmental load of a whole city district may be analyzed. If each factor is analyzed separately, different urban activities provide useful information for urban planning.

Evaluation findings of the initially developed areas of Hammarby Sjöstad as compared with a reference scenario are illustrated in figure 3.22. The results are positive: a 28 to 42 percent reduction in nonrenewable energy use, a 41 to 46 percent reduction in water use, a 29 to 37 percent reduction in global warming potential, a 33 to 38 percent reduction in photochemical ozone creation production, a 23 to 29 percent reduction in acidification potential, a 49 to 53 percent reduction in eutrophication potential, and a 27 to 40 percent reduction in radioactive waste. By monitoring the environmental loads from Hammarby Sjöstad, one may plan suitable societal and financial environmental measures to continue the development of the district, while offering guidance for similar projects.

Project management

The two municipal administrations responsible for planning and managing the project are the City Planning Administration and the Development Administration. These entities are under respective committees and the city council.

In the mid-1990s, Stockholm and its external stakeholders agreed to cooperate on planning objectives in the area. These stakeholders include the neighboring municipality of Nacka, the Stockholm Local Transport Authority, and the National Road Administration. After negotiating, the stakeholders agreed on a set of common planning features and infrastructure projects (1994–95). During this period, there was a political steering group and an official management group composed of representatives of key stakeholders. An organization was established to manage the project. All departments responsible for the planning, development, implementation, and maintenance of the area were involved from the beginning.[1] The city's Waste Collection Administration and the city's associated companies—the energy company and the water company—participated in preparing the project's environmental program; moreover,

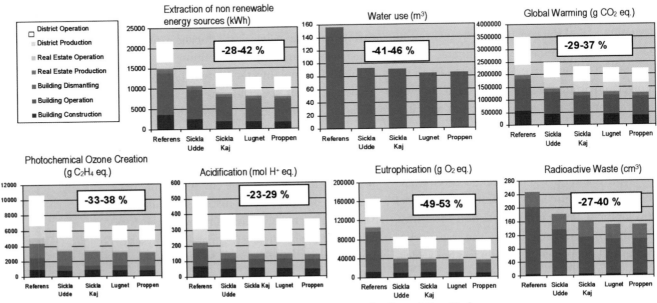

Figure 3.22 Monitoring Major Reductions in Environmental Loads, Hammarby Sjöstad, Stockholm

Source: Grontmij AB.

these companies had a vested interest because a power station and a wastewater treatment plant were located in the area.

A steering group composed of the executive officers of the departments involved and a cross-sector official management group have been active in the development of the project.[2] As a landowner, the city may initiate agreements and undertake contracts with developers. The city may specify various requirements depending on the issues that are important in each phase. Developers have contractual obligations to participate in the planning process (on the detailed development plan), the process of defining and implementing quality and design standards, and the implementation of relevant aspects of the environmental program.

The national level

The Hammarby Sjöstad project was partially supported by a national subsidy program that aimed to encourage municipalities to become part of an ecologically sustainable society, while providing project-related jobs in municipalities (Bylund 2003). The program lasted from 1998 to 2002 and allocated SKr 6.2 billion (€671 million) to 211 local investment programs involving 1,814 projects in 161 municipalities (figure 3.23). This national investment leveraged SKr 27.3 billion (almost €3 billion) from municipalities, businesses, and other organizations. Of this amount, SKr 21 billion (about €2.3 billion) were investments directly related to sustainability and the environment. It has been estimated that 20,000 full-time short-term or permanent jobs were created (Swedish EPA and IEH 2004).

A report of the United Nations (2004: 4) states that, according to local authority estimates, "the grants awarded to local investment programmes for the period 1998–2002 will lead to annual reductions in energy use by 2.1 TWh [terawatt hour] while carbon dioxide emissions will be reduced by 1.57 million tonnes per year (equalling 2.8 percent of Sweden's emissions) and landfill refuse deposits will be reduced by about 500,000 tonnes per year. Emissions to water will be reduced by 2,460 tons of nitrogen and 180 tons of phosphorous per year, which correspond to two percent and four percent, respectively, of the current total emissions to the sea."

The next phase

Lessons and experiences from Hammarby Sjöstad will be considered in planning and implementing Stockholm's new eco-profiled city districts. These new areas will use the latest environmental technology with a view to serving as a model of the sustainable city concept. Energy, transportation, lifestyle, and behavioral issues will be particularly important variables determining whether the project objectives will be met.

For instance, the Stockholm Royal Seaport is a new urban development with a unique environmental profile (figure 3.24). Developing a new ecologically sustainable district places extra demands on technology in the construction of houses, the use of efficient materials, and methods for handling energy. This urban development contains plans for 10,000 new residences and 30,000 new workspaces. Phase 1 started in 2009, and about 5,000 units will be developed over the next decade. The first residents will arrive in 2011.

The vision for the area may be summarized in three comprehensive objectives:

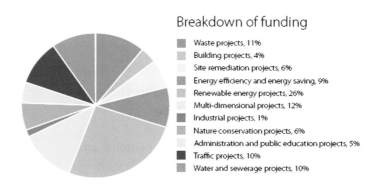

Breakdown of funding

- Waste projects, 11%
- Building projects, 4%
- Site remediation projects, 6%
- Energy efficiency and energy saving, 9%
- Renewable energy projects, 26%
- Multi-dimensional projects, 12%
- Industrial projects, 1%
- Nature conservation projects, 6%
- Administration and public education projects, 5%
- Traffic projects, 10%
- Water and sewerage projects, 10%

Figure 3.23 Local Investment Subsidy Program Funding across Types of Projects in Sweden

Source: Swedish EPA and IEH (2004).

Figure 3.24 Stockholm Royal Seaport: Vision of a New City District

Source: Lennart Johansson, Stockholm City Planning Administration.

1. By 2030, the area is to be a fossil-fuel-free city district.
2. By 2020, CO_2 emissions are to have been cut to 1.5 tons per person per year (CO_2 equivalent).
3. The area is to become adapted to the expected effects of climate change.

The project's focus areas are energy consumption and efficiency, sustainable transportation, climate change adaptation, ecocycle modeling, and the maintenance of high-quality lifestyles. Other important goals include implementation of a holistic and integrated process; constant evaluation and follow-up; and assessment and cooperation among private, public, and academic stakeholders.

Lessons Learned in the Stockholm Case

Great leadership in planning and implementing sustainable urban development strategies demonstrates Stockholm's strong commitment to sustainable development. Success in a project such as the one in Hammarby Sjöstad depends on good coordination among key stakeholders. For the project, Stockholm's various departments have been integrated into a single fabric led by a project manager and an environmental officer whose responsibilities have included guiding and influencing all stakeholders, public as well as private, to realize the environmental objectives of the project (Johansson and Svane 2002). Integrated planning and management through systematic stakeholder collaboration can lead to significantly greater life-cycle benefits.

After a few modifications, the ELP may serve as a decision-making tool in cities in developing countries in a fashion similar to the use of this ELP in the Swedish context. The ELP provides a systematic and standardized methodology to quantify the costs and benefits of the steps in development. For the application of an ELP in developing countries, one might propose the following:

1. Expanding the ELP to include assessments of other input variables, such as the impacts that efficient spatial planning, integrated land use, and the improved management of solid waste may have on output indicators.

2. Improving and fine-tuning the existing program by filling in gaps and streamlining the inclusion of the inputs. Moreover, the complete model needs to be adapted to large-scale use and adjusted to fit developing country contexts.

3. The outputs in the current ELP area are associated with environmental indicators, such as carbon emissions. Converting these indicators from environmental indicators to economic and fiscal indicators is necessary to help policy makers reach better decisions.

Notes

1. The departments included the City Planning Administration; the Real Estate, Streets, and Traffic Administration (now split into the Development Administration and the Traffic Administration); the City District Administration; and the Environment and Health Protection Administration.

2. The departments and companies involved included the City Planning Administration, the Development Administration, the Traffic Administration, the City District Administration, the Environment and Health Protection Administration, the Water Company, and the Housing Service Company.

References

Bylund, Jonas R. 2003. "What's the Problem with Non-conventional Technology? The Stockholm Local Investment Programme and the Eco-cycling Districts." In *ECEEE 2003 Summer Study Proceedings: Time to Turn Down Energy Demand,* ed. Sophie Attali, Eliane Métreau, Mélisande Prône, and Kenya Tillerson, 853–62. Stockholm: European Council for an Energy Efficient Economy. http://www.eceee.org/conference_proceedings/eceee/2003c/Panel_4/4214bylund/.

CABE (Commission for Architecture and the Built Environment). 2009. "Hammarby Sjöstad, Stockholm, Sweden." http://www.cabe.org.uk/case-studies/hammarby-sjostad.

Cervero, Robert. 1998. *The Transit Metropolis: A Global Inquiry.* Washington, DC: Island Press.

City of Stockholm. 2003. "Stockholm's Action Programme against Greenhouse Gas Emissions." City of Stockholm, Stockholm. http://www.stockholm.se/KlimatMiljo/Klimat/Stockholms-Action-Programme-on-Climate-Change/Downloads/.

——. 2007. "Vision 2030: A World-Class Stockholm." Executive Office, City of Stockholm, Stockholm. http://international.stockholm.se/Future-Stockholm/.

——. 2008. "The Stockholm Environment Programme." City of Stockholm, Stockholm. http://international.stockholm.se/Stockholm-by-theme/A-sustainable-city/.

——. 2009. "The City of Stockholm's Climate Initiatives." City of Stockholm, Stockholm. http://www.stockholm.se/vaxthuseffekten.

Fryxell, Stellan. 2008. "Planning Hammarby Sjöstad, Stockholm." Presentation at the Urban Land Institute, "Europe Trends Conference: Rethinking Tomorrow; Real Estate in a Fast Changing World," Stockholm, May 29.

Johansson, Rolf, and Örjan Svane. 2002. "Environmental Management in Large-Scale Building Projects: Learning from Hammarby Sjöstad." *Corporate Social Responsibility and Environmental Management* 9 (4): 206–14.

Swedish EPA (Swedish Environmental Protection Agency) and IEH (Swedish Institute for Ecological Sustainability). 2004. "Local Investment Programmes: The Way to a Sustainable Society." Investment Programmes Section, Swedish EPA, Stockholm. http://www.naturvardsverket.se/Documents/publikationer/91-620-8174-8.pdf.

United Nations. 2004. "Human Settlement Country Profile: Sweden." Division for Sustainable Development, Department of Economic and Social Affairs, United Nations, New York. http://www.un.org/esa/agenda21/natlinfo/countr/sweden/Sweden_HS.pdf.

USK (Stockholm Office of Research and Statistics). 2008. "Data Guide Stockholm 2008." USK, Stockholm.

CASE 3

Singapore

The One-System Approach: Integrated Urban Planning and Efficient Resource Use

Singapore is an island city-state at the southern tip of the Malay Peninsula (figure 3.25). With a limited land area of 700 square kilometers and a population of 4.8 million, Singapore has become developed because of innovative urban planning integrated with the efficient use of land and natural resources (CLAIR 2005; Statistics Singapore 2009).

Singapore's small size poses challenges related to the availability of land and natural resources. To optimize land use, Singapore promotes high-density development not only for businesses and commercial entities, but also for residential structures. High density lends itself to higher economic productivity per unit of land and facilitates the identification of green spaces and natural areas for preservation. Indeed, Singapore is known as the garden city. Furthermore, high-density development has translated into greater use of public transportation as major business, commercial, and residential areas are well connected to an integrated public transportation network. In 2004, public transportation as a share of all transportation modes during morning peak hours reached 63 percent. The significant use of public transportation helps reduce greenhouse gas emissions. High public transporta-

Figure 3.25 Singapore Cityscape
Source: Photo by Hinako Maruyama.

tion ridership also means Singapore has been able to recover all public transportation operating costs from fares, a feat achieved only by Hong Kong, China, and by Singapore among modern, highly developed cities (LTA 2008).

Singapore imports most of its natural resources, including food, water, and industrial materials. Careful resource planning is thus critical. For example, Singapore has adopted the comprehensive management of water resources by looping and cascading water, which represents a closed water cycle integrated into one system, rather than a water supply system based on once-through flows. Water efficiency is integrated into the activities in other sectors as a result of cross-sector coordination among

Profile of Singapore

Singapore

- An island city-state at the southern tip of the Malay Peninsula, 136.8 km north of the equator; located south of the Malaysian State of Johor and north of Indonesia's Riau Islands
- Population (2008): 4.84 million, including resident and nonresident population
- Land area: 700 km^2
- Population density (2008): 6,814 people per km^2
- GDP at current prices (2008): US\$181.9 billion
- Water and sewerage coverage: 100 percent
- Center of commerce and industry in Southeast Asia
- Global financial center and trading hub with one of the busiest seaports in the world

Map 3.5 Location of Singapore

Source: Map Design Unit, General Services Department, World Bank.

government departments and stakeholders. For example, new housing developments are equipped with efficient rainwater collection devices so that building roofs become water catchment areas.

Singapore has introduced various tools and incentives to manage the resource supply and demand. For example, it has implemented strategic water tariffs, creative energy policies, road pricing schemes, and a vehicle quota system. These measures discourage people and businesses from using resources beyond the city's capacity to supply them.

Singapore has demonstrated how a city may enhance economic productivity and growth, while minimizing ecological impacts and maximizing the efficiency of resource use. The strong leadership of the prime minister has been a major driver in the city-state's sustainable development, complemented by an integrated one-system approach and the active collaboration of stakeholders.

Approaches and Ecological and Economic Benefits

Singapore is committed to promoting sustainable development. The Inter-Ministerial Committee on Sustainable Development, which was established in 2008, enables integrated approaches across ministerial boundaries in the formulation of strategies for sustainable growth.

Integrated land use and transportation planning

Because of limited land resources, land use planning has been important in maintaining the quality of Singapore's environment and supporting its economic growth. Since independence in 1959, Singapore has actively expropriated land to obtain public land for public facilities, promote city redevelopment, and catalyze new development. Today, about 90 percent of land is owned by the city-state (Bertaud 2009). The city thus has strong authority over urban development plans and their implementation.

Singapore's Urban Redevelopment Authority within the Ministry of National Development is in charge of urban planning and promotes Singapore's policy of high-density development. For example, the central business district of Singapore has floor area ratios up to 13. Ongoing development near Marina Bay next to the central business district aims to produce high-density, mixed use development with floor area ratios up to 20 (URA 2009). Marina Bay will be more than a commercial center. It will also offer housing, shops, hotels, recreational facilities, and community zones such as green areas and open spaces.

Singapore's high-density, built-up areas have enabled the preservation of open spaces, natural parks, and greenery. Around 10 percent of all land is designated as green space, including natural reserves (figure 3.26). The share of green area in Singapore, including roadside greenery, was 36 percent in 1986, but increased to 47 percent in 2007. This gain was realized despite population growth of 68 percent.

Singapore's transportation plan is coordinated and well integrated with land use planning (Leitmann 1999). Recent high-density development, such as new towns, industrial estates, and commercial areas, is well connected to the city's mass rapid transit system. The mass rapid transit network runs underground in the city center and on the surface outside the city center and in other major areas. The network is the backbone of Singapore's public transportation system. Other transportation modes, such as buses and light rail transit, are well connected to network routes at interchange stations and serve local areas. To ease transfers, Singapore introduced a distance-based through-fare structure.

The integration of the mass rapid transit, light rail transit, and bus networks helped boost public transportation's share in all transportation modes (including taxis) to 63 percent in

2004, although this represented a decline from 67 percent in 1997 that was attributable to the growing use of private cars. In addition, among major cities in developed countries, the full recovery of the operating costs of public transportation through fares has been achieved only by Hong Kong, China, and by Singapore (LTA 2008). Because the transportation system has been integrated into high-density development areas with sizable populations, it has been possible to maintain the financial viability and high-quality service of the system. People are well satisfied with public transportation[1] (LTA 2008).

Transportation measures

Singapore's Land Transport Authority was established in 1995 by integrating four separate land transport departments comprehensively to plan, control, and manage relevant policies. The authority aims to provide a high-quality transportation system, enhance the quality of the lives of citizens, and maintain Singapore's economic growth and global competitiveness.

Singapore provides incentives to control the number of private cars. In 1990, the vehicle quota system was introduced by the government to limit the number of newly registered cars to 3 to 6 percent each year. A consumer wishing to purchase a new car must apply to the Land Transport Authority to conduct an open bidding process. Car owners must obtain certificates of enrollment that are valid for a decade following registration (Leitmann 1999, CLAIR 2005).

To respond to growing traffic and congestion, Singapore introduced an area licensing scheme in 1975 to manage cars entering the central business district during peak hours. In 1998, to boost effectiveness, the area licensing scheme was replaced by the current electronic road pricing system. The new system electronically collects fees from drivers through in-vehicle units installed in cars that enter designated areas of the city center during certain periods of peak

Figure 3.26 A Green Area in Singapore

Source: Photo by Hinako Maruyama.

traffic. The system has several price options depending on road types (arterial and highway) and periods. Higher prices are applied during the most congested times. In addition, Singapore uses several other demand control measures, such as encouraging off-peak driving or park-and-ride schemes through financial incentives (Leitmann 1999, CLAIR 2005).

Taken together, these road traffic, public transportation, and mobility measures mean that 71 percent of trips in Singapore may be completed in less than an hour (IMCSD 2009). Traffic congestion is alleviated, and average traffic speed is maintained. Unnecessary vehicular emissions are thus avoided. This translates into less greenhouse gas linked to climate change. However, travel demand is expected to increase from 8.9 million trips in 2008 to 14.3 million trips in 2020. Within Singapore, 12 percent of the land is dedicated to roads, and 15 percent to housing. Moreover, it is highly unlikely that more land can be dedicated to roads to accommodate travel demand (LTA 2008). Singapore must thus accommodate increased demand through public transportation services, not automobiles.

Water resource management

Singapore is considered a water-scarce city-state despite high annual precipitation of 2,400 millimeters per year (Tortajada 2006a).[2] Singapore imports water from neighboring Malaysia. To reduce dependency on external water sources, Singapore is taking steps to improve water security and establish an independent water supply within its own territory. The approach Singapore has developed and implemented to achieve this aim is considered successful because of the city-state's institutional effectiveness and its highly efficient control of water demand and supply. Singapore successfully lowered its annual water demand from 454 million tons in 2000 to 440 million tons in 2004, while its population and GDP grew by 3.4 and 18.3 percent, respectively (Tortajada 2006a). Singapore has shown that comprehensive water resource management is achievable using new approaches and that these approaches are financially viable.

The institutional framework that enabled the integrated approach

The Public Utilities Board (PUB), a statutory board under the Ministry of the Environment and Water Resources, manages the entire water cycle, including collection, production, distribution, and reclamation. It is the national water agency of Singapore. When PUB was established in 1963, it managed several utilities, including water, electricity, and gas. To reduce costs and improve services, PUB underwent institutional restructuring in 2001. The electricity and gas services were privatized, and sewerage and drainage functions were transferred to PUB. Since 2001, PUB has developed and implemented comprehensive and holistic approaches to the water system, rather than managing each water function individually (water supply, sewage, drainage, and so on). In this way, the water loop is closed, which enables PUB to implement the Four National Taps, a long-term strategy to ensure that Singapore has sustainable water supplies (figure 3.27). The Four National Taps are (1) water from local catchments, (2) imported water, (3) desalinated water, and (4) NEWater (reclaimed water from wastewater). By approaching the water system holistically, PUB is able to efficiently address various issues and activities, such as water resource protection, storm water management, desalination, demand management, water catchment management, private sector engagement, and community-driven programs, including public education and awareness campaigns. PUB also runs a research and development facility in which experts research water technology.

PUB's effective engagement of the private sector is a distinctive aspect. To lower costs, PUB harnesses the private sector in areas where it does not have competence or competitive

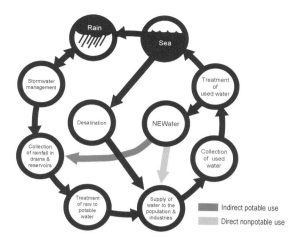

Figure 3.27 A Closed Water Loop in Singapore
Source: PUB (2008a).

Indirect potable use
Direct nonpotable use

advantage. For example, public-private partnerships are used in water desalination and wastewater reclamation.

Supply management

Because water is scarce, Singapore carefully manages its water supply. Sewerage covers 100 percent of the city-state area, and all wastewater is collected. Singapore has a separate drainage system to ensure wastewater and runoff do not mix. Wastewater and drainage water are recycled into the city-state's water supply.

The Four National Taps strategy considers the following as water sources (PUB 2008b):

1. *Water from local catchments* (catchment management): Rainwater is collected from rivers, streams, canals, and drains and stored in 14 reservoirs. Because storm drains are separated from the sewerage system, rainwater may be sent directly to rivers or reservoirs for later treatment to produce tap water. Reservoirs are linked via pipelines. Excess water may be pumped from one reservoir to another, thus optimizing storage capacity and preventing flooding during heavy rains. Catchment areas are protected, and polluting activities are prohibited in these areas by strict regula-

tions. By 2009, water catchment areas will have expanded from one-half to two-thirds of Singapore's land surface. Activities that generate pollution are allowed on only 5 percent of Singapore's land area; all other land is protected. Water catchments provide about half of Singapore's water needs (Tigno 2008).

To improve environmental and resource management, the government pays close attention to water catchment areas and to the locations of industrial sites. Singapore also pursues integrated urban planning. For example, PUB and the Housing and Development Board, which is under the Ministry of National Development, collaborate to enhance Singapore's water catchment areas. PUB considers rainfall an important resource, and rainfall collection and drainage systems are installed on the roofs of housing structures developed by the Housing and Development Board. Newly developed properties are equipped with rainfall collection and drainage systems. Collected water is stored in neighboring holding basins and transferred to reservoirs. This strategy allows built-up areas to participate in water catchment. Two-thirds of Singapore's land area is expected to participate in water catchment.

2. *Imported water:* Singapore will continue to import water from Malaysia under two bilateral agreements that expire in 2011 and 2061 respectively. Imported water accounts for about a third of the country's water needs (Tigno 2008).

3. *Desalinated water:* In September 2005, Singapore opened a US$200 million desalination plant, which was PUB's first public-private partnership project. The plant can produce 30 million gallons (136,000 cubic meters) of water a day; it is one of the largest seawater reverse osmosis plants in the region. In 2007, the plant provided about 10 percent of the country's water needs (Tigno 2008).

4. *NEWater:* Used water (wastewater) is also an important water resource. Wastewater is collected through an extensive sewerage system and treated at water reclamation plants. Wastewater is purified using advanced membrane technology to produce high-grade reclaimed water, known as NEWater, which is safe to drink. Because such water is purer than tap water, it is ideal for industry uses that require high-quality water, such as the manufacture of precision equipment and information technologies. Each day, PUB blends 6 million gallons (28,000 cubic meters) of NEWater with raw reservoir water, which is later treated to become tap water. The amount to be blended will increase to 10 million gallons a day (46,000 cubic meters) by 2011. Four NEWater factories operate in Singapore, and a fifth plant is being built under a public-private partnership agreement. In 2008, NEWater satisfied more than 15 percent of Singapore's total daily water needs, and the share is expected to rise to 30 percent by 2010 (PUB 2008c, 2008d).

The water supply may be optimized if nonrevenue water or water lost to leaks is reduced. The 4.4 percent share of nonrevenue water in Singapore's water supply in 2007 was low (Lau 2008), and there are no illegal connections (Tortajada 2006a).[3]

PUB has built a deep tunnel sewerage system as an integral part of the water loop. Though sewerage coverage is 100 percent, the aging sewerage network posed problems. The new system comprises deep sewerage tunnels that intercept water flows from existing sewerage pipes, pumping stations, and linked sewerage pipes. The designed lifespan of the system is 100 years. Because wastewater flows by gravity through the system to a centralized water reclamation plant (the Changi plant), intermediate pumping stations may be abolished. This removes the risks of surface water pollution caused by failures at intermediate pumping stations and the risks of damage to pumping mains. Water reclamation plants and pumping stations require about 300 hectares of land. New water reclamation plants in the deep tunnel system occupy only 100 hectares; 200 hectares of land may therefore be released for other uses. Building the system proved to be more cost-effective (by more than S$2 billion, or about US$1.35 billion) than expanding and upgrading existing infrastructure (Tan 2008). The system also enhances the closed water loop by collecting wastewater effectively for NEWater production.

Demand management

PUB has a well-planned, holistic policy for managing water demand. Tariffs rely on several rates depending on consumption level, not lump-sum proxies (table 3.2). If domestic use surpasses more than 40 cubic meters per month, the unit charge becomes higher than the nondomestic tariff. The basic water tariff has increased each year since 1997. The water con-

Table 3.2 Water Tariff in Singapore

SERVICE	CONSUMPTION BLOCK, M³/MONTH	TARIFF BEFORE GST, S$/M³	WATER CONSERVATION TAX BEFORE GST, % OF TARIFF	SANITARY WATERBORNE FEE AFTER GST, S$/M³	APPLIANCE FEE AFTER GST, S$/ CHARGEABLE FITTING/MONTH
Domestic	0 to 40	1.17 (US$0.81)	30	0.30 (US$0.21)	3.00 (US$2.07)
	above 40	1.40 (US$0.97)	45	0.30 (US$0.21)	3.00 (US$2.07)
Nondomestic	all units	1.17 (US$0.81)	30	0.60 (US$ 0.41)	3.00 (US$ 2.07)

Source: PUB Web site, http://www.pub.gov.sg/mpublications/FactsandFigures/Pages/WaterTariff.aspx (accessed May 2009)

Note: The U.S. dollar amounts shown in parentheses reflect the exchange rate of S$1.00 = US$0.69 as of June 4, 2009. Before GST (goods and services tax) and after GST indicate tariffs and fees excluding and including, respectively, the GST of 7 percent as of May 2009, rounded to the nearest cent. GST = goods and services tax; M³ = cubic meter.

Table 3.3 Water Consumption and Water Bills per Household in Singapore, 1995, 2000, and 2004

INDICATOR	1995	2000	2004
Population (1,000s)	3,524.5	4,028	4,167
GDP (US$ millions)	84,288.1	92,720.2	109,663.7
National water consumption (millions m³)	403	454	440
Average monthly water consumption (m³)	21.7	20.5	19.3
Average monthly water bill, including taxes (S$)	14.50	31.00	29.40

Source: Tortajada (2006b).

Note: m³ = cubic meter.

servation tax is levied to reinforce water conservation. In addition, a waterborne fee is charged to cover the costs of wastewater treatment and the maintenance and extension of the public sewerage system. This represents a financial disincentive on household water consumption. Consequently, as water bills are raised (inclusive of all taxes), water consumption decreases (table 3.3). The tariff system has significantly influenced water usage. Although annual water use in Singapore increased from 403 million cubic meters in 1995 to 454 million cubic meters in 2000, these demand control policies helped lower demand to 440 million cubic meters in 2004 (Tortajada 2006b).

Social considerations and awareness raising

To ensure equity, the government provides direct subsidies to lower-income families. Lifeline tariffs subsidize all water consumers, not only those who cannot afford to pay high tariffs. As such, Singapore provides subsidies only to targeted poor households. Targeted subsidies are widely considered more efficient in socioeconomic terms relative to subsidies on the initial amount of water consumed by all households irrespective of economic status. The tariff system makes clear that those who consume more water will be penalized (through basic tariffs and taxes) even more heavily than commercial and industrial uses.

Other environmental approaches

Singapore supports intense economic activity in the small island-state. Maintaining a quality environment is thus a critical issue. The Ministry of the Environment and Water Resources issued the Singapore Green Plan 2012 in 2002 and updated it in 2006. The plan addresses six main areas: air and climate change, water, waste management, nature, public health, and international environmental relations (MEWR 2006). The plan builds on the 1992 Singapore Green Plan. Since 1992, local officials have actively tackled environmental issues by implementing various activities involving a range of stakeholders, including citizens and public and private sector entities. In 2009, the Sustainable Singapore Blueprint—"A Lively and Livable Singapore: Strategies for Sustainable Growth"—was launched by the Inter-Ministerial Committee on Sustainable Development to ensure that Singapore not only met the target set in the Green Plan, but also would go beyond this to achieve economic growth and a good living environment in an integrated way (IMCSD 2009).

Energy: To avoid overconsumption, Singapore does not subsidize energy. Electricity supplies are established by market demand and competition, and industries are encouraged to find better solutions and to be energy efficient. For improvement of cost-effectiveness, natural gas–based electricity generation has recently surpassed oil-based generation. The share of electricity produced using natural gas rose from 19 percent in 2000 to 79 percent in 2007. In addition, energy consumption per unit of GDP has been reduced, and the efficiency of electricity generation has been enhanced (IMCSD

2009). To raise public awareness about energy concerns, the government has introduced E² Singapore, a national energy efficiency plan. The government has also made investments in energy research and technologies. For example, to capitalize on Singapore's tropical location, the government promotes solar energy research with a view to reinforcing the clean energy sector.

Air pollution measures: To minimize air pollution, land use plans locate industrial facilities outside the urban area. Car emissions are another source of air pollution. The vehicle quota system and the electronic road pricing system help reduce traffic congestion, and the integrated public transportation system encourages public transportation ridership. Additional car emissions are avoided, including airborne particulate matter and greenhouse gases. In 2008, 96 percent of the days in the year showed good air quality according to the Pollutant Standards Index (IMCSD 2009).

Waste management: Rapid economic and population growth has resulted in increased waste. Because it has limited land for landfills, Singapore incinerates wastes that cannot be recycled or reused. Incineration reduces the weight and volume of waste by, respectively, 10 percent and 20 percent, and has proven to be an efficient waste treatment process (CLAIR 2005). Electricity produced from incineration provides 2 to 3 percent of the city's electricity needs (IMCSD 2009). Singapore has only one remaining landfill site, which is located 8 kilometers south of the mainland and is the city-state's first constructed offshore landfill. There is no more land available for landfills or the disposal of residue from incineration. It is expected that the life of this offshore landfill will surpass its 2040 closure because of the recycling efforts of citizens (SG Press Centre 2009). However, the city is facing waste management challenges, especially as daily waste increased by a factor of 6, to 7,600 tons, between 1970 and 2000 owing to economic growth, population increases, and improved living standards (CLAIR 2005).

To promote recycling and waste reduction, Singapore's National Recycling Program encourages various activities, and per capita domestic waste has fallen despite economic growth. In 2008, the recycling rate reached 56 percent. Additionally, government-industry collaboration has promoted reduced waste from packaging (IMCSD 2009).

River cleanup: Singapore has successfully cleaned and restored the environmental conditions of its once deteriorated rivers. In 1977, Singapore and the prime minister supported a major project to clean the Singapore River and Kallang Basin, which covers about one-fifth of the city-state's land area. Uncontrolled waste and wastewater from farms, houses not on the sewerage system, and squatters were being discharged directly into the rivers. In response, houses and other polluting activities were relocated, and efforts were undertaken to improve the physical condition of the rivers. The riverbeds were dredged; waterfront facilities were upgraded; and greenery was added to riverbanks. Government agencies, grassroots communities, and nongovernmental organizations contributed to the cleanup. The rivers were revitalized in 10 years at a cost of S$200 million (Best Policy Practices Database). Today, the river waterfronts, including canals and reservoirs, are well preserved and maintained. These river zones act as water catchments and flood prevention areas, while providing community recreational space (for example, see PUB 2008e).

Singapore's waterways, including its rivers and reservoirs, are designed to be people friendly. The designs complement Singapore's vision as a city of gardens and water. Waterways and embankments are often recreational sites; moreover, people are reluctant to contaminate a resource they eventually drink. PUB provides educational opportunities through a visitor center and learning courses. PUB also encourages water conservation by providing tips and devices for saving water in households.

Greening: Singapore's Garden City campaign has been promoted since the 1970s to green the country by planting trees along roads, in vacant plots, on reclaimed land, and in new developments. Flowers are added, too. Since Singapore's independence in 1959, more than a million trees have been planted, and a high standard has been achieved in landscaping in the country (Leitmann 1999).

Housing

The government aims to supply affordable housing to its citizens. The Housing and Development Board plans and develops public housing and facilities in new towns. Because land is limited, high-density development and high-rise buildings are promoted for commercial, business, and residential uses. Urban renovation and the development of new and satellite towns are encouraged; 20 such towns have been constructed. New towns are connected to public transportation and Singapore's city center. In 2003, 84 percent of Singaporeans resided in publically built housing, and 92.8 percent had their own housing (CLAIR 2005). Since 1989, the Housing and Development Board has implemented an ethnic integration policy to ensure a balanced mix of ethnic groups in public housing (HDB 2009). Singapore has myriad ethnic groups, including Chinese, Indians, and Malays. The policy prevents the establishment of racial enclaves and promotes diverse communities and social integration.

Lesson Learned in the Singapore Case

Singapore faces challenges related to the scarcity of land and natural resources amid strong economic and population growth. Singapore shows that innovative and comprehensive management of land and other resources is achievable. Singapore has capitalized on its understanding of local conditions to develop a high-density city that preserves green spaces and open spaces. Public transportation works efficiently and is financially viable and integrated with land uses. Because of Singapore's comprehensive and integrated management of resources, the city-state is successfully addressing ecological, economic, and social concerns, while ensuring sustainability and productivity.

Notes

1. According to the Land Transport Authority (LTA 2008), 86.5 percent of the population are satisfied with the bus and rail services. About 80 percent are satisfied with the overall travel times on buses and trains. About 85 percent are satisfied with the accessibility and the locations of bus stops and mass rapid transit stations.

2. For comparison, data on annual rainfall among major cities in the world show the following: Bangkok, 1,530 millimeters; Beijing, 575 millimeters; Jakarta, 1,903 millimeters; Kuala Lumpur, 2,390 millimeters; London, 751 millimeters; Manila, 1,715 millimeters; New York, 1,123 millimeters; Shanghai, 1,155 millimeters; Tokyo, 1,467 millimeters (see Statistics Bureau 2008).

3. In most urban centers in Asia, nonrevenue water accounts for around 40 to 60 percent of water supplies.

References

Bertaud, Alain. 2009. "Urban Spatial Structures, Mobility, and the Environment." Presentation at "World Bank Urban Week 2009," World Bank, Washington, DC, March 11.

Best Policy Practices Database. Asia-Pacific Forum for Environment and Development. http://apfed-db.iges.or.jp/dtlbpp.php?no=23 ("Cleaning up of Singapore River and Kallang Basin").

CLAIR (Council of Local Authorities for International Relations). 2005. "Singapore no Seisaku" シンガポールの政策 [Policies of Singapore]. Tokyo: CLAIR.

HDB (Housing Development Board). 2009. "Ethnic Group Eligibility." HDB, Singapore. http://www.hdb.gov.sg/fi10/fi10004p.nsf/ECitizen/SELLING/$file/Selling_HDBEnq_FAQB.htm.

IMCSD (Inter-Ministerial Committee on Sustainable Development). 2009. "A Lively and Liveable Singapore: Strategies for Sustainable Growth." Ministry of the Environment and Water Resources and Ministry of National Development, Singapore. http://app.mewr.gov.sg/web/contents/ContentsSSS.aspx?ContId=1034.

Lau, Yew Hoong. 2008. "Sustainable Water Resource Management in Singapore." Presentation at the United Nations Economic and Social Commission for Asia and the Pacific, "1st Regional Workshop on the Development of Eco-Efficient Water Infrastructure in Asia Pacific," Seoul, November 10–12. http://www.unescap.org/esd/water/projects/eewi/workshop/1st/asp.

Leitmann, Josef. 1999. *Sustaining Cities: Environmental Planning and Management in Urban Design.* New York: McGraw-Hill.

LTA (Land Transport Authority). 2008. "LTMasterplan: A People-Centred Land Transport System." LTA, Singapore. http://www.lta.gov.sg/ltmp/LTMP.html.

MEWR (Ministry of the Environment and Water Resources). 2006. *The Singapore Green Plan 2012.* Singapore: MEWR.

PUB (Public Utilities Board). 2008a. "About Us." PUB, Singapore. http://www.pub.gov.sg/about/Pages/default.aspx.

——. 2008b. "Four National Taps Provide Water for All." PUB, Singapore. http://www.pub.gov.sg/water/Pages/default.aspx.

——. 2008c. "NEWater Wins Its Second International Award at Global Water Awards 2008." Press release, April 22, PUB, Singapore. http://www.pub.gov.sg/mpublications/Pages/PressReleases.aspx?ItemId=176.

——. 2008d. "Plans for NEWater." PUB, Singapore. http://www.pub.gov.sg/newater/plansfornewater/Pages/default.aspx.

——. 2008e. "Explore Bedok Reservoir." Brochure, PUB, Singapore. http://www.pub.gov.sg/abcwaters/Documents/Bedok_reservoir_nov25.pdf.

SG Press Centre. 2009. "National Environment Agency Launches a Commemorative Book to Celebrate Semakau Landfill's 10th Anniversary." Press release, Singapore Government, Singapore. http://www.news.gov.sg/public/sgpc/en/media_releases/agencies/nea/press_release/P-20090808-1.

Statistics Bureau. 2008. "Sekai no toukei 2008" 世界の統計 2008 [*World Statistics 2008*]. Tokyo: Statistics Bureau, Ministry of Internal Affairs and Communications. http://www.stat.go.jp/data/sekai/pdf/2008al.pdf.

Statistics Singapore. 2009. "Statistics: Time Series on Population (Mid-Year Estimates)." Singapore Department of Statistics, Singapore. http://www.singstat.gov.sg/stats/themes/people/hist/popn.html.

Tan, Yok Gin. 2008. "Managing the Water Reclamation Infrastructure for Sustainability: The Singapore Water Story." Presentation at the International Water Association, "World Water Congress 2008," Vienna, September 9.

Tigno, Cezar. 2008. "Country Water Action, Singapore; NEWater: From Sewage to Safe." Asian Development Bank, Manila. http://www.adb.org/Water/Actions/sin/NEWater-Sewage-Safe.asp.

Tortajada, Cecilia. 2006a. "Water Management in Singapore." *International Journal of Water Resources Development* 22 (2): 227–40.

——. 2006b. "Singapore: An Exemplary Case for Urban Water Management." Additional Paper, Human Development Report, United Nations Development Programme, New York.

URA (Urban Redevelopment Authority). 2009. "Embrace the World at Marina Bay." URA, Singapore. http://www.marina-bay.sg/index.html.

CASE 4
Yokohama, Japan

Waste Reduction by Engaging Stakeholders in the Private Sector and Civil Society

For Eco[2] cities, the case of Yokohama offers information on ways to realize significant environmental and economic benefits by engaging private sector and civil society stakeholders. Yokohama is the largest city in Japan (figure 3.28).[1] It reduced waste by 38.7 percent between fiscal years[2] 2001 and 2007, despite the growth of 165,875 people in the city's population. This reduction in waste is attributable to the city's success in raising public awareness about environmental issues and the active participation of citizens and businesses in Yokohama's 3Rs program (reduce, reuse, and recycle).[3]

Yokohama has been able to shut down two incinerators because of the significant reduction in waste. The incinerator closures have saved US$6 million[4] in annual operating costs and US$1.1 billion that would have been needed to renovate the incinerators (table 3.4) (City of Yokohama 2006). Around 5 percent of the fiscal year 2008 budget of the Resources and Wastes Recycling Bureau, the city's waste management entity, was derived from the sale of recycled material (US$23.5 million). In addition, the city raises US$24.6 million annually by selling the electricity generated during the incineration process (City of Yokohama 2008a).

Yokohama's success demonstrates that a city may achieve waste reduction through the cooperation of its stakeholders, particularly citizens. Reducing waste also results in significant cuts

Figure 3.28 The Yokohama Waterfront

Sources: Yokohama Convention and Visitors Bureau (left) and City of Yokohama (right).

Profile of Yokohama

Yokohama

- The largest city in Japan after Tokyo
- Population (2009): 3.65 million
- Land area: 435 km²
- Population density (2009): 8,409 persons per km²
- The Port of Yokohama was opened to international trade in 1859, the year the government of Japan decided to abandon its isolationist policy and initiate modernization and the opening to foreign cultures. The city is celebrating the 150th anniversary of the port's opening in 2009.
- In 2005, about 21 percent of the population was commuting out of the city for purposes of employment or education.
- The population becomes quite involved in participatory civil activities.
- In 2008, the city was selected as one of Japan's Eco-Model Cities.

Map 3.6 Location of Yokohama

Source: Map Design Unit, General Services Department, World Bank.

Table 3.4 The Power of Stakeholder Engagement in Yokohama, Fiscal Years 2001–07

Total waste reduction	623,000 tons (–38.7 percent)
Economic benefit	US$1.1 billion in capital costs saved because of two incinerator closures
	US$6 million in operating costs saved because of two incinerator closures
	Life of landfill sites was extended
CO_2 reduction	840,000 tons

Source: Author compilation (Hinako Maruyama).

in greenhouse gas emissions. In addition, a city may cut expenditures by reducing waste, while generating revenue from the recyclables and by-products resulting from waste treatment. Encouraged by these achievements, Yokohama now aims to reduce greenhouse gas emissions to lead Japan toward the national reduction target and demonstrate its place as one of the Eco-Model Cities.[5]

Background and Approaches to Waste Reduction

Yokohama's population has increased slowly by 0.5 to 1 percent per year. Population growth and the associated economic activities have gener-

ated more waste, and this has put pressure on the city's landfill sites, which have limited capacity. In 2000, the city had seven incinerators (of which six were in operation) and two landfill sites (an inland site and a sea reclamation site). To reduce the environmental impact of incineration and landfill disposal and to nudge Japanese society toward a zero waste cycle, Yokohama started the G30 Action Plan in 2003. The G30 plan aims to reduce waste by 30 percent by fiscal year 2010, using fiscal year 2001 waste quantities as baselines.

The G30 plan identifies the responsibilities of all stakeholders—households, businesses, and the city government—to reduce waste through the 3Rs based on polluter pay schemes and extended producer responsibility principles (City of Yokohama 2003). The plan provides integrated approaches to reduce waste that are supported by detailed action programs. For example, Yokohama citizens must separate waste into 15 categories and properly dispose of each category of waste at designated places and times. Businesses are requested to provide products and services that produce less waste and to implement the 3Rs actively. The city, which is one of the largest entities producing waste, is committed to decreasing waste and to working with citizens and businesses as a model player.

To disseminate the G30 approach, the city conducts environmental education and promotional activities that enhance public awareness and foster collaborative action to achieve the G30 goal. To promote adequate waste separation, the city has undertaken public activities, including more than 11,000 seminars among neighborhood community associations—80 percent of Yokohama's population participates in neighborhood community associations—to explain waste reduction methods such as the segregation of waste (City of Yokohama 2008b; see figure 3.29). In addition, about 470 campaigns have been held at railway stations; about 2,200 awareness campaigns have also been organized in the mornings at local waste disposal points; and so on (City of Yokohama 2006). Campaign activities have been initiated along local shopping streets, at supermarkets, and at various events (figure 3.29). The G30 logo is posted in all city publications, on city-owned vehicles, and at city events.

As a result, the waste reduction target of 30 percent was achieved in fiscal year 2005, five years earlier than expected (fiscal year 2010). By fiscal year 2007, waste had fallen 38.7 percent relative to 2001, despite the growth in the population by 165,875 people over the period (table 3.5, figure 3.30).

Figure 3.29 Public Awareness Campaigns for Waste Reduction and Separation in Yokohama

Source: City of Yokohama.

Table 3.5 Waste in Yokohama, Fiscal Years 2001–07

INDICATOR	FY2001	FY2002	FY2003	FY2004	FY2005	FY2006	FY2007
Population (millions)	3.46	3.50	3.53	3.56	3.58	3.60	3.63
General waste, excluding recyclables (1,000s of tons)	1,609	1,586	1,532	1,316	1,063	1,032	987
Waste from households (1,000s of tons)	935	928	919	855	651	652	628
Waste from business activities (1,000s of tons)	674	658	613	461	412	380	359
Collected recyclables, including compost waste (1,000s of tons)	50	50	53	72	166	162	160

Sources: City of Yokohama (2008a); City of Yokohama statistics portal, http://www.city.yokohama.jp/me/stat/.

Note: FY = fiscal year.

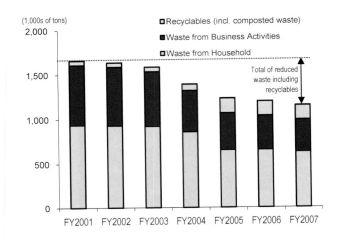

Figure 3.30 Waste Reduction in Yokohama, Fiscal Years 2001–07

Sources: Author compilation (Hinako Maruyama) based on City of Yokohama (2008a); City of Yokohama statistics portal, http://www.city.yokohama.jp/me/stat/.

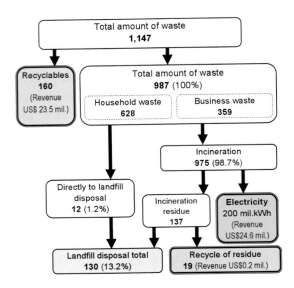

Figure 3.31 The Waste Flow in Yokohama, Fiscal Year 2007

Source: Author compilation (Hinako Maruyama) based on City of Yokohama (2008a, 2008c).

Note: The amounts in bold are thousands of tons. kWh = kilowatt-hour; mil.= million.

The Environmental Benefits of Waste Reduction

In Yokohama, almost 99 percent of nonrecyclable waste is brought to incinerators for treatment (figure 3.31). Waste treatment is the largest contributor to carbon dioxide (CO_2) emissions among the city's public works activities, which include office work, waste treatment, water provision, sewage treatment, and public transportation. For instance, CO_2 linked to waste treatment comprised 54.8 percent of total CO_2 emissions from city public works in fiscal year 2000.

According to Yokohama's life-cycle assessment, the waste reduced between fiscal years 2001 and 2007 was equivalent to avoiding 840,000 tons of CO_2 emissions. This included 760,000 tons of avoided CO_2 emissions from obviated waste collection, incineration, and landfill disposal and 110,000 avoided tons of CO_2 from recycling waste. Incinerators produce electricity from heat and steam generated by burning waste, then reuse this electricity for their own operations or sell it to electricity companies or other facilities. However, because reduced waste results in less incineration and electricity production, the electricity company that purchased electricity from incinerators

must produce additional electricity. In Yokohama, this additional supply of electricity was equal to 30,000 tons of CO_2; thus, the balance of avoided CO_2 is 840,000 tons (table 3.6), which is equivalent to the amount of CO_2 that 60 million Japanese cedar trees are able to absorb in one year. Planting that many cedar trees would require an area of approximately 600 square kilometers, an area 27 percent larger than the city (City of Yokohama 2009).

The Economic Benefits of Reduced Waste

In 2000, the city had seven incinerators, but, by 2006, two incinerators had been shut down owing to significant reductions in waste. This closure represented a savings of US$1.1 billion in capital expenditures that would have been needed to reconstruct and renovate the two incinerators. It also saved annual operating expenditures of US$6 million (that is, US$30 million

Table 3.6 CO$_2$ Reduction through Waste Reduction, Fiscal Years 2001–07

tons

INDICATOR OF REDUCTION	QUANTITY OF CO$_2$
CO$_2$ reduction because of reduced waste collection, incineration, and landfill	760,000
CO$_2$ reduction because of recycling	110,000
CO$_2$ increase caused by the additional supply of electricity by the electricity utility	(–30,000)
Total CO$_2$ reduction	840,000

Source: City of Yokohama (2009).

in savings from obviated annual operating costs, minus US$24 million in expected annual expenditures for intermediate waste treatment and separation costs, recycling, contracting, and so on) (City of Yokohama 2006).

Yokohama has two landfill sites. When the G30 was planned in 2003, it was forecast that the landfills would have 100,000 cubic meters of capacity remaining in 2007 and would be full by 2008. However, owing to the waste reduction achieved, the two sites had 700,000 cubic meters of capacity remaining in 2007. The value of the additional capacity of 600,000 cubic meters is equivalent to US$83 million (City of Yokohama 2006). In addition, the development of a new landfill site or reclamation area in the sea has been postponed.

The Economic Benefits of the Efficient Use of Resources

The city's five incinerators produce heat and steam during the incineration of waste. The heat and steam are used to operate the incinerators, including the heating, cooling, and generation of hot water, and to power adjacent public facilities, including an indoor pool and elder care facilities. Turbines in the incinerators produce electricity from the steam. In fiscal year 2007, the incinerators produced 355 million kilowatt-hours of electricity. Of this power, 42.2 percent was reused by the incinerators; 55.4 percent was sold to electricity companies

under competitive tendering; and 2.4 percent was harnessed by nearby public facilities, such as a sewage treatment plant, sludge recycling facility, and seaside line railway. In fiscal year 2007, US$24.6 million was earned by selling 200 million kilowatt-hours of electricity, which is equivalent to one year of electricity for 57,000 households (City of Yokohama 2008a).

Yokohama began earning revenue by selling recyclables, such as cans, bottles, paper, furniture, and electronic appliances, as well as reusable metal and material produced from incinerated ash. Collected recyclables are sold to private companies for additional treatment and reuse. Incineration ash is recycled into construction materials. About US$23.5 million in revenue is secured by selling recyclables to treatment companies (City of Yokohama 2008a).

As a result of these measures, about 10 percent of the US$480 million budget of the Resources and Wastes Recycling Bureau in fiscal year 2008 came from selling recyclables (US$23.5 million) and electricity generated from incineration (US$24.6 million) (City of Yokohama 2008a, 2008c).

To promote efficient waste management, the city also began contracting key activities (such as waste collection and transportation) to the private sector, which often provides higher-quality services at lower cost. Between 2003 and 2005, the city saved US$26.4 million in operating costs by contracting services to the private sector (City of Yokohama 2006).

Lesson Learned in the Yokohama Case

The Yokohama case shows that the cooperation of stakeholders, particularly citizens, is important in achieving city targets. Of course, substantial and consistent efforts are needed at the grassroots level to raise the awareness of citizens and businesses and to attempt to change behaviors. However, the measures in Yokohama have not required new technology or huge investments. Moreover, cities can count on citizen power to make headway once people understand relevant issues, change their behavior, and become active players in implementing plans.

Encouraged by the achievements of the G30, Yokohama now aims to continue to reduce greenhouse gas emissions to lead Japan and demonstrate its qualities as one of the country's Eco-model Cities. In Yokohama's 2008 Climate Change Action Policy, CO-DO 30, the city aims to reduce greenhouse gas emissions by more than 30 percent by fiscal year 2025 and by more than 60 percent by fiscal year 2050 (relative to the levels in fiscal year 2004) (City of Yokohama 2008d). Action plans are being established on the basis of seven approaches to realize the plan's targets.[6] In addition, Yokohama aims to increase its use of renewable energy by a factor of 10 relative to fiscal year 2004 baselines. Citizens are actively participating in these activities, including by purchasing city-issued bonds to fund the construction of a new wind power generator. Finally, in light of Yokohama's reduced waste and the need to soon undertake costly renovations of an aged incinerator, the city is planning to close one more incinerator by fiscal year 2010 and, henceforth, use only four incinerators. More reductions in CO_2 emissions and operational savings are expected.

Notes

1. In Japan, there are several hierarchies and categories of administrative areas defined with such terms as *prefecture, city, county, ward, town,* and *village.* Among those areas in Japan categorized as *city,* Yokohama has the largest population.
2. Yokohama's fiscal year runs between April and March of the following year.
3. In this case study, waste denotes waste produced by households or businesses (commercial and services). Industrial waste is not included. See also City of Yokohama (2008a) and Yokohama statistics portals, http://www.city.yokohama.jp/me/stat/ and http://www.city.yokohama.jp/me/stat/index-e.html.
4. In this case study, $ = ¥100 was used for currency calculation.
5. The government launched the Eco-model Cities Initiative in 2008. A total of 13 cities were selected to serve as model cities. The selection was based on (a) achievement of a difficult target in the reduction of greenhouse gases, (b) a comprehensive and original approach that may be replicated by other cities, (c) appropriate local conditions and features, (d) the feasibility of the target and the plans and wide stakeholder participation, and (e) long-term and sustainable implementation. Apart from Yokohama, the cities of Iida, Kita-Kyushu, Kyoto, Minamata, Miyakojima, Obihiro, Sakai, Toyama, and Toyota; the towns of Kasihara and Shimokawa, and the Tokyo Ward of Chiyoda were selected.
6. The seven approaches are (a) living: to change society with anti-climate-change actions among individuals; (b) business: to change society with anti-climate-change business styles; (c) building: to plan and develop a city through energy-efficient building construction; (d) transportation: to promote city planning and development to create an attractive city where people may travel on foot, by bicycle, or on public transportation and to promote anti-climate-change measures with regard to automobiles; (e) energy: to increase recyclable energy ten-fold; (f) city and green areas: to plan and develop a green city through urban heat island measures and so on; and (g) city hall: to develop an anti-climate-change city hall.

References

City of Yokohama. 2003. "Yokohama shi ippan haikibutsu shori kihon keikaku, Yokohama G30 plan" 横浜市一般廃棄物処理基本計画、横浜G30プラン [City of Yokohama, Master Plan for Management of General Waste: Yokohama G30 Plan]. City of Yokohama, Yokohama, Japan. http://www.city.yokohama.jp/me/pcpb/keikaku/kei1.html.

———. 2006. "Yokohama G30 Plan—Kenshou to kongo no tenkai ni tsuite" 横浜G30プラン「検証と今後の展開」について [Yokohama G30 Plan: Verification and next steps]. Resources and Wastes Recycling Bureau, City of Yokohama, Yokohama, Japan. http://www.city.yokohama.jp/me/pcpb/keikaku/G30rolling/.

———. 2008a. "Heisei 20 nendo jigyou gaiyou" 平成20年度事業概要 [Operation Outline for Fiscal Year 2008]. Resources and Wastes Recycling Bureau, City of Yokohama, Japan, Yokohama. http://www.city.yokohama.jp/me/pcpb/keikaku/jigyo_gaiyou/20gaiyou/.

———. 2008b. "Kankyou model toshi teian sho" 環境モデル都市提案書 [Proposal for Eco-Model Cities]. Climate Change Policy Headquarters, City of Yokohama, Yokohama, Japan. http://www.city.yokohama.jp/me/kankyou/ondan/model/.

———. 2008c. "Heisei 20 nendo yosan gaiyou" 平成20年度予算概要 [Budget Outline for Fiscal Year 2008]. Resources and Wastes Recycling Bureau, City of Yokohama, Yokohama, Japan. http://www.city.yokohama.jp/me/pcpb/keikaku/yosan/20yosan.pdf.

———. 2008d. "CO-DO 30: Yokohama Climate Change Action Policy." Leaflet, Climate Change Policy Headquarters, City of Yokohama, Yokohama, Japan. http://www.city.yokohama.jp/me/kankyou/ondan/plan/codo30/leaf_english.pdf.

———. 2009. ごみの分別による効果 - 二酸化炭素削減効果 [Effect of segregation of garbage—reduction of carbon dioxide]. Resources and Wastes Recycling Bureau, City of Yokohama, Japan. http://www.city.yokohama.jp/me/pcpb/shisetsu/shigenkai/lca/ (accessed March 2009).

Brisbane, Australia

Actions on Climate Change in a Rapidly Growing City in a Subtropical Region

With 2 percent population growth in 2006–07, Brisbane, the capital of the State of Queensland, was one of the most rapidly growing capital cities in Australia (ABS 2008). The population of Brisbane in 2007 was approximately 1.01 million, making it the first local government area in Australia to exceed the milestone of 1 million people (ABS 2008). Brisbane is among the top 10 most rapidly growing cities in the countries of the Organisation for Economic Co-operation and Development and the second most rapidly growing city in the western world (Brisbane City Council 2006). Brisbane's population is expected to continue to grow over the next two decades (Brisbane City Council 2006).[1]

Since 2000, Brisbane has experienced increased electricity consumption and annual growth in peak electricity loads (Brisbane City Council 2007a).[2] Because the city has a subtropical climate, increased domestic air-conditioning has been a major factor prompting higher demand for electricity, along with poor housing design, an energy-intensive economy, and growth in population and disposable income (Brisbane City Council 2007a). The demand for electricity is expected to rise consistently through 2030. Brisbane is also experiencing a shortage of potable water during a period of growth and climate change that is straining water resources

and highlighting the need to shift to a new form of water management.

In 2007, the Brisbane City Council issued Brisbane's Plan for Action on Climate Change and Energy, which delineates the selected actions to be achieved in the short term (about 18 months) and the long term (more than five years) (see Brisbane City Council 2007b). Brisbane has three major challenges: climate change, high peak oil demand, and greenhouse gas emissions (see Brisbane City Council 2007c). Analyses suggest that, if Brisbane responds intelligently to these challenges, the city may generate significant economic benefits by developing sustainable industries, while saving resources. Brisbane is actively introducing various approaches to sustainable development. In addition, in the city's "Our Shared Vision: Living in Brisbane 2026" policy document, authorities have committed to cutting greenhouse gas emissions in half, reusing all wastewater, and restoring 40 percent of the natural habitat by 2026 (Brisbane City Council 2006).

The Ecological and Economic Benefits of the CitySmart Program

To implement actions in Brisbane's Plan for Action on Climate Change and Energy, officials have initiated the Green Heart CitySmart Program (Brisbane City Council 2009a). The program introduces residents and businesses to

Profile of Brisbane

Brisbane

- Capital city of the state of Queensland, Australia
- Population (2007): 1.01 million
- Population increase (2006–07): 2.0 percent
- The largest populated local government area in Australia
- Brisbane is located on a coastal plain in southeast Queensland. The eastern suburbs line the shores of Moreton Bay, and the city's central business district is only 27 km from the mouth of the bay.
- Brisbane is a subtropical river city and has hot, humid summers and dry, mild winters.

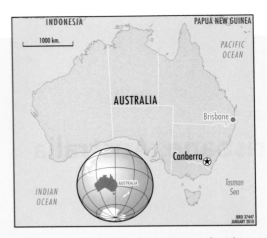

Map 3.7 Location of Brisbane

Source: Map Design Unit, General Services Department, World Bank.

practical and affordable ways to implement actions indicated in the climate change action plan. These practical tips help residents and businesses become energy and resource efficient, thus improving the environment and saving money (box 3.2).

For instance, residents are offered tips on hot water use, heating and cooling, waste disposal, lighting and electronic appliances, bathroom and laundry facilities, house renovations, urban gardening, the installation of rainwater tanks, and so on. Moreover, Brisbane aims to reduce the annual carbon footprint of an average household from 16 tons of carbon dioxide (CO_2) in 2006 to 4.5 tons by 2026. To encourage household participation, the city offers rebates and grants supporting environmentally sustainable projects (box 3.3). The city recommends that homes reduce their greenhouse gas emissions, particularly by installing solar hot water systems (rebates available) to reduce up to 3 tons of CO_2, undertaking energy audits and monitoring (rebates available) to reduce up to 3 tons of CO_2, and connecting to GreenPower (renewable energy from government accredited sources) to save up to 9 tons of CO_2.

Brisbane's trees are vital in protecting and improving the urban environment. Trees pro-

---

BOX 3.2

The Measures in the CitySmart Program in Brisbane

- Shifting to energy-efficient light fittings
- Installing rainwater tanks in homes
- Using more efficient air-conditioners
- Continuing to recycle and preserve water
- Installing solar panels and solar hot water systems
- Signing up for green energy
- Thinking about alternative public transportation solutions
- Reducing vehicle emissions
- Implementing the 2 Million Trees Project

Source: Brisbane City Council (2009b).

BOX 3.3

Examples of Grants and Rebates for Environmentally Sustainable Home Projects in Brisbane

- $A 50 rebate on the installation of home energy monitors
- $A 400 rebate on solar hot water systems
- Rebates for installing rainwater tanks with internal connections to toilets and cold water washing machine taps
- Funding up to $A 50,000 to local nonprofit community groups for the installation of devices to save energy and water

Source: Brisbane City Council (2009c).
Note: The information is current as of May 2009.

vide shade and transpire water to cool the air and surface temperatures. In subtropical cities, it is important to identify ways to become less dependent on air-conditioners to reduce energy use and carbon emissions. Shade allows more people to enjoy outdoor activities. Trees absorb greenhouse gases, including CO_2, and remove pollutants from the air. In addition, trees reduce storm water runoff and evaporation, an important outcome in cities in which water resources need protection. Brisbane city officials have provided 133,000 free plants to residents to maintain the city's unique subtropical landscape. Furthermore, the city is committed to planting two million trees between 2008 and 2012. People involved in this effort will restore bushland on a large scale, cultivate new trees along streets, and support the greening of landfill and infrastructure sites (Brisbane City Council 2009d).

The Brisbane City Council aims to be carbon neutral in its daily operations by 2026 by adhering to sustainability principles in its offices and facilities. As a result, public sector electricity use and greenhouse gas emissions have already decreased (table 3.7). The city council also actively engages residents and businesses to promote actions that reduce negative environmental impacts.

Urban Development in Brisbane

As in many other cities in Australia, most of Brisbane's citizens reside in detached homes built in low-density suburbs outside the city boundaries (Dingle 1999). The suburban lifestyle in Australia is highly dependent on private motor vehicles because, for the past 50 years, suburbs have been built on the assumption that most people will not need public transportation services (Newman 1999). The shape of Brisbane demonstrates this dependence. Peak oil prices have multiple implications for Brisbane's economy and society and increase the need for fuel-efficient vehicles and public transportation options. For many years, the problem of urban sprawl has been addressed for reasons other than peak oil prices. Local and regional planning has incorporated the principles of transport-oriented development, which aims to promote the development of mixed residential and employment zones to maximize the efficient use of land through a high level of access to public transportation (Brisbane City Council 2009f). However, the results are still mixed; economic structures and traditional housing preferences do not always coincide with these planning initiatives (Brisbane City Council 2009b).

Table 3.7 Greenhouse Gas Emissions and Electricity Use by the Brisbane City Council, Fiscal Years 2005–2008

INDICATOR	2005	2006	2007	2008
Net greenhouse gas emissions (tons of equivalent carbon dioxide)	—	441,850	376,471	—
Direct emissions[a]	—	199,284	180,255	—
Indirect emissions arising from the consumption of electricity, heat, and steam	—	218,988	205,669	—
Other indirect emissions	—	30,148	40,864	—
GreenPower[b]	—	(6,570)	(53,317)	—
Offsets	—	—	(95,000)	—
Electricity use (megawatt hours)	224,603	209,357	200,719	—
GreenPower purchased (percent)	6	6	25	50

Source: Brisbane City Council 2009e.

Note: — = not available; FY = fiscal year.

a. Direct emissions are from transport (trucks, buses, ferries); manufacturing (for example, asphalt production); and the on-site generation of energy, heat, steam, electricity, and fugitive emissions from landfill and wastewater treatment.

b. GreenPower is renewable energy that comes from the sun, wind, and waste. GreenPower produces no greenhouse gas emissions; the energy must be supplied by government-accredited sources.

Urban Renewal Brisbane is a US$4 billion program to revitalize specific areas of the inner city (Brisbane City Council 2009g). The program has been implemented in several urban areas, including Brisbane City Center (the central business district). It has incorporated innovative principles and practices, such as high-quality urban designs, modern construction, mixed land use, higher-density development, diverse transit options, and enhanced accessibility.

The Brisbane City Council is working with the development industry to promote sustainable living and working environments. The city council has developed guidelines to help architects, engineers, planners, developers, and builders incorporate principles that promote sustainability in development applications. While such principles offer broad markers for sustainable development, the guidelines explain ways to apply them practically. For example, the buildings in Brisbane used to be designed to be open to breezes, with overhead ceiling fans, shaded areas, and good circulation. However, recent designs depend on air-conditioners that are energy dependent. Today, Brisbane is promoting new approaches to urban construction and spatial designs that create attractive living environments and walkable areas in this subtropical city.

The Water Cycle and Water Catchment Management

Brisbane's growing population is increasing pressure on the city's supply of potable water. The average annual rainfall in southeast Queensland is about 1,200 millimeters (compared with 2,400 millimeters in Singapore). Although higher than in other Australian cities, Brisbane's rainfall is less predictable, and careful water resource management is required. In recent years, drought has become a serious national problem. States with authority in water management may undertake measures to conserve water, including by applying water use restrictions (with

penalties for overuse) and subsidizing rainwater tanks. Brisbane has also pursued integrated water cycle management encompassing water provision, wastewater treatment, storm water management, and strategic land management. Poor land management in water catchments results in lower-quality water and higher water treatment costs. As a subtropical city, Brisbane is endowed with creeks, waterways, and rich biodiversity. The city is working to restore the health of its waterways and creeks through various means, including removing weeds, encouraging communities to plant native seedlings, and reducing illegal dumping by sponsoring community campaigns (for example, see Brisbane City Council 2007d).

Public Transportation: Bus Rapid Transit Systems

Brisbane has two bus rapid transit systems: the Brisbane South East Busway, which opened in 2001, and the Brisbane Inner-Northern Busway, which opened in 2004. These systems fall under the jurisdiction of the Queensland government and Queensland Transport, which is committed to public transportation provision to support growth and connectivity in greater Brisbane. They are designed to provide public transportation services to areas that existing rail lines (Queensland Rail) do not cover. The Brisbane South East Busway connects Brisbane's central business district to the city's sprawling southeastern suburbs. The busways are two-lane, bidirectional roads used exclusively by buses and emergency vehicles. This permits buses to bypass congestion. The system also provides high-quality, well-designed bus stations with good pedestrian access (Queensland Transport 2008).

Busways reduce the growth of car traffic on roadways mainly because of their greater carrying capacity. One motorway lane may accommodate 2,000 passengers per hour, but one busway lane is able to carry 15,000 passengers per hour. In addition, busways significantly reduce travel time. For example, a typical route that takes 60 minutes on a Brisbane motorway is reduced to 18 minutes by riding a bus on the South East Busway. Fewer cars and less travel time decrease vehicular emissions, which helps mitigate climate change and improve air quality. In general, less time commuting translates into greater urban productivity and economic activity. The bus rapid transit systems also affect land development. Along the South East Busway, property values within six miles of bus stations have risen as much as 20 percent; moreover, the rates of growth in property values have been two to three times higher in these areas than in areas farther from stations (Currie 2006).

Lesson Learned in the Brisbane Case

Brisbane has responded to its unique local situation as a subtropical city under growth pressures. Climate change has already started to affect the city. Water is scarce, and temperatures are higher. Responding to its natural conditions, Brisbane protects water resources, plants trees to improve its urban ecology, and promotes a sustainable built environment. These actions save money for the city and its residents. Many developing-country cities are in tropical and hot climates and may be vulnerable to climatic change. Some cities may be highly dependent on air-conditioning, which is relatively energy consuming compared with other viable strategies. In this context, Brisbane's measures and actions may provide good examples for how cities might respond to such challenges, while remaining ecologically and economically vibrant.

Notes

1. The State of Queensland will have to accommodate 1 million new residents over the next two decades, 25 percent of whom will arrive in Brisbane.

2. The State of Queensland experienced a 53 percent increase in electricity consumption and an 8 percent annual increase in peak load growth over the 10 years between 1997 and 2007.

References

ABS (Australian Bureau of Statistics). 2008. "Regional Population Growth, Australia, 2006–07." Catalogue 3218.0, ABS, Canberra, March 31.

Brisbane City Council. 2006. "Our Shared Vision: Living in Brisbane 2026." Brisbane City Council, Brisbane, Australia. http://www.brisbane.qld.gov.au/bccwr/about_council/documents/vision2026_final_fulldocument.pdf.

———. 2007a. "Brisbane Long Term Infrastructure Plan." Brisbane City Council, Brisbane, Australia. http://www.brisbane.qld.gov.au/bccwr/plans_and_strategies/documents/brisbane_long_term_infrastructure_plan.pdf.

———. 2007b. "Brisbane's Plan for Action on Climate Change and Energy." Brisbane City Council, Brisbane, Australia. http://www.brisbane.qld.gov.au/bccwr/environment/documents/brisbane_climate_change_and_energy_action_plan.pdf.

———. 2007c. "Climate Change and Energy Taskforce Report; Final Report: A Call for Action." Maunsell Australia Pty Ltd, Milton, Queensland, Australia, March 12.

———. 2007d. "Know Your Creek; Moggill Creek: Improving Our Waterways from Backyard to Bay." Brisbane City Council Information, Brisbane City Council, Brisbane, Australia. http://www.brisbane.qld.gov.au/bccwr/environment/documents/know_your_creek_moggill_2008.pdf.

———. 2009a. "Green Heart CitySmart Home." Brisbane City Council, Brisbane, Australia. http://www.brisbane.qld.gov.au/BCC:CITY_SMART::pc=PC_2796.

———. 2009b. "Message from the Lord Mayor." Brisbane City Council, Brisbane, Australia. http://www.brisbane.qld.gov.au/BCC:CITY_SMART::pc=PC_2803.

———. 2009c. "Grants and Rebates." Brisbane City Council, Brisbane, Australia. http://www.brisbane.qld.gov.au/BCC:CITY_SMART::pc=PC_5014.

———. 2009d. "2 Million Trees Project." Brisbane City Council, Brisbane, Australia. http://www.brisbane.qld.gov.au/BCC:CITY_SMART::pc=PC_2645.

———. 2009e. "What Council Is Aiming For." Brisbane City Council, Brisbane, Australia. http://www.brisbane.qld.gov.au/BCC:CITY_SMART::pc=PC_5475.

——. 2009f. "Urban Renewal Glossary." Brisbane City Council, Brisbane, Australia. http://www.brisbane.qld.gov.au/BCC:BASE::pc=PC_1745.

——. 2009g. "Urban Renewal Brisbane." Brisbane City Council, Brisbane, Australia. http://www.brisbane.qld.gov.au/BCC:BASE::pc=PC_1727.

Currie, Graham. 2006. "Bus Rapid Transit in Australasia: Performance, Lessons Learned, and Futures." *Journal of Public Transportation* 9 (3): 1–22.

Dingle, Tony. 1999. "'Gloria Soame': The Spread of Suburbia in Post-War Australia." In *Changing Suburbs: Foundation, Form and Function,* ed.

Richard Harris and Peter J. Larkheim, 189–201. London: Routledge.

Newman, Peter. 1999. "Transport: Reducing Automobile Dependence." In *The Earthscan Reader in Sustainable Cities,* ed. David Satterthwaite, 173–98. London: Earthscan Publications.

Queensland Transport. 2008. "South East Busway: Planning to Springwood; Project Guide." Queensland Transport, Queensland Government, Brisbane, Australia. http://www.transport.qld.gov.au/resources/file/eb6b7c0e3065e66/Pdf_seb_project_guide.pdf.

CASE 6

Auckland, New Zealand

Regional Collaboration, Including a Planning Framework

The Auckland metropolitan area is New Zealand's largest and most populous urban area (figure 3.32, map 3.8). The Auckland Region is home to over 1.3 million people, about one-third of the national population. The region's population grew by 12.4 percent between the 2001 and 2006 censuses. Auckland is characterized by ethnic diversity; 37.0 percent of the region's residents were born overseas. In the region, there are four cities and three districts, each with its own council; there is also one regional council.

Currently, each council develops its own plans and strategies. This results in areas of overlap and competing priorities. Collective regional strategies for growth, the urban form, economic development, and transportation planning have been devised. However, they do not have common goals or principles to ensure their alignment.

The lifestyle typical of the Auckland Region and the employment opportunities there continue to attract new inhabitants, but drawbacks have also become evident, namely, a lack of a cohesive and effective approach to ongoing transportation problems and concerns about the pattern and nature of urban growth. The Auckland Regional Growth Forum was therefore established in 1996 as a cooperative meeting place for political representatives of the Auckland Regional Council and the local terri-

Figure 3.32 Auckland Harbor Viewed from the East

Source: Photo by Sebastian Moffatt.

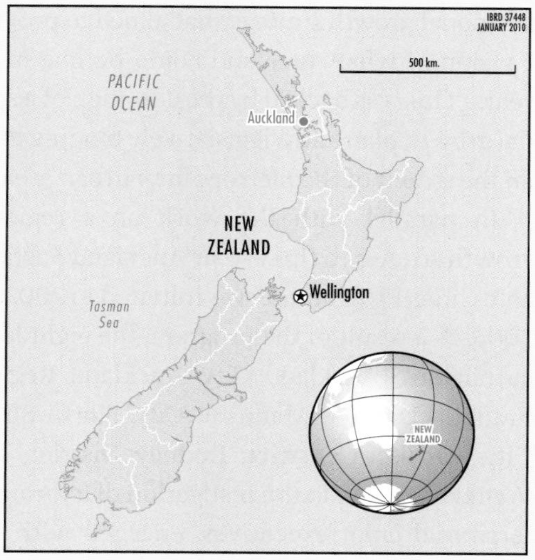

Map 3.8 Location of Auckland

Source: Map Design Unit, General Services Department, World Bank.

torial authorities in the region. The aim of the forum is to develop and implement a strategy for managing the effects of growth.

Governments at every level recognize the need for a collaborative, regional process

The interconnectedness of national and local Auckland issues (such as housing and education) with growth and innovation and the major required investments (particularly in land transport) have created complex and difficult issues among multiple authorities. Despite Auckland's importance to the New Zealand economy and the areas of common interest, such as transportation and energy provision, the national government did not initially play a close role in directing regional and local government planning. Concern emerged that, without agreement on an overarching regional strategy and framework, decision making in the region could become ad hoc and adversarial if each stakeholder tried to have a say from a narrow perspective and without viewing the region as a whole. As a result, there was a clear need for coordinated strategic planning across the Auckland Region to ensure that Auckland would be able to remain competitive in today's globalized world. The response involved a process undertaken in 2001 to prepare a regional growth strategy that aimed to provide a vision of what Auckland could be like in 50 years. This was backed by the adoption of a spatial growth plan and a legislatively binding limit on the extent of the metropolitan urban area.

In parallel with the work on a regional growth strategy, a three-year Auckland Sustainable Cities Programme was initiated in 2003. In 2006, as a result of the program, the eight local authorities (Auckland City, Auckland Region, Franklin District, Manukau City, North Shore City, Papakura District, Rodney District, and Waitakere City), at the instigation of a forum of territorial chief executives, engaged with the central government to develop a long-term sustainability framework. Initially called START

(Sustaining The Auckland Region Together), the approach represented an attempt to evaluate how forces of change (such as climate change, global resource depletion, and changing demographics) might impact Auckland and how the local and regional councils and the central government might align their efforts and create strategic directions to ensure the region's long-term success (figure 3.33).[1] The engines of START included the need to develop resilient and adaptive systems able to respond (1) to persistent pressures over short and long time horizons with no obvious alternative solutions and (2) to many vested interests with apparently irreconcilable demands.

Making a START: Gathering information

The START working group developed a prototype framework with a cascading set of deliverables, including a vision, goals, initial foundation, process principles, initial themes, and some potential responses (which included catalyst projects, long-term sustainability goals, and the development of indicators to measure progress). Critical to progressive development was consideration of the forces that would shape Auckland's future over the next 100 years. Also significant to the development of the framework was the involvement of expert groups that included academics and experts from the business and community sectors, who, through facilitated

Figure 3.33 The START Logo in the Auckland Region
Source: ARC (2006).

workshops, developed theme papers on key issues identified in the prototype framework—the built environment, urban form and infrastructure, energy, economic transformation, social development, cultural diversity and community cohesion, and environmental quality. Each group deliberated around four sustainability principles—resilience, prosperity, livability, and ecology—and considered how these might be influenced by the forces that would shape the future.

In a linked, but parallel process, a working group representing all Maori tribes (New Zealand's indigenous people) of the Auckland Region developed its own collective long-term framework, the Mana Whenua Framework. The working groups involved in these processes built links between the two frameworks, including a basic common structure; common analysis via the forces and theme papers; a Maori goal in the overall framework; and an indigenous concept of sustainability, which fed into the definition of sustainability in the overall framework.[2] Meanwhile, the overall framework acknowledges Mana Whenua as the first peoples of the region and as an intimate part of the region's ecological and cultural fabric.

In August 2006, a three-day START design workshop enabled 120 representatives of local authorities and the central government, academia, and the community and business sectors to contribute expertise and perspectives to the development of the draft 100-year framework. The methodology drew heavily on the Vancouver CitiesPlus model, which progressed from a high-level vision to responses and indicators through an adaptive management approach to the development of a resilient urban planning framework able to address future challenges (CitiesPlus 2002). The workshop relied on a charrette format, a process whereby new design ideas emerge and evolve quickly (figure 3.34). The process is interactive and harnesses the talents of a range of parties to resolve planning challenges. The charrette format is particularly

Figure 3.34 Strategic Planning among Many Stakeholders at a Three-Day Regional Charrette, New Zealand

Source: ARC (2006).

successful in helping local government authorities engage communities in planning. The product is usually a tangible plan ready for immediate implementation.

Stakeholder consultations and interagency coordination

As a result of feedback and wider strategic discussions following the START workshop, it was decided that the framework should include the following:

- A shift from business as usual as a key component of the framework
- The addition of integrated goals, key directions, leadership goals, and Maori goals
- The adoption of a revised version of a regional vision developed by a youth contingent
- The development of a draft set of indicators
- The development of a process and tools for the application of the framework

A governance and reporting structure was set up whereby the project was overseen by a steering committee of the council officers that was sponsored by the chief executives forum responsible for final approval of the framework.

Consultation with stakeholders and the public took place from February to May 2007 through 19 workshops involving around 200 participants, plus written submissions from several individuals, four organizations, and two regional councils.

A revised version, the Auckland Sustainability Framework (ASF), was endorsed in September 2007 by the Auckland Regional Growth Forum after it had been endorsed by all member local authorities and government agencies. It also received high-level support within the central government. The ASF goals and visions were consistent with central government priorities, especially in the substantive shifts that would be required (box 3.4). In turn, the ASF was expected to provide a tool to review the effect of national policies on Auckland. However, it was also clear that a better understanding was needed of the methods for achieving goals and of the proper indicators for assessing progress.

The ASF is also intended to guide and align regional strategies (such as the Regional Growth Strategy, the Regional Land Transport Strategy, and the Auckland Regional Economic Development Strategy). The process of developing a framework was therefore highly inclusive, with many conversations feeding into the framework and the emerging responses. The Auckland Regional Growth Forum, for example, facilitated regionwide discussions, joint political decision making, and the establishment of a reference group of council members to provide direction and support. Similarly, local authorities and the central government formed a senior officers steering group and an officers working group. Key collaborative elements were the relationship between the central government and local governments and the common governance elements, primarily because of the involvement of the Government Urban and Economic Development Office, including a joint commitment to the development of a shared long-term view of a sustainable Auckland.[2]

The final framework that was adopted consists of the following (figure 3.35):

- The identification of key challenges to sustainability that the region will need to address
- A 100-year vision
- The eight long-term goals
- Eight shifts from current practice that are required to meet the goals
- Suggested strategic responses
- A measurement framework and monitoring process
- A toolkit to apply the framework to strategies, significant decisions, and plans and to integrate regional planning

The framework's role consists of the following:

- To align existing regional strategies and projects, for example, the Regional Growth Strategy, the Regional Land Transport Strategy, and the Auckland Regional Economic Development Strategy
- To align future regional strategies and projects
- To guide the development of a single regional plan (the One Plan; see the following section)

BOX 3.4

Eight Goals Direct the Auckland Sustainability Framework

The ASF is built around eight interrelated long-term goals that will enable the region to take a sustainable development approach:

Goal 1   A fair and connected society
Goal 2   Pride in who we are
Goal 3   A unique and outstanding environment
Goal 4   Prosperity through innovation
Goal 5   Te puawaitanga o te tangata (self-sustaining Maori communities)
Goal 6   A quality, compact urban form
Goal 7   Resilient infrastructure
Goal 8   Effective, collaborative leadership

Source: RGF (2007).

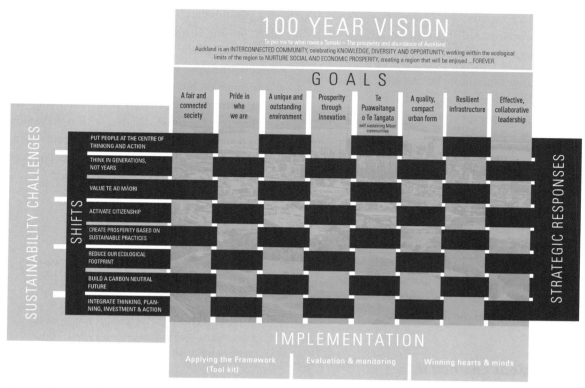

100 YEAR VISION
Te pai me te whai rawa a Tamaki – The prosperity and abundance of Auckland.
Auckland is an INTERCONNECTED COMMUNITY, celebrating KNOWLEDGE, DIVERSITY AND OPPORTUNITY, working within the ecological limits of the region to NURTURE SOCIAL AND ECONOMIC PROSPERITY, creating a region that will be enjoyed ...FOREVER.

GOALS

| A fair and connected society | Pride in who we are | A unique and outstanding environment | Prosperity through innovation | Te Puawaitanga o Te Tangata (self sustaining Maori communities) | A quality, compact urban form | Resilient infrastructure | Effective, collaborative leadership |

SUSTAINABILITY CHALLENGES

SHIFTS
- PUT PEOPLE AT THE CENTRE OF THINKING AND ACTION
- THINK IN GENERATIONS, NOT YEARS
- VALUE TE AO MĀORI
- ACTIVATE CITIZENSHIP
- CREATE PROSPERITY BASED ON SUSTAINABLE PRACTICES
- REDUCE OUR ECOLOGICAL FOOTPRINT
- BUILD A CARBON NEUTRAL FUTURE
- INTEGRATE THINKING, PLANNING, INVESTMENT & ACTION

STRATEGIC RESPONSES

IMPLEMENTATION

| Applying the Framework (Tool kit) | Evaluation & monitoring | Winning hearts & minds |

Figure 3.35 The Auckland Sustainability Framework
Source: RGF (2007).

- To provide methods to adapt business-as-usual scenarios, for example, the 10-year community investment plans of a local council
- To identify strategic responses that must be undertaken to achieve sustainability goals

The ASF "will provide direction so that our local authorities and central government agencies can work together with a common purpose to embrace the opportunities and face the challenges associated with developing a truly sustainable region" (ARC 2008).

Keys to Success

Extended peer communities
The overall process created considerable buy-in at political and administrative levels, and the resulting framework is owned by all parties. However, there has been a considerable change in political representation at the local and national levels since the adoption of the ASF. Many new council members have not been involved in the development of the framework, and the national government has redefined sustainability into the narrower concept of natural resource management.

Nonetheless, the ASF has been used to develop a collective investment plan, which is referred to as the One Plan, as well as a number of local council plans, including the Manukau City Council's 2060 Strategic Framework and the Waitakere City Council's social strategy.

Stretched thinking
The framework and, especially, the participatory process have stretched the thinking of many participants with regard to the following topics:

- Recognizing that the world and Auckland are going to experience exponential change over the next 50 years and that they have limited time to prepare for this change

- Recognizing that many business-as-usual practices will have to be altered or abandoned

- Understanding the meaning of sustainable development, especially by bringing in a Maori perspective

- Developing the Mana Whenua Framework

The development of a separate, but linked Maori framework has ensured that the long-term planning for Maori is being undertaken by Maori. The depth of indigenous understanding of generational thinking and the holistic and spiritual understanding of the relationship between the environment and people are fully realized in the Mana Whenua Framework and have challenged and stretched the thinking on the ASF.

Lessons Learned in the Auckland Case

Two groups appear to have been less well represented in the process of the development of the ASF: business representatives and the developers who would eventually implement the strategies and activities based on the ASF. A special process may be needed to engage these groups because they are typically reluctant to attend open meetings and because they require a process that is especially efficient.

After the ASF was adopted, the region quickly focused on new priorities. As a consequence, one component of the framework—winning hearts and minds—did not achieve progress (see figure 3.35). Winning hearts and minds acknowledged the importance of the social learning that council members, key staff members, and stakeholders experienced through the development of the ASF. Continued dialogue and education on the challenges and solutions involved in achieving sustainability are required among these key decision makers and the public.

While the ASF has been adopted as a guiding framework, no hard targets have yet emerged for planning and strategy making. Likewise, no bottom-line thresholds for public sector decision making have appeared. Without these elements, the ASF may become a useful tool for some parties, but may be ignored by others. The new national government is restructuring the eight local government bodies within the region into a single unitary council, and it remains to be seen whether this new council will adopt the ASF as the guiding regional framework.

Notes

1. The government of New Zealand is in the process of restructuring the Auckland local government and plans to replace the existing seven local councils and one regional council with one super council and 20 to 30 local community boards.

2. See Frame (2008) for a critical analysis of the regional planning process and outcome.

References

ARC (Auckland Regional Council). 2006. "A Workshop to Design the Auckland Region's Future: Summary of Proceedings." Auckland Regional Council, Auckland, New Zealand. http://www.arc.govt.nz/albany/fms/main/Documents/Auckland/Sustainability/START%20workshop%20report.pdf .

———. 2008. "Auckland Sustainability Framework." Auckland Regional Council, Auckland, New Zealand. http://www.arc.govt.nz/auckland/sustainability/auckland-sustainability-framework.cfm.

CitiesPlus. 2002. "Canada's 100-Year Plan for a Sustainable Region." CitiesPlus, Vancouver, Canada. http://www.citiesplus.ca/index.html.

Frame, Bob. 2008. "'Wicked,' 'Messy,' and 'Clumsy': Long-Term Frameworks for Sustainability." *Environment and Planning C: Government and Policy* 26 (6): 1113–28.

RGF (Regional Growth Forum). 2007. "Auckland Sustainability Framework. An Agenda for the Future." Auckland, New Zealand. http://www.aucklandoneplan.org.nz/auckland-sustainability-framework/sustainability-concepts-and-challenges/sustainability-concepts-and-challenges_home.cfm.

Eco^2 Sector Notes

A Sector-by-Sector Lens on Urban Infrastructure

We will now look at important urban infrastructure issues in more detail through the lens of each sector. Ideally, this leads to a kaleidoscopic view of the city that recognizes the interrelationships of energy, water, transportation, and solid waste. These interrelationships apply across sectors and with respect to the built form of the city. In this context, the final note in part 3, "Managing the Spatial Structure of Cities," provides important lessons on how spatial planning and land use regulations may powerfully influence mobility and affordability.

It is clear that many of the operational and jurisdictional boundaries of sectors impede innovation and creativity in the effort to achieve better outcomes. It is also clear that investments made in one sector may result in savings in another sector (for example, investments in water efficiency usually result in large energy cost savings). Pooling scarce resources to invest in multifunctional and multipurpose common elements may also benefit urban residents (for instance, through single-purpose underground infrastructure corridors).

What emerges from a closer analysis is an understanding of how these infrastructure systems interact with a city's spatial form. Infrastructure investments trigger and enable urbanization. However, urban planning and spatial development establish the locations, concentrations, distributions, and nature of demand nodes for sector infrastructure systems. Urban and spatial planning also identifies the physical and economic constraints and parameters of infrastructure systems, including capacity limits, service delivery technologies, and cost recovery requirements. Good urban planning and spatial development provide proactive demand-side management and improve resource efficiency by identifying and assessing the viability of technology and infrastructure options. For instance, public transportation is financially viable only at certain threshold urban densities and forms and under good coordination of urban land use.

In addition to illustrating the opportunities and strategies for realizing benefits within and across sectors, the following notes shed light on critical sector-specific issues that are not under the direct control of city authorities, but nonetheless influence city sustainability. These issues may need to be addressed on a sector-by-sector basis. Moreover, identifying critical pressure points beyond the direct control of city authorities is important in devising an expanded platform for collaboration.

Cities and Energy

Overview

Cities and urban areas account for about two-thirds of the world's annual energy consumption. In the coming decades, urbanization and income growth in developing countries are expected to push this urban consumption even higher.[1] As the main consumers of energy and as implementers of national and regional sustainable energy policies and programs, cities may play a crucial role in improving our energy and environmental futures by making smart choices in urban development, energy demand management, and energy supplies. In return, cities stand to become more livable, affordable, and sustainable.

Traditionally, urban energy planning and management have aimed to improve access, security, reliability, and affordability. These efforts have focused on developing network-based energy systems (on which cities have become dependent), such as electricity grids, district heating networks, and natural gas pipelines. These efforts remain essential because modern cities simply cannot function without such networks. However, the potential for dire environmental impacts of traditional urban energy use persists, as exemplified in the London smog disaster of 1952 that killed 12,000 people. Today, heavy urban air pollution in many developing countries is a sober reminder that growing cities cannot always cope with the serious health-related impacts of the consumption of

fossil fuels. The first oil crisis of 1973 highlighted the importance of energy efficiency, conservation, and renewable energy. However, 35 years later, achieving progress in energy efficiency and renewable energy remains a tough challenge in both developed and developing countries. The emergence of climate change as a global development constraint, much of it related to the energy consumption habits and infrastructure in cities, also calls for fundamental changes in how countries and cities approach urban development, manage energy demand, and secure energy supplies.

How can cities address their multidimensional energy challenges, which affect their success and long-term development prospects? The evolution of urban energy agendas—from access, security, reliability, and affordability to environment and public health concerns and, more recently, to climate change mitigation and adaptation—has challenged cities and national and regional governments to break away from supply-centric practices and strengthen environmental rules in planning and management. The largely successful control of local and regional air pollution in cities in developed countries is encouraging and suggests possibilities for the expansion of efforts in developing countries. This success relied mainly on relocating factories, switching to cleaner fuels, and adopting stringent national and regional emissions regulations for industries and motor vehicles. As a result, many cities have become more attractive and competitive.

Controlling the carbon footprints of cities represents the greatest current energy challenge, but urban planners may tap this challenge to strengthen energy security and enhance energy access, affordability, and reliability. To be successful, cities must manage energy demand by promoting energy efficiency across sectors and the uptake of efficient and renewable energy supplies. It is also important, particularly in developing countries, for urban planners to support solutions incorporating energy efficiency and renewable energy in urban land use planning and land development. These efforts require that cities be actively involved in energy planning and management and that they adopt a long-term vision of urban development and redevelopment.

Visionary cities are adopting a new paradigm of integrated urban energy planning and management. Recent examples include PlaNYC 2030 of New York City and the Paris Climate Protection Plan (City of New York 2007; Mairie de Paris 2007). However, implementation hurdles remain, and the real test of turning visions into realities lies ahead. City governments are often faced with urgent tasks and competing interests and must prioritize actions against the constraints in human and financial resources. City administrations often lack a single department with adequate authority to spearhead a cross-cutting agenda, with the exception of the mayor's office, which commonly cannot sustain efforts because of mayoral term limits. In addition, urban energy planning and management are not entirely within the jurisdiction of city governments. In fact, prevailing urban energy infrastructures, with the exception of district heating systems, are usually not under the direct purview of local governments.[2] If cities are to succeed, they need strong support from their national and regional governments.

Why should a city government care about being assertive and making and implementing sustainable energy decisions? The short answer is that it pays. Most energy efficiency and conservation measures are not high-technology applications or expensive solutions, and the initial costs may usually be quickly recovered. The Municipality of Emfuleni in South Africa, for example, initiated an energy and water efficiency project that cost US$1.8 million and achieved annual savings of about 7 billion liters of water and 14 million kilowatt-hours. This equated to annual monetary savings of over US$4 million; thus, the project paid for itself in under six months. Because the contract was financed and implemented by an energy services company, the municipality saved money not only from reduced water losses and pumping costs, but also through less investment up front. The energy services company, however, recouped its investment quickly by sharing part of the cost savings (USAID 2005). The Växjö Municipality in Sweden began, in 1994, to replace its streetlights with high-efficiency lamps, which reduced energy use by 50 percent. After a project investment of about US$3.6 million, the city saved US$0.75 million per year, which meant the project paid for itself in less than five years (C40 Cities 2009a). Cities facing budget shortfalls are well advised to consider mining current expenditures for energy savings in their facilities and operations.

Energy efficiency and cleaner energy in developing-country cities with serious air pollution promote productivity and reduced medical bills, which improves urban livability and competitiveness. A recent joint study by the Chinese government and the World Bank has estimated that the cost of ambient air pollution in China's urban areas in terms of air pollution–related premature deaths and illnesses amounted to US$63 billion in 2003, equivalent to 3.8 percent of China's GDP (World Bank 2007). In fact, China's efforts in the past two decades to modernize energy infrastructure and improve energy efficiency have aimed to reduce the health impacts of air pollution. This is evident in the rapid penetration of gaseous cooking fuels and the rapid expansion of district heating systems in northern Chinese cities, which are also implementing national energy efficiency standards for buildings.

Among rapidly growing cities in developing countries, shifting to a new paradigm of urban energy planning and management is as much about contributing to global welfare as enhancing capacity to serve growing energy needs at lower costs and with greater security. Good environmental stewardship in energy planning and management is essential to mitigate regional and global environmental impacts that affect the long-term well-being of cities (for example, acid rain, climate change–induced storms, and rising sea levels). Making cities more energy efficient and more accessible to renewable energy supplies also helps hedge the risks of higher energy costs if a global agreement is reached to reduce anthropogenic greenhouse gas emissions drastically. This does not mean that developing cities should necessarily address all sustainable energy options at the same time. Pursuing actions on sustainable energy, however cost-effective they may be, requires public and private investment, efforts from city governments and citizens, and the strong support of regional and national governments. Cities should, moreover, tailor their efforts to the available resources and pursue initial steps toward sustainable energy that generate significant and immediate local benefits.

Where should a city start? In general, there are three areas where actions and interventions at the city level are critical and where city governments are in the driver's seat:

1. *Investing in sustainable energy supply and retrofits in city government facilities and operations.* Cities might start by adopting a range of energy efficiency and conservation measures in government-owned buildings and municipal services, such as water supply and wastewater treatment facilities, public lighting, transportation, and solid waste management. Large government complexes are often good candidates for distributed energy supply options such as the cogeneration of heat and power using natural gas. Local governments might also expand renewable energy supplies by purchasing green electricity and accommodating renewable energy technologies, such as photovoltaic systems and solar water heating, in their own buildings and facilities.

2. *Promoting energy efficiency and the application of renewable energy technologies in the urban built environment.* City governments may promote energy efficiency and renewable energy options in nonmunicipally owned or operated sectors by harnessing their dominant role in shaping the urban built environment. One of the most critical and effective interventions involves enforcing national or regional energy efficiency standards in new building construction and building renovations.[3] A more ambitious green building agenda may also include additional requirements for water efficiency and conservation, the adoption of renewable energy technologies, incentive programs for industry and residential users, and other measures to reduce the environmental impact of buildings (see CBSC 2009).

3. *Promoting energy efficiency and renewable energy through land use planning and land development policies.* Within their jurisdictions, city governments may shape or reshape land use and development patterns in ways that minimize carbon footprints, while ensuring lower overall operating costs. In this area, energy planning encounters and integrates with transportation planning and other urban infrastructure planning to serve a city's growth ambitions and environmental aspirations effectively.

Cities in developing countries face much tougher challenges than do their counterparts in developed countries. Technical capacity is often lacking. Competition for resources is fierce. Because of growth pressures and capital constraints, compromises are often reached to

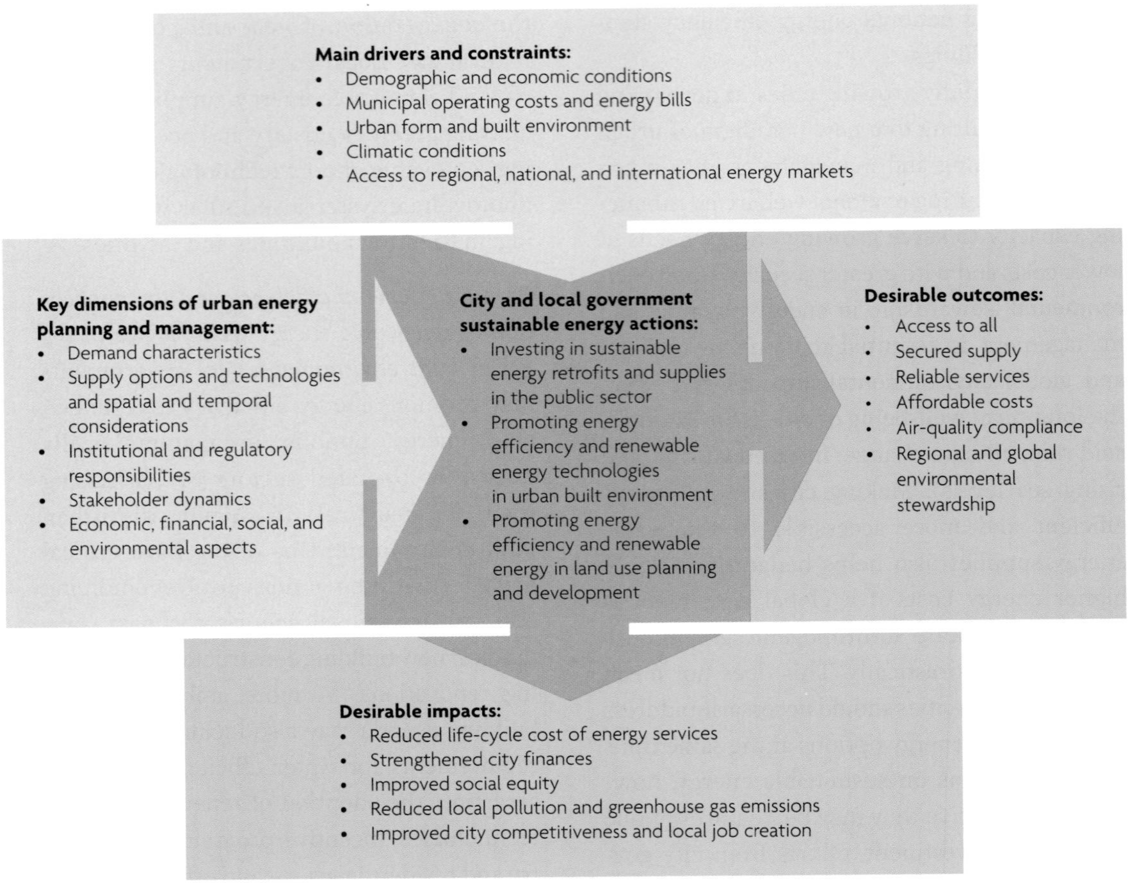

Main drivers and constraints:
- Demographic and economic conditions
- Municipal operating costs and energy bills
- Urban form and built environment
- Climatic conditions
- Access to regional, national, and international energy markets

Key dimensions of urban energy planning and management:
- Demand characteristics
- Supply options and technologies and spatial and temporal considerations
- Institutional and regulatory responsibilities
- Stakeholder dynamics
- Economic, financial, social, and environmental aspects

City and local government sustainable energy actions:
- Investing in sustainable energy retrofits and supplies in the public sector
- Promoting energy efficiency and renewable energy technologies in urban built environment
- Promoting energy efficiency and renewable energy in land use planning and development

Desirable outcomes:
- Access to all
- Secured supply
- Reliable services
- Affordable costs
- Air-quality compliance
- Regional and global environmental stewardship

Desirable impacts:
- Reduced life-cycle cost of energy services
- Strengthened city finances
- Improved social equity
- Reduced local pollution and greenhouse gas emissions
- Improved city competitiveness and local job creation

Figure 3.36 A Stylized Framework for Urban Energy Planning and Management

Source: Author compilation (Feng Liu).

serve more interests rather than to serve more people more effectively.[4] While cities must actively engage in the promotion of sustainable energy solutions, urban leaders need the support and cooperation of regional and national governments to be successful. Substantial donor support, knowledge, and finance are also required to encourage cities to enact sustainable energy actions in these three areas.

This chapter reviews the general urban energy landscape, particularly in cities in developing countries. It reviews activities linked to basic energy consumption; the options in energy services and supplies; factors affecting urban energy planning and management; and good practices, lessons, and challenges in urban energy planning and management. Figure 3.36 illustrates the aspects of sustainable urban

energy planning and practice that the chapter examines.

Energy Use in Cities

A city's energy profile—the level of use, mix of energy types, and patterns of use by sector or end use activity—is determined by many factors, including population, income, economic structure, energy prices, end use efficiencies, climate conditions, urban forms, built environments, and access to regional and national energy markets. Understanding the dynamics or constraints imposed by these factors is the starting point of sustainable urban energy planning. The amount of energy used is not a good indicator of the level of energy service supply or demand (for example, in lighting, cooling, heating, or refrigeration). The critical factor is *energy efficiency,* which refers to the adoption of improved technologies and practices to reduce the energy required to provide a similar level of energy output or service provision. In the urban context, it is important to assess the amount of useful energy that may be extracted from the primary source, delivered to end users, and turned into energy services.[5] In buildings, energy efficiency also implies reducing energy needs by improving the structural design and use of materials.[6]

A recent accounting of urban energy use conducted by the International Energy Agency delineated all energy-consuming activities within a city (IEA 2008). Based on this accounting, urban energy applications may be lumped into four broad categories: industry, transportation, municipal services, and buildings. A breakdown of these categories is presented in table 3.8.

Buildings that do not fit in the first three categories include a broad spectrum of structures ranging from single-family houses and apartment buildings to schools, hospitals, offices, and shopping malls. Factory buildings are excluded. For statistical purposes, buildings are usually divided into residential and commercial build-

ings. Residential buildings, which account for most of the urban building stock, are well defined as owner- or renter-occupied houses or apartments. Commercial buildings are diverse and usually include office buildings, shopping malls, supermarkets, hotels, and other buildings that host commercial or public entities. Government buildings are separately identified in table 3.8 because they represent special opportunities for sustainable energy interventions by city governments.

Typically, urban energy use in service-oriented cities in developed countries is dominated by buildings and transportation, which account for two-thirds or more of energy consumption. In rapidly industrializing developing countries, such as China, industrial energy use is often predominant in large cities. Even in Beijing, one of the most modern, high-income cities in China, manufacturing still accounted for about half of all energy consumption in 2006 (IEA 2008). In general, buildings and transportation are the most rapidly growing energy sectors in cities in developing countries. They are also the sectors in which sustainable energy measures may have the greatest impact. Countries with a growing middle class typically show explosive growth in the use of electricity for

Table 3.8 Energy Consumption in Cities: Main Sectors and Clusters

SECTOR/CLUSTER CATEGORY	SUBCATEGORY	CITY GOVERNMENT SUSTAINABLE ENERGY INTERVENTION, POTENTIAL LEVERAGE
Industry	Manufacturing	Indirect, relatively weak
	Construction	Indirect, relatively week
Transport	Private motor vehicles	Indirect, relatively weak
	Commercial motor vehicles	Indirect, relatively weak
	Public transit systems	Direct, strong
	Government motor vehicles	Direct, strong
Municipal services	Water supply and wastewater treatment	Direct, strong
	Solid waste management	Direct, strong
	Public lighting and traffic lights	Direct, strong
Buildings	Government buildings	Direct, strong
	Commercial buildings (nongovernment)	Indirect, strong in new construction
	Residential buildings	Indirect, strong in new construction

Source: Author compilation (Feng Liu).

residential air-conditioning and larger appliances. Although cities generally do not control appliance efficiency, and equipment standards are usually under the purview of national governments, cities may adopt incentive programs to encourage the use of more efficient appliances.

Though industries form part of the urban landscape, including industries in urban energy accounting may skew the understanding of city energy consumption and performance because the type and significance of industries vary across cities. For consistency in cross-city energy comparisons, it may be necessary to exclude (or separate) industrial energy consumption from the typical urban energy-consumption sectors indicated in table 3.9.

For urban energy planners, it is also necessary to separate urban energy demand and consumption into key end use activities, often within the four main sector categories outlined earlier. End use activities are more or less similar across cities, although the energy type supporting specific end uses may vary even within a city (see table 3.9).

Excluding industrial consumption, end use energy patterns in developing-country cities, especially cities in low-income provinces or states, are skewed toward the most basic energy services, such as lighting and cooking (and space heating in cold climates). The direct use of solid fuels, such as coal and firewood, is common in developing-country cities and is often the main cause of indoor and ambient air pollution. This is particularly true in low-income urban areas and slums in which access to cleaner cooking fuels is limited.

Electricity is the form of energy used most extensively in cities. The share of electricity in total energy use and the amount of electricity per capita often indicate the modernity and wealth of a city. Satisfying rapidly growing electricity needs often dominates the energy agenda of developing-country cities. (At the other extreme, gasoline is exclusively used for transport.)

Energy costs are critical to understanding energy use in cities and are often a primary energy-related concern of city officials. Decisions on sustainable energy must be economic and financial. However, the data on the costs according to energy type and on aggregate energy costs in urban sectors are often inadequate. Adequate cost information on individual end use activities and even simple data on common energy indicators are also rare (for example, kilowatt-hours per cubic meter of water

Table 3.9 Energy Consumption in Cities: Key End Use Activities and Energy Types

MAIN ENERGY END-USE ACTIVITIES	COMMON ENERGY TYPES USED						
	ELECTRICITY	NATURAL GAS[a]	LPG[b]	KEROSENE	GASOLINE, DIESEL	COAL	FIREWOOD, CHARCOAL
Lighting	■			■			
Cooking	■	■	■			■	■
Water heating (domestic hot water)	■	■	■			■	
Appliances (refrigerators and so on)	■						
Home and office electronics	■						
Air conditioning	■						
Space heating (cold climate)	■	■				■	■
Motorized transportation	■				■		
Motive power (stationary)	■						
Processing heat or steam	■					■	

Source: Author compilation (Feng Liu).
a. In some cities, gas supplies are still provided by coal-gasification or coking facilities, but in general, town gas is no longer an attractive energy supply option in cities.
b. LPG = liquefied petroleum gas.

delivered, tons of oil equivalent per person per mode of transport, or watts per square meter of building lighting).

Few cities in developing countries systematically track energy consumption patterns and costs. Without adequate energy consumption and cost information, cities will not be able to plan and implement sustainable energy measures effectively. Recent efforts to establish an international protocol and tools to inventory urban greenhouse gas emissions are helping build a platform to facilitate improved urban decision making on sustainable energy approaches (ICLEI 2008). Besides basic accounting, a critical element of urban energy planning is the provision of information to stakeholders about the opportunities for demand management through investments in energy efficiency, conservation programs, and alternative supplies. Simple benchmark data, such as quantifiable measures of energy use in lighting and heating, may help city managers to identify sectors that exceed norms and to plan remedial interventions. Additional supply options such as cogeneration in wastewater treatment plants or methane capture in landfills may also be assessed. The evaluation of such options requires tools to help cities compare their energy performance with good or best practice and understand the cost and benefit implications. Practical decision support tools and methods for sustainable urban energy planning and management help cities quickly identify and prioritize sustainable energy actions grounded on local capacities and conditions.

Energy Supply Options and Spatial and Temporal Considerations

Modern cities are highly dependent on network-based electricity and, to a lesser extent, natural gas supplies that are connected to regional or national networks. Power plants are often located within city boundaries, but these are frequently owned or operated by regional or national electricity utilities or independent power producers.[7] Developing-country cities generally aim to ensure secure and reliable access to energy supplies based on regionally integrated networks. District heating systems represent another network-based energy service common in cold climate cities, especially in China and Europe. However, they are limited to areas of a city with sufficient building density. Supplies of solid and liquid fuels, such as coal and petroleum products, are usually decentralized; thus, users may buy fuels from different producers or local distributors. The supply of transport fuels is usually vertically controlled by oil companies. In low-income countries, cities with significant periurban and slum populations often rely heavily on firewood and charcoal as cooking fuels and, in cold climates, also as heating fuels. The firewood is typically supplied locally and is often collected by individual households; charcoal is usually supplied by informal service providers. As a city grows in wealth, there is a progression among households and other dispersed service points toward greater dependence on network-based energy supplies and away from the use of solid fuels (coal and firewood). In general, cities and urban areas are almost entirely dependent on external energy supplies; even the power plants located in cities need to import fuel.

It is possible to conceive of a city's energy supply options and technologies along the three main energy delivery channels depicted in figure 3.37. In mildly cold and mildly warm climates, centralized heat supply is generally not an economically viable option and is not considered. In cold climates, electricity and centralized heat are often the focus of urban energy optimization because they may be produced together in combined heat and power plants. Cooling may be provided by using heat energy to drive a cooling system based on absorption chiller technology. Thus, district heating systems may provide cooling services in

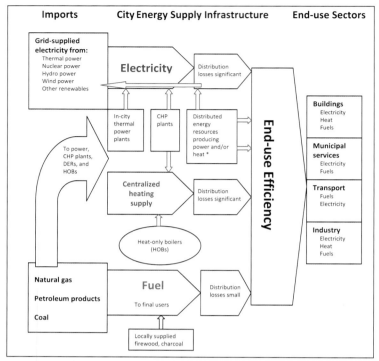

Figure 3.37 Urban Energy Supply Sources and Systems: A Stylized Sketch

Source: Author compilation (Feng Liu).
Note: Many gas-fired microgeneration facilities produce electricity and provide heating and cooling (using absorption chillers).

the summer if it is economically justified. Distributed energy resources often produce electricity, while offering heating and cooling services. Natural gas not only represents a cleaner alternative to oil and coal, but also adds more flexibility to urban energy services through distributed generation facilities. For a large city, fitting all the pieces together to optimize sustainable energy outcomes is not an easy undertaking. This is especially challenging in developing-country cities in which energy supplies are less well organized or streamlined relative to developed-country cities in which energy supplies are primarily network based.

Advances in centralized and distributed renewable energy supply technologies, such as wind towers, solar water heaters, biomass, and photovoltaic systems, enable cities to source a small, but increasing amount of renewable energy. Heat pumps and shallow geothermal energy sources also provide additional ways to reduce

reliance on purchased energy. Considering the energy saved from efficiency and conservation measures as a valid source of energy supply has become a compelling concept in demand-side management and energy supply planning.

The consumption of solid fuels by households and other dispersed end use points, such as restaurants, tends to decline as gaseous fuels—liquefied petroleum gas or natural gas—become available or electricity becomes more abundant. Such a transition may take decades and often requires the construction of regional and national energy infrastructure. In China, the dispersed use of solid fuels in urban areas has decreased dramatically over the last 20 years. Solid fuels have been largely eliminated from cooking and are now mainly used in a falling number of cold climate urban households that have no access to centralized heating or natural gas. This trend has been generated because of the strong support of the national government for boosting the supply of liquefied petroleum gas and expanding natural gas transmission networks.

Spatial and temporal concerns are important in developing network-based urban energy infrastructure. Spatial planning entails the layout of networks within existing and planned built-up areas to achieve the most efficient routing and siting of generation and distribution facilities based on demand and load distribution. Temporal planning addresses system size based on current and anticipated demand and load and, most critically, the size of mains and trunk lines that are difficult to rehabilitate once built. This is especially important in rapidly growing cities and has significant financial implications. Owing to uncertainty in predicting demand, using the proper size in infrastructure is part science and part luck. However, decisions on size become more reliable if planners understand urban energy demand patterns and trends and have access to knowledge developed in other cities confronting similar situations.

Urban planners should also consider the constraints of overlapping energy supply networks (for example, covering the same urban area with both natural gas and district heating networks, which has occurred in China and some Eastern European cities). In China, the scarcity of natural gas, which is piped into households for cooking and water heating, means that natural gas distribution represents a relatively expensive investment for gas companies. Space heating is therefore normally provided by separate district heating systems. In Eastern European cities, natural gas has been introduced more recently and is competing with established district heating systems. While competition is normally good in markets, it is not necessarily beneficial in this case because it is undermining the capital investment in district heating systems. In Germany, many cities do not allow utilities to provide district heating and natural gas services in the same area because, to a large extent, both energy carriers provide the same service—space heating (box 3.5).

The future of urban energy supply will still lie in network-based systems that facilitate the adoption of distributed energy generation and decentralized renewable power systems. Thus, if long-term gas supplies are secure, urban energy infrastructure investments should focus on developing modern power grids and natural gas networks. In densely populated cold climate cities in which natural gas is scarce or unavailable, the development of district heating systems is the key to reducing air pollution and improving space heating services. The planning and engineering of specific network-based systems (that is, electric, gas, or heat) have become sophisticated, and technologies are still advancing. For urban planners, the real challenge and the essential task involve fostering the integration and adaptation of network-based energy infrastructure to enhance the efficiency of energy supply and facilitate the uptake of distributed energy resources and other local low-carbon energy sources (for example, methane from landfills and wastewater treatment plants).

Policies, Legislation, and Regulations

In general, national and regional legislators and governments are responsible for energy sector policies and regulations. Cities have limited influence on policy and legislative processes except as regards locally based energy services that require government interventions, such as

---

BOX 3.5

Energy Planning in the City of Mannheim

To strengthen energy planning, Mannheim, Germany has been divided into zones based on the type of energy network. A utility owned by the municipality supplies natural gas, electricity, and district heating. Electricity is universally supplied. Space heating is supplied using natural gas, district heating, or electricity. In areas with greater heat loads, district heating is provided and is the least costly option. In areas with medium -size heat loads, natural gas provides decentralized heating. Areas with low heat demand are supplied using off-peak electrical heat storage devices. Large customers such as department stores, hotels, and office buildings are cooled using absorption chillers linked to the district heating system.

Source: Bernd Kalkum.

By avoiding parallel gas and district heating networks, least cost energy provision is achieved. In zones served by district heating, gas is no longer offered. Electricity and district heat are produced by a combined heat and power plant in the city. The same utility operates public transportation and supplies water. In this way, energy demand and production may be optimized to meet the most important needs of the city.

An important result of Mannheim's plan has been the conversion to cleaner energies. In 1983, 37 percent of all residential buildings were heated by coal or oil-fired heating units. In 1995, this share had dropped to less than 10 percent. In addition, sulfur dioxide emissions have been reduced by about 85 percent, mono-nitrogen oxides by 40 percent, and carbon dioxide by about 30 percent.

district heating systems. The degree of regulation and government oversight in the energy sector varies by country. In many large economies, the energy sector is governed by numerous policies and regulations and is influenced by a mix of government institutions because of concerns about energy security, market competition, social and environmental issues, and other considerations. The fees and charges for network-based energy services are usually regulated so that they respond to social concerns—for example, unduly high energy costs for the poor—and protect against monopolistic price gouging. The pricing of solid and liquid fuels is also often subject to government intervention through taxes and subsidies. Energy sector policies and regulations used to be supply-centric, but this has changed substantially since the first oil crisis in 1973. Many countries now implement regulations and standards requiring minimum energy efficiency levels in energy-consuming equip-

ment, appliances, and building components. The standards are commonly called minimum energy performance standards. Governments may also initiate special policies and programs to create incentives for the adoption of renewable energy and energy-efficient equipment. Table 3.10 summarizes the general elements of energy policies and regulations and the ways in which cities are affected or involved.

Institutions

The multitiered and multifaceted nature of energy sector management and regulation favors complicated institutional interactions. Box 3.6 provides an example of one of the more elaborate institutional and regulatory settings for urban energy planning and management.

The role of national and regional governments is critical. National and regional energy

Table 3.10 Energy Policies and Regulations and Links to Cities

POLICIES AND REGULATIONS	EXAMPLES	CITY GOVERNMENT ROLE
General legislation	The Energy Policy Act (United States)	Local enforcement
	Energy Conservation Law (China)	
Supply-side measures		
Sector-specific measures	Power sector regulations	Interactions only in local distribution or retail
	Oil and gas sector regulations	
	Coal sector regulations	
District heating	Pricing and billing regulation	Strong involvement or autonomy
Renewable energy	Renewable Energy Law (China)	Local implementation
	Mandatory market share policies	Beneficiary
	Feed-in tariffs	
Demand-side measures		
Minimum energy performance standards	Appliance energy efficiency standards	Local programs to replace existing and inefficient equipment
	Industrial motor energy efficiency standards	
Automobile fuel economy standards	Corporate Average Fuel Economy (United States)	Beneficiary
Building construction and renovation	Building energy efficiency standards	Local enforcement
Utility demand-side management	Electricity rate decoupling	Beneficiary
National and regional financial and fiscal incentives	Subsidies for hybrid cars	Beneficiary
	Tax credit for photovoltaic systems	
Environmental protection	Air pollutant emissions standards	Local enforcement
		Beneficiary

Source: Author compilation (Feng Liu).

policies, legislation, and regulations influence the transparency, consistency, and predictability of modern energy supply systems in individual cities and address common social and environmental issues. National and regional governments also establish general provisions that incentivize cities to adopt sustainable energy practices. These provisions include, for example, renewable energy feed-in tariffs that mandate electricity utilities to purchase wind- or solar-generated electricity at set prices and energy performance standards that set minimum energy efficiency levels for new appliances and new buildings. In contrast, national and regional regulations may also hinder the sustainable energy measures of cities. For example, in most countries, the prevailing regulations on electricity utilities discourage utility demand-side management and the installation of distributed generation facilities, including renewable technologies.

The role of city government in setting broad energy sector policies and regulations is limited and is likely to remain so because of the nature of modern energy systems. However, this does not prevent cities from planning and deciding what, where, and how urban energy infrastructure should be built. Cities may take steps to influence national policies, while seeking to influence local behavior through voluntary programs and initiatives. Because city governments are intimately involved with every aspect of urban development and management and wield power and influence over urban energy demand, they are uniquely able to tie urban energy supply and demand together. This means that cities are one of the most effective actors in pursuing sustainable energy. Nonetheless, most

---

BOX 3.6

Public Agencies with Significant Influence on Electricity Production, Distribution, and Use, California

Federal

- *Federal Energy Regulatory Commission:* wholesale rates, interstate and international transmissions, and hydropower licensing
- *U.S. Environmental Protection Agency:* compliance with the Clean Air Act and the Clean Water Act, overseeing enforcement and regulatory actions delegated to the states
- *U.S. Department of Energy:* technology research, development, and promotion; energy efficiency programs; national standards for appliances and end use

State

- *California Energy Commission:* licensing for thermal generators of 50 megawatts or greater; end use efficiency standards; systems analysis, planning, and forecasting; planning of intrastate electricity transmission infrastructure; energy research, development, and demonstration in the public interest
- *California Public Utilities Commission:* rates on investor-owned utility services for retail customers; systems analysis, planning, and forecasting; monitoring of electricity market; public and private sector efficiency and education programs; representation of the state before the Federal Energy Regulatory Commission; transmission delivery infrastructure

- *California Independent System Operator:* monitoring and planning for system reliability; systems analysis, planning, and forecasting; planning of electricity transmission infrastructure
- *California Air Resources Board:* emission standards for distributed generation resources and diesel backup generators

Regional

- *Regional water quality control boards:* issuance and enforcement of permits under the Clean Water Act and California regulations for power generator discharges into and use of regulated bodies of water
- *Regional air quality management districts:* issuance and enforcement of permits under the Clean Water Act and California regulations for air emissions from power generators

Local

- *Cities and counties:* long-term land use planning, enforcement of energy efficiency standards for buildings, approval of site plans and urban designs in private development, permits and site authorizations for all power plants under 50 megawatts

Source: Lantsberg (2005).

cities have not yet become effectively organized to pursue sustainable energy planning and management. In the traditional supply-driven and network-oriented urban energy landscape, the role of cities is limited. Even in a sophisticated city such as New York, officials realized that the New York City Energy Planning Board was needed to link supply and demand effectively as part of PlaNYC 2030, the city's integrated energy strategy (City of New York 2007).

Stakeholder Dynamics

Urban energy planning and management are shaped by several principal stakeholders: local, regional, and national governments and their relevant agencies or authorities; public and private energy utilities, companies, vendors, and investors; customers; and public interest entities. Other stakeholders include financiers, equipment and service providers (for example, energy service companies), and city service users. The relationships among these stake-

holders are relatively straightforward. Governments regulate urban energy services to ensure quality, safety, environmental controls, and fairness to customers and investors; energy providers produce, transmit, transport, distribute, and retail energy to customers; and customers pay for energy to sustain services and reward investors. The public interest entities or organizations advocate on behalf of disadvantaged social groups, such as low-income households, to improve access and affordability. These entities and groups also inform and educate other stakeholders about sustainable energy solutions and press for relevant action. In particular, climate change has mobilized many international and domestic public interest entities.

Traditionally, city governments are most concerned about the needs and interests of consumers in their jurisdictions and strive to safeguard reliable and affordable energy services, especially in electricity (and heating service in cold climate cities). But the means of intervention are limited, as illustrated by the example of New York City (figure 3.38). In this case, the city government has been only marginally involved

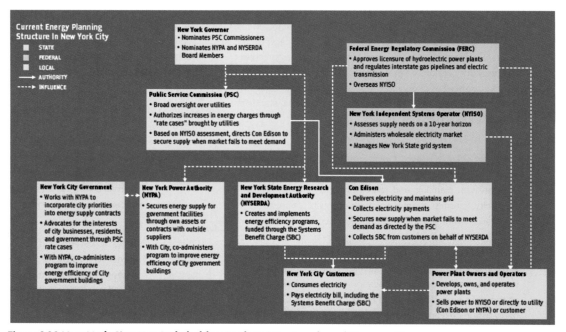

Figure 3.38 New York City: Key Stakeholders in Electricity Supply and Consumption

Source: City of New York (2007).

in planning and managing electricity supply and demand, which officials are seeking to improve through PlaNYC 2030.

City governments are uniquely positioned to influence the stakeholder dynamics in favor of sustainable energy because they are significant energy consumers themselves and are able to affect the behavior of energy consumers in the city. Cities also determine how urban areas are built, including energy supply infrastructure. However, intracity consultations are often challenging. Energy use cuts across many agencies, but stimulating interagency collaboration is a big challenge, particularly if energy costs and benefits are being borne unevenly. The functional areas within agencies—represented by technical staff members, environmental officers, budget teams, procurement personnel, and so forth—also bring unique biases, expertise, incentives, and constraints to efforts to improve energy efficiency. Some of these issues may be addressed through policies and programs, but the strong leadership of mayoral offices is often needed to push the parties to work together.

Economic, Financial, Social, and Environmental Aspects

Sustainable urban energy planning and practice should be economically justifiable, financially viable, socially equitable, and environmentally sensible. These considerations form the basis for the proper selection and design of sustainable energy actions by cities.

Economic justification requires cities clearly and consistently to account for and evaluate the costs and benefits of alternative urban energy solutions so as to facilitate robust comparisons. This is often challenging because evaluating environmental externalities, such as health benefits or hazards, is difficult. A critical aspect of economic analysis is the calculation of the *life-cycle cost* of alternative energy solutions. Many sustainable energy actions, especially energy efficiency measures, have life-cycle costs lower

than the business-as-usual alternatives. For example, Council House 2, a visionary building developed by the City of Melbourne, Australia, involved features reducing electricity use by 82 percent, gas by 87 percent, water by 72 percent, and corresponding carbon dioxide emissions by 87 percent. While the initial costs were high, it was possible for the city to envision a financial payback period of only about 10 years because of the energy savings (C40 Cities 2009b). Though the commercial sector normally considers projects viable if payback periods are less than five years, city governments tend to have longer investment horizons because their built environments last for decades. Other sustainable energy options may have longer payback periods, but may yield benefits that are more difficult to quantify, such as local investment, job creation, improved competitiveness, and enhanced quality of life (for example, reduced commuting times, improved air quality and health, more green space, and more community space).

Financial viability requires city actors to obtain sufficient funds to implement sustainable energy solutions, sustain outcomes, and maintain a positive return on investment within prevailing and projected financial cash flows. For a city to acquire and sustain modern energy services (for example, electricity, natural gas, or district heating), prices need to ensure cost recovery. For energy efficiency measures to be viable, the saved energy has to be as reliable as and cheaper than conventional supply options. In other words, a viable sustainable energy solution must be a viable business proposition. Because market valuations often fail to account for environmental externalities, some renewable energy solutions, such as wind electricity and solar photovoltaic systems, may still require government subsidies or regulations (feed-in tariffs) to be viable. The recent expansion of carbon financing markets should improve the financial attractiveness of sustainable energy investments by providing a new and sometimes more secure revenue stream for such projects. But many financially viable opportunities for greater energy efficiency remain unimplemented because of various market barriers.[8]

The investments most likely to be undertaken will be partly driven by economic considerations (table 3.11). In this context, serious analysis may be able to identify short- and medium-term payback measures that a city may wish to pursue. The key challenge is the development of the best overall package of investments that is below an acceptable payback or other investment threshold. Access to special funding, such as concessional donor funds or carbon finance revenues, may increase overall returns, while maximizing the effect of the measures in the investment package.

Social equity requires cities to address issues of access and affordability among the poor. Artificially suppressing energy prices or providing universal subsidies is not an effective way to approach these challenges. City governments should target subsidies only on people who can-

Table 3.11 Indicative Economics of Sustainable Energy Options

SECTOR	SHORT-TERM PAYBACK, UNDER 5 YEARS	MEDIUM-TERM PAYBACK, 5–10 YEARS	LONG-TERM PAYBACK, 10+ YEARS
Public buildings	Equipment retrofits Labeling building performance Energy service company contracting Solar water heating	Building envelope measures Green roofs Training in good practices in building operations and maintenance	Building codes Certification of building materials Building integrated photovoltaics Equipment standards
Public lighting	Lighting retrofits using high-pressure sodium vapor or metal halide Redesign of lighting systems Control systems and sensors	Retrofits using light-emitting diodes	Street and traffic lighting standards
Transport	Optimization of traffic signals Fuel efficiency vehicle standards Congestion taxes, tolls	Alternative fuels for public buses, taxis Bus rapid transit systems	Modal shifts Vehicle inspection and maintenance programs Changes in land use patterns to promote densification
Water, wastewater	Pumping retrofits Correct sizing of pumps Leak reduction Load management Energy service company contracting	System redesign and optimization Methane recovery for power generation from wastewater Water demand-side management (low-flow outlets, drip irrigation)	n.a.
Solid waste	n.a.	Methane recovery for power generation from landfills Recycling programs	n.a.
Electricity, heating	Supply-side loss reduction Power factor correction measures Improved metering and pricing Renewable energy portfolio standards Retrofits of boiler and piping systems	Combined heat and power provision Load management Energy storage systems Promotion of distributed generation with feed-in tariffs	n.a.
Cross-cutting	Bulk purchase of efficient products Awareness raising on energy issues to public sector staff Agency awards and contests for energy efficiency	Procurement standards for product procurement	Improved city design and planning systems

Source: Author compilation (Jas Singh).
Note: n.a. = not available.

not afford to pay full cost recovery prices for energy services. Social equity also means that cities should link sustainable energy actions with energy equity objectives. A good example is promotion of the use of compact fluorescent lamps in low-income or slum areas.

Environmental sensibility requires that cities be mindful of the local, regional, and global environmental impacts of energy practices and adjust energy plans to mitigate impacts. For example, in the City of Rizhao in Shandong Province, China, the adoption of sustainable urban energy solutions, while addressing social equity issues, has made good business and environmental sense (box 3.7).

Indicators and Benchmarks

Sustainable urban energy planning and practice are elusive without realistic metrics to quantify performance (using indicators) and measure progress (using benchmarks). For cities in developing countries, indicators and benchmarks do not merely reveal gaps; they also inspire actions to achieve better energy services without reducing affordability or compromising the environment, as exemplified in the case of the City of Rizhao. Developing metrics is a worthy, but difficult task because every city is unique in its energy uses and levels of energy service. We should therefore focus on a small set of key indicators that allow meaningful cross-city comparisons. Industrial energy consumption and related indicators should not be included and need to be addressed separately. One must also bear in mind that many developing-country cities are underserved in energy (lack of access or lack of affordability) compared with their developed-country counterparts. Thus, the indicators that are sensitive to distortion (for example, per capita–denominated indicators) should be carefully considered. In general, cities should have two sets of metrics on sustainable energy:

An Extensive Solar Water Heating Program in Rizhao, China

Rizhao, a city in northern China with a population of 350,000, is using solar energy for water heating and lighting. In the early 1990s, a municipal government retrofit program mandated that all buildings install solar water heaters. After 15 years, 99 percent of the households in the central district had obtained solar water heaters. Solar water heating is now ubiquitous. The city uses more than 500,000 square meters of solar panels to heat water. This is equivalent to the number of electric water heaters necessary to produce about 0.5 megawatts of power. Most traffic signals and street and park lights are powered by solar cells, reducing the city's carbon emissions and local pollution. Using a solar water heater for 15 years costs about US$1,934 (Y 15,000), which is less than the cost of a conventional electric heater. This shift has generated annual household savings of US$120 in a part of China where per capita incomes are lower than the national average.

This achievement is the result of a convergence of three factors: a regional government policy that promotes the solution and provides financial support for the research, development, and deployment of solar water heating technologies; a new industry that capitalizes on fresh opportunities; and city officials who have the vision and the leadership to implement changes in the attitudes of other stakeholders.

How does it work? The municipal government, the community, and local solar panel producers have had sufficient political will to adopt and apply the technology.

The provincial government of Shandong provided subsidies and funded the research and development of the solar water heater industry.

The cost of a solar water heater was reduced to the cost of an electric water heater, about US$190. This represented about 4 or 5 percent of the annual income of an average household in Rizhao and about 8 to 10 percent of an average rural household income.

Panels are simply attached to the exteriors of buildings. The city helps install the panels.

The city raised awareness through community campaigns and education. Rizhao held public seminars and supported advertising on television.

The city mandated that all new buildings incorporate solar panels and oversaw the construction process to ensure proper installation.

Source: Bai (2006).

Table 3.12 Sustainable Urban Energy Indicators and Benchmarks: Preliminary Proposal

SECTOR	INDICATORS[a]	BENCHMARKS[b]
Long-term and strategic goals	• Share of renewable energy supply in final energy consumption • Carbon content of final energy consumption (kilogram equivalent CO_2 per megajoule) • Urban density indicator • Energy cost and affordability indicator	Benchmarks should draw on a group of comparable cities in terms of climate conditions and indicate the medium-level practice and best practice, respectively.
Municipal services	• Electric distribution losses • Energy used for delivering and treating one cubic meter of water • Technical and nontechnical water losses • Public lighting energy efficiency • Methane recovery from landfills and wastewater treatment plants	See above.
Buildings	• Residential buildings: cooling, heating, and lighting efficiency • Office buildings: cooling, heating, and lighting efficiency • Government buildings: cooling, heating, and lighting efficiency • Energy efficiency of key appliances	See above.
Transport	• Carbon emissions of passenger traffic (kilogram equivalent CO_2 per person-kilometer)	See above.

Source: Author compilation (Feng Liu).
Note: In the table, urban energy does not include industrial energy consumption. CO_2 = carbon dioxide
a. Indicators represent the current performance of a city.
b. The benchmarks and indicators are the same set of metrics, but the benchmarks represent the medium-level practice and best practice, respectively, among a set of cities that are comparable in terms of climate conditions.

one set reflecting the long-term strategic goals of sustainable urban energy planning and practice and the other highlighting the performance and efficiency of energy consuming sectors in the cities. Table 3.12 represents a preliminary list of sustainable energy metrics or categories of metrics proposed for cities.

Barriers to Investing in Sustainable Energy in the Public Sector

Many sustainable energy actions may be justified solely on the basis of cost-effectiveness. However, for various reasons, many investments are unrealized because of administrative, policy, and market barriers. Key issues include the following: (1) government agencies are typically unresponsive to price signals because they lack a commercial orientation; (2) public procedures for equipment and service procure-

ment are generally inflexible to new approaches; and (3) constrained annual budgets restrict funding for capital upgrades, while restrictions on public financing and typical one-year budget appropriations mean that the amortization of costs is difficult. A list of typical barriers categorized by stakeholder is provided in table 3.13.

Sustainable Energy Actions of City Government

The development of modern interconnected energy systems over the past century or so has gradually reduced the capacity of cities to understand and plan for their energy needs. Cities have become passive participants in the urban energy agenda, leaving most responsibilities to regional and national governments and the private sector. To pursue a sustainable urban energy agenda, cities need to become more

Table 3.13 Typical Barriers to Public Sector Sustainable Energy Investments

POLICY AND REGULATORY BARRIERS	PUBLIC END USERS	EQUIPMENT AND SERVICE PROVIDERS	FINANCIERS
• Low-energy pricing or collections • Procurement policies (lowest cost, defined project, unbundled services) • Annual budget cycles may not allow multiyear contracting • Ad hoc planning	• No incentive to change or take risk • No discretionary budget for upgrades or special projects • Unclear about ownership of cost and energy savings • Weak technical ability to assess options • Behavioral biases	• Higher transaction costs for public sector projects • Concerns over late or no payment • High project development costs • Limited technical, business, and risk management skills • Low track record in the market for new contractual models	• High perceived public credit risk • New technologies • New contractual mechanisms • Small sizes and high transaction costs • High perceived risks • Behavioral biases

Source: Author compilation (Jas Singh).

assertive and involved in decisions that affect energy demand and supply options. City governments need to become stronger partners of regional and national governments and to guide and mobilize private sector participation. Most important, cities need to act within their own authority to implement sustainable energy solutions.

Energy efficiency and renewable energy solutions in the public sector

Energy costs often constitute a significant portion of the operating budget of city governments. In the State of California, for instance, energy is the second largest expenditure item in city government operations, after employee salaries and benefits (Lantsberg 2005). The share of public sector consumption is particularly high in electricity and heating. The public sector accounts for 9 percent of Brazil's electricity use. Public agencies account for 20 percent of Eastern Europe's electricity and heating loads, and about 10 percent of the European Union's electricity and heating demand arises from the public sector.[9] As a first step, city governments should consider initiating sustainable energy solutions within city boundaries because this may produce rapid benefits and may be implemented more easily. Common targets for improvements include government-owned buildings and facilities; water supply and wastewater treatment; public lighting and traffic lights; and municipal services such as solid waste management, public transportation, and, in cold climates, district heating.[10]

Government-owned buildings and facilities: Buildings consume about one-third of global energy and present significant potential for energy savings. Government buildings, particularly those in developing countries, tend to be older and use more inefficient equipment, underlining the potential for energy efficiency gains. Measures to realize gains may focus on building envelopes (windows and insulation), electrical appliances (lighting, pumping, and heating and cooling) and office equipment (computers, copiers, and printers). Though measures are beneficial, public facilities are often subject to rigid procurement practices that focus heavily on costs and lack discretionary budgets with which to make meaningful improvements. In addition, principal-agent relationships or split incentives complicate investments. For example, a parent budget agency may determine a subsidiary's capital budget and even specify equipment, despite the subordinate agency's responsibility for paying monthly energy bills.

Energy efficiency programs often initially support relatively low-cost modulated measures, such as lighting retrofits, or the replacement of old equipment, such as heating, ventilation, and air-conditioning systems. In public building complexes, such as city halls, schools, and hospitals, a whole-building approach may be needed to achieve cost-effective control of the energy budget of a building (annual energy consumption). Moreover, buildings are complex energy systems, and trade-offs are often made to optimize energy efficiency. For example, planners must evaluate the efficiency of a heating, ventilation, and air-conditioning system against the thermal pass-through of a building envelope because each option reduces the effectiveness of the other. For new government buildings, the adoption of best practice in sustainable design and construction reduces life-cycle costs and serves as an example for the private sector. A comprehensive analysis of the financial costs and benefits of LEED-certified office and school buildings in the United States has found that a minimal up-front investment of about 2 percent of construction costs typically yields (20-year) life-cycle savings of over 10 times the initial investment (Kats 2003).[11]

Recently, some governments in developing countries have experimented with retrofitting multiple municipal facilities under common authority. Though this may be more complex, it may also substantially reduce transaction costs and allow for scaled-up investments. In Hungary, for example, the Ministry of Education issued a tender in 2006 for a single consortium to finance and retrofit all the schools in the country under an energy service company contract. The International Finance Corporation provided a portfolio credit guarantee to the winning bidder for up to US$250 million. To date, about US$22 million has been invested in about 200 projects.

Water supply and wastewater treatment: The operation of water and wastewater systems is often the largest outlay in municipal energy budgets. For example, cities in California spend over 50 percent of their energy budgets on water and wastewater pumping (Lantsberg 2005). Estimates suggest that 2 or 3 percent of the world's energy consumption is devoted to pumping and treating water and that the potential exists for related energy savings of more than 25 percent. In many cities, energy and water are scarce resources, and cities often introduce efficiency programs to save energy and water simultaneously in light of links between these sectors. In developing countries, water and wastewater systems are often poorly designed, rely on outdated equipment, and suffer from high nonmetered water losses owing to inadequate investment and expertise. Many systems operate without adequate commercial incentives to be efficient. In light of these obstacles, the Alliance to Save Energy launched Watergy, a program that demonstrates the significant benefits of increasing clean water access by reducing energy costs and water losses.[12] In Fortaleza, in northeast Brazil, the Alliance to Save Energy worked with the local utility, the Companhia de Água e Esgoto do Ceara, to develop and implement measures to improve water distribution and access to sanitation services, while reducing operating costs and environmental impacts. The local utility invested about R$3 million (about US$1.1 million) in various activities such as the installation of an automatic control system. The utility saved 88 gigawatt hours and US$2.5 million over four years. More important, the utility established an additional 88,000 new connections, while decreasing overall energy costs.

Efforts to improve energy efficiency should consider both supply- and demand-side measures and the relevant links. For example, as water leakage and waste are reduced, additional efficiency gains may be realized by downsizing pumping stations. Other measures should also be considered to boost efficiency, such as system redesign, pressure management, pump impeller reduction, installation of low-friction pipes and variable speed pumps, load management, power factor improvements, improved maintenance procedures, improved metering, and water recycling. Wastewater treatment plants might also be made more efficient by recovering waste heat, capturing methane for power generation, and improving pumping systems.

Many cities in developing countries have a pressing need to expand water supply and wastewater treatment capacities. The reclamation of runoff water and the on-site treatment of domestic wastewater are increasingly prevalent in new real estate development projects. Such practices, if properly configured within urban water and wastewater networks, may enhance overall energy efficiency, while relieving the pressure on scarce freshwater resources.

Public lighting: Public lighting is often considered an essential public service that enhances economic activity and improves the quality of life (for example, by reducing crime and vehicular accidents). Streetlights may be provided more effectively and extensively by using energy-efficient lighting technologies, which are now cheaper and more plentiful. However, the procurement of lamps is often based on consideration of the initial costs without taking into account the impact on recurring energy bills. To varying degrees, municipal governments possess limited capital budgets to replace lighting; lack credible information on alternatives; and, in some cases, fail to pay electricity bills for streetlighting regularly. To illustrate the range of options available in streetlighting, table 3.14 examines the cost-effectiveness of alternative systems that were considered by the State of New York in 2002.

Streetlamp retrofits may potentially save 30 to 40 percent of typical energy costs and, depending on cost structures and available lamps, may have payback periods of less than three years. Installing time clocks and automated control systems and redesigning systems (to eliminate overlighted and underlighted areas) may achieve additional energy savings. In India, the State of Tamil Nadu issued a tender for seven municipalities to be retrofitted to reduce energy use in public lighting and water pumping. Through an urban infrastructure development fund, bids were solicited that required a minimum of 30 percent energy savings. Several competitive bids were received, an award was made, and the project has been in operation since 2008 (Singh and others 2010).

Other municipal services: There are other opportunities to realize energy savings through municipal services, such as solid waste (waste recycling, methane recovery in landfills for power generation, and so on) and transportation (alternative fuel vehicles, maintenance of the public transit bus fleet, establishment of rapid transit systems and congestion tolls, for example). An especially important aim in cold climates is to improve the efficiency of district heating systems (box 3.8).

Beyond the public sector: Focusing on the built environment

As enforcers of national, regional, and local regulations, city governments substantially influ-

Table 3.14 A Comparative Economic Analysis of Selected Streetlighting Systems

COMPARED ELEMENT	MERCURY COBRAHEAD, CONVENTIONAL	METAL HALIDE COBRAHEAD, ENERGY-EFFICIENT	HIGH PRESSURE SODIUM CUTOFF, ENERGY-EFFICIENT
Lamp type	400-watt mercury vapor	250-watt metal halide	250-watt high-pressure sodium
Number of luminaries	12	12	11[a]
Installed cost, US$	36,672	36,240	35,618
Annual energy cost, US$	2,391	1,551	1,419
Annual operating cost, US$[b]	2,536	1,677	1,601
Total annualized cost, US$[c]	6,271	5,368	5,229

Source: NYSERDA (2002).

a. Assumes a 10 percent reduction in the number of poles needed because of the higher luminous efficacy of high-pressure sodium.

b. Includes energy and maintenance costs.

c. Includes initial capital investment, energy, and maintenance costs annualized over 20 years.

Improving Energy Efficiency, Reducing Energy Costs, and Releasing Municipal Budgets

With partial support from a World Bank loan in 1991–99, the cities of Gdansk, Gdynia, Krakow, and Warsaw, Poland, undertook renovations in heat supply systems distributed heating meters for buildings, and reformed heat pricing from a tariff based on square meters of serviced area to a two-part tariff calculated per building.

The government of Poland implemented energy sector reforms requiring that payment for heat gradually become the responsibility of households. Households subsequently began to use heat more efficiently. Households or companies operating as the agents of households invested in thermostatic radiator valves, heat allocation meters, better windows, and insulation. A key result was that the costs of heating a given apartment area fell by 55 percent because of consumer-driven efficiency improvements and technical, operational, and management improvements in the heat supply companies. This reduction helped to make the removal of subsidies less burdensome to households.

Results in Four Cities

INDICATOR	1991/92	1999	CHANGE, %
Household heat bill subsidy (%)	67	<5 (1994)	n.a.[a]
Heat bill charged to households, 1999 (US$ per square meter)	13.7	6.2	−55
Heated floor area, square meter (millions)	63.8	68.6	7
Heat energy sold (gram calories per square meter)	0.27	0.22	−18
Energy savings	n.a.	n.a.	22

Note: n.a. = not applicable.
a. Nationwide, household heat subsidies, provided by municipal governments, fell from 78 percent in 1991 to zero by the end of 1997.

Source: World Bank (2001).

Nationwide, household heating subsidies, provided by municipal governments, were reduced from 78 percent in 1991 to zero by the end of 1997. The installation of building heat meters has been mandatory in all buildings since 1999. The use of heat allocation meters has become a popular method to allocate heat bills within buildings; 5.5 million such meters had been installed as of 1997, covering about 30 percent of dwellings nationwide. More than 10 companies have been formed and compete in the market for billing services, including allocation meter installation, meter reading, billing, and maintenance. The energy savings reflected in customer heating bills stemming from the reform, including savings from private investments spurred by the reform, typically range from 20 to 40 percent.

ence the nature of the sustainable energy solutions adopted in the urban built environment. This is especially important in rapidly growing cities in developing countries, where inaction leads only to future energy waste. Officials should focus primarily on the features and functions of new buildings that affect energy consumption, especially heating and cooling systems. Other factors include site plans, building layouts, building envelopes, lighting fixtures, and water heaters. Much relevant experience has been accumulated in developed countries over the past 30 years. Matured technologies and efficient materials may be applied to create buildings with low or near zero energy use for heating and cooling (Rosenthal 2008). Implementing energy efficiency standards requires coordination among national, regional, and city efforts, but local enforcement is critical. China provides a good example of the success of energy efficiency programs in rapidly urbanizing developing countries. In 1995, the first mandatory energy efficiency standard was introduced for new residential buildings in cold climates in China. Among large northern Chinese cities, the compliance rate was a meager 6 percent in 2002. Since then, the national government has increased the assistance to local governments for enforcement and inspections. The rate of compliance rose to about 40 percent in 2005 and 70 percent in 2007. Compliant buildings, on average, lose 35 percent less heat than conventional buildings. The national government will soon promulgate a revised energy efficiency standard for new residential buildings in cold climates that will cut heat losses by an additional 30 percent. This time, many cities are ahead of the national government. For example, in 2005, Beijing and Tianjin adopted energy efficiency standards for buildings similar to the pending revised national standards. In 2007, the provinces of Hebei and Liaoning did the same.

Many developed countries have broadened their efforts to promote sustainable buildings by incorporating other conservation strategies, such as the improved management of water and waste and steps to enhance the quality of indoor environments. For example, in 2008, the State of California adopted the first green building standards in the United States. Developing countries should take note, however, that it takes years to create adequate capacity to enforce energy efficiency and green standards. Moreover, it is important to sequence sustainable building interventions in ways that suit local capacity and priorities.

The big picture: Urban spatial development

Ultimately, individual cities and regional urban clusters must become more efficient in using natural resources, including energy. In cities, sustainable urban energy planning and practices should be integral parts of the implementation of resource-efficient growth, which, it is hoped, complements sustainable development agendas at regional and national levels. To achieve intelligent resource-efficient growth, cities may need to drop expansionary urban spatial development linked to motorized transportation and refocus development in neighborhoods to ensure that key services are within walking distance or the range of bicycle travel and public transportation. Details on the impacts of urban spatial development on urban energy efficiency are discussed in chapter 5 and in sector note 3. In essence, the key message is that urban energy requirements may be reduced by increasing urban densities, which reduces the extent of major municipal infrastructure, such as roads, water and wastewater systems, power lines, and gas pipelines. Infrastructure capital and operations and maintenance costs also fall under condensed systems. Figure 3.39, for example, illustrates the general relationship between urban density and transportation fuel consumption. Density also has drawbacks and limits, however, and must be

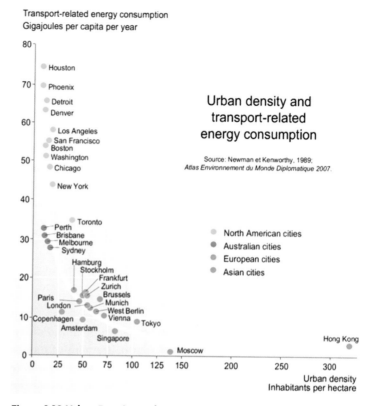

Figure 3.39 Urban Density and Transportation-Related Energy Consumption

Source: Adapted from Kirby (2008).

planned on the basis of existing physical, socio-economic, and natural conditions.

Conclusions

Because energy cuts across multiple sectors, the planning and implementation of sustainable energy measures in urban settings are complex. Though many energy investments may be justified on the basis of the financial or economic returns, environmental concerns should be factored into project assessments. Some general recommendations for promoting sustainable energy and increasing energy efficiency and clean energy include the following:

- *Ensure that the energy sector works properly.* Energy sector restructuring, utility commercialization, pricing reform, and other measures may reduce energy costs, while reducing energy waste. These efforts are most effectively led at the national level.

- *Explore options to retrofit the existing stock of infrastructure.* This may be accomplished by auditing energy sources and organizations, changing procurement guidelines, contracting energy service companies, devising public agency targets for energy efficiency, and so on. Access to financing is key to realizing these gains.

- *Consider the options in addressing the new built environment.* This might entail adopting energy efficiency standards for buildings and equipment, improving city planning and design processes, strengthening land use schemes, and so forth.

- *Seek options to bundle city programs.* For example, combine the procurement of equipment to negotiate better prices, combine similar services across cities, and boost the city's influence at the national level.

- *Seek ways to incentivize public agencies and staff on sustainable energy options.* Offer environmentally sustainable awards, publish agency energy and environmental performance records, provide incentive grants, and so on.

- *Create mechanisms for sharing cities' experiences across the country.* This could be done through associations, case studies, newsletters, and so forth.

Notes

1. This sector note reflects the International Energy Agency's definition of cities as a general and interchangeable reference for urban areas, which may be large metropolitan city-regions, such as New York City, or small urban settlements that have only a few thousand people (see IEA 2008). The exact definition of urban areas varies by country.

2. For example, the supply and prices of grid-based electricity are generally regulated by regional or national governments.

3. In general, building energy codes are regulated at the regional, provincial, state, or national level, but compliance depends on local enforcement.

4. Energy-efficient alternatives often require greater expense in the short term, but save money in the long term. They require capital investment at the start-up, but their overall life-cycle costs are lower. Less efficient alternatives are often less expensive in the short term (requiring a smaller capital investment), and cities may choose them to provide a less expensive and easier solution for a wider population in a shorter time frame even though this may not be the optimum solution in the longer term.

5. Electric lighting is a good example. By the time the electricity has reached a light bulb, an average of about 70 percent of the energy content of coal has already been lost through conversion, transmission, and distribution. A compact fluorescent lamp delivers the same amount of lighting service (that is, brightness per square meter) using about 20 percent of the electricity of an incandescent lamp.

6. Passive houses using ultralow energy for space cooling and heating have been successfully demonstrated in Europe and the United States (Rosenthal 2008).

7. Distributed energy resources also exist in urban areas. These are based on parallel, stand-alone electricity generation units within electricity distribution systems. The units are located at or near end users. Examples include gas microturbine systems, wind turbine systems, fuel cells, and rooftop photovoltaic systems. Distributed generation may be beneficial for electricity consumers and, if properly integrated, the electricity utility.

8. Market barriers to investment in energy efficiency refer to factors, usually social and institutional, that prevent the realization of the full economic potential of energy efficiency opportunities. The barriers help explain the difference between observed energy efficiency choices and decisions and the corresponding choices and decisions predicted by economic theory. Some common market barriers include misplaced incentives, lack of access to financing, high transaction costs, regulatory price distortions, lack of information, and misinformation.

9. For additional information on those and other cases, see European Commission (2003) and Taylor and others (2008).

10. District heating systems are the only modern urban energy infrastructure that is entirely bound

to cities. The ownership structure has undergone significant changes, but city governments still exert great influence on the development and management of these systems.

11. LEED (Leadership in Energy and Environmental Design) is a green building rating system developed by the U.S. Green Building Council. It provides a suite of criteria for environmentally sustainable construction. The main financial benefits of meeting the criteria include the lower costs of energy, water, and waste disposal.

12. See the Alliance to Save Energy's 2007 Watergy Handbook (Barry 2007) for a discussion on the barriers and opportunities for tapping into water and energy efficiency in water utilities.

References

Bai Xuemei. 2006. "Solar-Powered City: Rizhao, China." In *State of the World 2007: Our Urban Future*, ed. Worldwatch Institute, 108–09. Washington, DC: Worldwatch Institute. http://www.worldwatch. org/taxonomy/term/467.

Barry, Judith A. 2007. "Watergy: Energy and Water Efficiency in Municipal Water Supply and Wastewater Treatment; Cost-Effective Savings of Water and Energy." Handbook, Alliance to Save Energy, Washington, DC. http://www.watergy.net/ resources/publications/watergy.pdf.

C40 Cities. 2009a. "Lighting: Växjö, Sweden." C40 Cities Climate Leadership Group. http://www. c40cities.org/bestpractices/lighting/vaxjo_ streetlight.jsp.

———. 2009b. "Buildings: Melbourne, Australia." C40 Cities Climate Leadership Group. http://www. c40cities.org/bestpractices/buildings/melbourne_ eco.jsp.

CBSC (California Building Standards Commission). 2009. "2008 California Green Building Standards Code." CBSC, Sacramento, CA. http://www. documents.dgs.ca.gov/bsc/2009/part11_2008_ calgreen_code.pdf.

City of New York. 2007. *PlaNYC: A Greener, Greater New York*. New York: City of New York. http:// www.nyc.gov/html/planyc2030/downloads/pdf/ full_report.pdf.

European Commission. 2003. *Harnessing the Power of the Public Purse: Final Report from the European PROST Study on Energy Efficiency in the Public Sector*. Stockholm: Borg and Co. AB. http://ec.europa. eu/environment/gpp/pdf/harnessing_power_ prost_study.pdf.

ICLEI (ICLEI–Local Governments for Sustainability). 2008. "International Local Government GHG Emissions Analysis Protocol." Release version 1.0, ICLEI, Toronto. http://www.iclei.org/fileadmin/ user_upload/documents/Global/Progams/GHG/ LGGHGEmissionsProtocol.pdf.

IEA (International Energy Agency). 2008. *World Energy Outlook 2008*. Paris: IEA.

Kats, Greg. 2003. "The Costs and Financial Benefits of Green Buildings: A Report to California's Sustainable Building Task Force." California Integrated Waste Management Board, Sacramento, CA. http://www.cap-e.com/ewebeditpro/ items/O59F3259.pdf.

Kirby, Alex. 2008. *Kick the Habit: A UN Guide to Climate Neutrality*. Nairobi: United Nations Environment Programme.

Lantsberg, Alex. 2005. "Sustainable Urban Energy Planning: A Road Map for Research and Funding." Consultant report, CEC-500-2005-102, California Energy Commission, Sacramento, CA. http:// www.energy.ca.gov/2005publications/CEC-500- 2005-102/CEC-500-2005-102.PDF.

Mairie de Paris (Mayor's Office of Paris). 2007. "Paris Climate Protection Plan" [Plan climat de Paris]. Mairie de Paris, Paris. http://www.paris.fr/portail/ english/Portal.lut?page_id=8118&document_type_ id=2&document_id=66812&portlet_id=19237.

NYSERDA (New York State Energy Research and Development Authority). 2002. "NYSERDA How-to Guide to Effective Energy-Efficient Street Lighting: For Planners and Engineers." Nyserda, New York. http://www.rpi.edu/dept/lrc/nystreet/ how-to-planners.pdf.

Rosenthal, Elisabeth. 2008. "The Energy Challenge: No Furnaces but Heat Aplenty in 'Passive Houses.'" *New York Times* December 27: A1.

Singh, Jas, Dilip R. Limaye, Brian J. Henderson, and Xiaoyu Shi. 2010. *Public Procurement of Energy Efficiency Services: Lessons from International Experience*. Washington, DC: World Bank.

Taylor, Robert P., Chandrasekar Govindarajalu, Jeremy Levin, Anke S. Meyer, and William A. Ward. 2008. *Financing Energy Efficiency: Lessons from Brazil, China, India, and Beyond*. Washington, DC: World Bank.

USAID (U.S. Agency for International Development). 2005. "Watergy Program Pioneers Performance Contract to Save Water, Energy in S. Africa." *Energy Update* 2 (April/May): 6–7.

World Bank. 2001. "China: Opportunities to Improve Energy Efficiency in Buildings." Report, Asia Alternative Energy Programme and Energy and Mining Unit, East Asia and Pacific Region, World Bank, Washington, DC.

———. 2007. *Cost of Pollution in China: Economic Estimates of Physical Damages*. Washington, DC: World Bank. http://siteresources.worldbank.org/ INTEAPREGTOPENVIRONMENT/Resources/ China_Cost_of_Pollution.pdf.

Worldwatch Institute. 2006. *State of the World 2007: Our Urban Future*. Washington, DC: Worldwatch Institute. http://www.worldwatch.org/taxonomy/ term/467.

Cities and Water

Overview

Water is indispensible to human activity. Ancient civilizations flourished around water sources, including ancient China, Egypt, and Rome. Water has shaped the destinies of great cities, such as Beijing, Cairo, Frankfurt, London, New York, Paris, Rome, and Sydney. However, many pioneering cities, such as Babel and Sheba in the ancient Middle East, diminished or disappeared because water sources dried up. Water plays an important role in economic growth, quality of life, and environmental sustainability. Some people define water as a divine gift, while others view it as an economic commodity. In any case, water is a limited resource that must often be processed to become usable, and there are costs associated with its transportation, distribution, and management. Water has social value, and access to sufficient water to survive is a human right. In this context, politicians and managers typically take steps to guarantee that the poor have access to an equitable share of water services, particularly in developing countries. Water is a shared resource that plays a vital role in the development of other economic sectors.

Given the importance of water, there is a need for integrated management at the sectoral level and at the macrolevel to ensure that the resource is used in optimal and sustainable ways (that is, integrated water resources management). To ensure resource optimization and sustainability, governments must address key aspects of integrated water sector management and cross-cutting issues among various sectors. These aspects and issues involve policies, regulations, planning activities, sector investments, financing methods, service provision, and institutional factors.

The input-output model for the water sector is shown in figure 3.40, which specifies input parameters, desired outputs, relevant interventions, and undesired outputs that must be minimized. All interventions in a city should lead to desired objectives, which include (1) accessibility for all residents, including the poor; (2) adequate service quality; (3) high operational efficiency; (4) service reliability; (5) supply security and sustainability; (6) environmental preservation; and (7) service affordability. These objectives are interlinked, and trade-offs must be recognized. Interventions may be related to planning, water resource protection and enhancement, infrastructure, service delivery, and management. These interventions are subject to relatively unchangeable input constraints (independent inputs) such as the characteristics of water resources, hydrology and hydrogeology, climate and atmospheric conditions, demographic and economic conditions, and social norms and historical rights. Parameters that are manageable include policy, legislation, regulations, institutions, physical systems technology, spatial planning, stakeholders, and economic and financial aspects. Undesirable impacts should be mitigated or eliminated. These

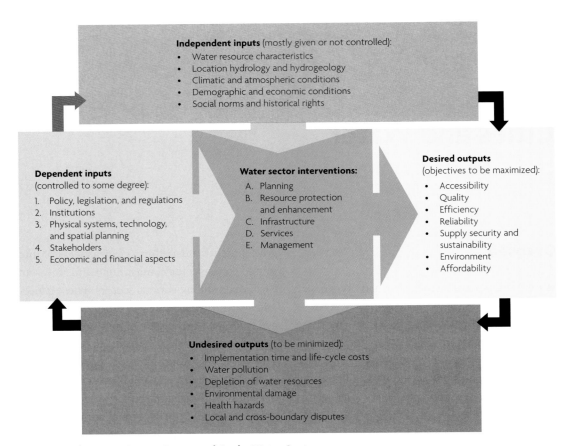

Independent inputs (mostly given or not controlled):
- Water resource characteristics
- Location hydrology and hydrogeology
- Climatic and atmospheric conditions
- Demographic and economic conditions
- Social norms and historical rights

Dependent inputs
(controlled to some degree):
1. Policy, legislation, and regulations
2. Institutions
3. Physical systems, technology, and spatial planning
4. Stakeholders
5. Economic and financial aspects

Water sector interventions:
A. Planning
B. Resource protection and enhancement
C. Infrastructure
D. Services
E. Management

Desired outputs
(objectives to be maximized):
- Accessibility
- Quality
- Efficiency
- Reliability
- Supply security and sustainability
- Environment
- Affordability

Undesired outputs (to be minimized):
- Implementation time and life-cycle costs
- Water pollution
- Depletion of water resources
- Environmental damage
- Health hazards
- Local and cross-boundary disputes

Figure 3.40 The Input-Output Framework in the Water Sector

Source: Author compilation (Khairy Al-Jamal).

potential impacts include, but are not limited to implementation time and life-cycle costs, water pollution, depletion of water resources, environmental damage, health hazards, and local and cross-boundary disputes.

This sector note sheds lights on strategies for water sector management. It is designed to assist urban decision makers in putting together an optimal and well-coordinated set of programs. A key challenge is the high degree of nonlinear interconnections among urban sectors, including water, energy, solid waste, telecommunications, and transportation, which share many economic, environmental, and political constraints. Nonetheless, optimization of the net benefits amid these sectors and links is the defined objective.

Water Sector Policy, Legislation, and Regulations

The policy, legislative, and regulatory framework defines the rules for managing the water sector at the national and local levels. The framework may expand beyond national boundaries and address cross-country issues if water resources are shared or water management and protection require international cooperation. The framework strives to satisfy myriad aims, including ensuring the adequate protection of water resources, developing and promoting sustainable water services, ensuring equitable distribution and access, improving health and environmental conditions, enabling economic growth, and promoting efficiency and optimization in the use of water resources to enhance the viability of the sector.

As suggested in table 3.15, water sector management systems are linked, and systems and subsystems need to be connected. The three main interconnected systems are water resources, water services, and governance systems. Each of these main systems is composed of downstream systems and subsystems. Though the management and implementation of these systems and subsystems may intersect within the water sector, some systems and subsystems may also intersect with other sectors, such as land, construction, and infrastructure operations, particularly in decision making on the allocation of water resources.

The high degree of connectivity and interaction in these systems and subsystems calls for well-developed policies, legislation, and regulations at the national and local levels.

At the national level, the framework influences the minimum service standards for the quality of drinking water and wastewater treatment and disposal in accordance with quality regulations and the technical standards of water systems. Standards in the allocation of water resources are normally based on environmental regulations, and prices are usually set on the basis of economic regulations, water laws, and the modalities of service providers. At the local level, the framework addresses equity and access to services, community participation in decision making, distribution and collection system layout, physical planning, the spatial configuration of water works, and acoustic and odor control and management (table 3.16).

Though policies and legislation are normally formulated at the central level, regulations may be imposed centrally or locally. In some cases, regulations may be imposed through contracts.

Table 3.15 Water Sector Management Systems

WATER RESOURCES SYSTEMS	WATER SERVICES SYSTEMS	GOVERNANCE SYSTEMS
Water resource planning system Financial planning subsystem Organization planning subsystem Physical planning subsystem	**Water service planning system** Financial planning subsystem Organization planning subsystem Physical planning subsystem	**Policy-making system** **Regulatory system** Environmental regulation subsystem Economic regulation subsystem Drinking water quality regulation subsystem
Water resurce operations system Construction management subsystem Operations and maintenance subsystem	**Water service operations system** Construction management subsystem Water operations and maintenance subsystem Wastewater operations and maintenance subsystem Systems hardware and software management subsystem	**Accountability system**
Water resource management system Abstraction licensing subsystem Allocation subsystem Supply and demand management subsystem	**Water services management system** Services quality compliance subsystem Effectiveness and efficiency subsystem Pricing subsystem	
	Commercial system Customer services subsystem Billing subsystem Collection subsystem	
	Human resources management system	
	Management information system	

Source: Author compilation (Khairy Al-Jamal).

Table 3.16 The Policy, Legislative, and Regulatory Framework Affecting the Water Sector

LEVEL	POLICIES, LEGISLATION, AND REGULATIONS FRAMEWORK
National	• Water laws
	• Allocation of water resources to the domestic sector and the share with provinces and cities
	• Drinking water quality standards
	• Wastewater treatment and disposal standards
	• Water works and systems standards
	• Tariff structure and pricing policy
Local	• Physical planning and spatial distribution
	• Metering and usage charges
	• Billing and collection
	• Equity and access
	• Affordability
	• Efficiency of operations
	• Local environmental impacts such as noise, appearance, and odor
	• Participation and community empowerment

Source: Author compilation (Khairy Al-Jamal).

The Institutional Context

A strong and appropriate institutional setup ensures smooth and successful compliance with sector policies, legislation, and regulations. Institutions ideally execute interventions to optimize gains despite sectoral constraints and boundary conditions. Institutions should prioritize the achievement of sectoral targets, but also interface with other sectors to ensure optimal development on a larger scale. The key institutional entities include policy makers, regulators, service providers, and customers (figure 3.41). Regardless of the institutional arrangements, the integrity of water sector systems must be preserved (see table 3.15). An institutional setup should show integrity, comprehensiveness, a sound division of roles and responsibilities, and representation in other developmental forums, as follows:

- *Policy maker:* The policy-making function resides mainly at the central level, and national policies are normally imposed on cities.

- *Regulator:* The regulatory system is responsible for enforcing rules to guarantee compliance with service standards and other sector policies to ensure sustainability. This requires providing adequate services at affordable prices. The three subsystems under the regulatory function are the environmental regulation subsystem (issuing licenses for abstraction and disposal), the quality regulation subsystem (ensuring compliance with standards for drinking water, wastewater treatment, and the quality of works), and the economic regulation subsystem (reviewing prices to ensure tariffs are proportionate to real costs, promoting efficiency and conservation, and enabling sustainability and affordability among the poor).

In some cases, municipalities may propose and implement regulations. However, it is important to ensure separation between regulators and service providers to avoid conflicts of interest. The regulator ensures that customers receive services up to agreed standards to mitigate the risk that service providers underperform. Both bodies should not be under the jurisdiction of the same entity. It is equally important that the policy-making body and regulator be separate institutions.

In the United Kingdom, regulations are extensive. Environmental regulations are executed by the Environment Agency, which is separate from the independent economic and quality regulator (the Water Services Regulation Authority). This separation enhances the internal transparency and accountability of the regulatory system because decisions are made in the open. However, regulatory agencies should coordinate closely to manage trade-offs. The prime interest of the environmental regulator is to minimize the abstraction from water resources and to enforce stringent standards on disposed sewage. However, the main interest of the economic regulator is to ensure that levied tariffs cover costs, which often means supporting relaxed disposal standards and maximizing the use of natural water resources before considering more expensive and nonconventional options, such as seawater desalination.

- *Service provider*: Service providers are responsible for providing water services in the city, including water treatment and distribution and associated customer relations. Water sources may be located within the city and managed by the service provider or outside the city and managed by a different water provider. The same service provider should provide storm water collection, flood management, and wastewater collection and treatment services. Consolidating these services will help improve the control over all services and promote accountability and more efficient operations. For instance, a water supplier normally encourages customers to reduce sewage if it also handles the sewage. Suppliers also promote the protection of water resources if they bear the cost of water treatment. Service providers may be private companies (France, Germany, the United Kingdom, and so on), public utilities (Australia, Germany, and South Africa), or municipalities (France, Egypt, Germany, and Jordan). In some cases, multifunction utilities offer water, electricity, and other services. The decision to combine utility services should depend on the scale of related industries and potential cost savings. Moreover, public utilities may outsource some or all of their operations to increase efficiency.

- *Civil society*: Civil society should be institutionalized by establishing user associations and appropriate participation channels. This helps ensure public participation in industry development and related decision making. Policy makers and regulators often consult user associations to assess and ensure the adequacy of policies, legislation, regulations, and service levels. Service providers should recognize users as genuine customers who drive industry revenue and sustainability.

Physical Systems Technology and Spatial Arrangement

Water systems are composed of four main systems: water supply, wastewater, storm water, and reclaimed water. The storm water and reclaimed water systems are similar in configuration and operation to the wastewater and water supply systems. Figure 3.42 illustrates a typical water system layout. The system includes water and wastewater treatment facilities, distribution

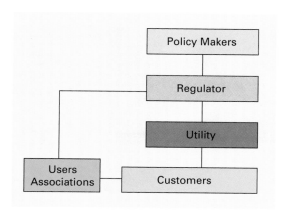

Figure 3.41 The Institutional Setup in the Water Sector

Source: Author compilation (Khairy Al-Jamal).

and collection networks, control valves, pumping stations, storage tanks, and disinfection facilities. The system serves distributed demand nodes.

The following sections highlight technological factors in designing water systems and the merits of the proper spatial distribution of demand nodes, which are governed by land use planning.

Water supply systems

The spatial distribution of demand centers: Cities should strive to limit urban sprawl to ensure that the demand for water is not overly dispersed and that the extent of distribution and collection systems is minimized. (The costs related to distribution networks typically account for 70 percent of the overall costs of a water supply system.) Denser land developments help minimize the costs of capital and operations. Energy consumption also typically declines because consumption is linearly proportionate to pipe length. In addition, dense developments promote environmental protection.

Distribution system spatial configuration: It is important to achieve highly reliable water supply services. This often entails some redundancy in the network, such as loop networks that supply a demand zone through more than one main pipe. This may involve trading higher capital costs for improved reliability and reduced energy costs. Proper system reconfiguration leads to reduced total distribution costs (box 3.9). To avoid substantial increases in network costs, loop networking may be operated at lower pressures and through smaller pipes made of material that is less expensive than ductile iron.

Water treatment plant: The water treatment process may be simple. Biological treatment is preferred over chemical-intensive processes. Treatment plants must be close to water resources and, it is hoped, close to urban demand centers. To ensure the security of water supplies, cities must build the treatment plant with space to expand to meet growing demand and should consider building more than one plant associated with different sources (if possible).

Groundwater and water wells: If the groundwater is viable, well fields should be developed and distributed near demand centers. This proximity leads to a simpler network and lower energy and capital costs. Distribution systems may often operate from minimal storage tanks because the aquifer represents a robust and feasible storage source. Moreover, many cities commonly use aquifers as natural storage receptacles for the surface water that infiltrates from basins along river embankments and other areas (for example, Paris). Infiltration is a natural treatment process that helps purify water at minimal cost.

Water pumps: Water pumps are the main energy consumers in water systems. The energy consumed is proportional to the set efficiency of the pump motor. Normally, pumps are most efficient at their designed operating points. However, owing to load changes, pumps are often operated outside peak efficiency, and significant energy is wasted. Variable speed pumps may be modified to address this waste. Pump speed is adjusted to maximize efficiency given any particular load. Pumps may also be noisy, but this may be mitigated by introducing acoustic insulators.

Storage Tanks: Depending on the mode of operation, storage tanks may be important com-

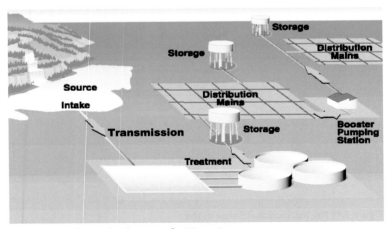

Figure 3.42 Schematic Diagram of a Water System

Source: Walski, Chase, and Savic (2001).

BOX 3.9

The Effect of Distribution System Configuration on Energy Consumption

The schematic represents a small town with hourly water demand of 450 cubic meters. The demand is split among nodes 2, 3, 4, and 5. The town is served through a fixed speed pump from a reservoir at an elevation of 10 meters. For illustrative purposes, two scenarios have been considered. In scenario 1, link 1–5 does not exist. In scenario 2, link 1–5 has been constructed.

DISTRIBUTION SYSTEM DATA						ANALYSIS RESULTS		
Node	Demand (m³/hr)	Elevation (m)	Pipe	Diameter (mm)	Length (m)	Node	Pressure[a] (m)	Pressure[b] (m)
1	0	10	1–2	500	1,000	1	58.33	43.75
2	100	30	2–3	400	1,000	2	37.02	23.39
3	50	30	3–4	400	1,000	3	34.58	23.03
4	100	30	4–5	300	1,000	4	27.13	22.47
5	200	25	5–1	400	1,000	5	23.58	27.66

Note: m = meter; m³/hr = cubic meters per hour; mm = millimeter.
a. Node pressure, scenario 1.
b. Node pressure, scenario 2.

Results:

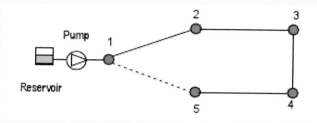

1. In scenario 2, the pump has been replaced by a smaller one, and the power demand has dropped to 71.5 kilowatts from 95.3 kilowatts (a 25 percent drop).
2. The annual energy savings is 209 megawatt hours, which may be equivalent to a savings of US$20,000 per year.
3. The capital investment to complete the loop and construct link 5–1 is less than US$100,000 and may be paid back in about five years using the gain from energy savings.
4. Additional improvement may be achieved if a full optimization analysis is conducted to target the sizes of the pipes (diameters), while maintaining other hydraulic parameters such as flow velocities and node pressures within the recommended hydraulic design parameters.

Note: The hydraulic analysis has been conducted using EP-Net (enhanced prioritized Petri net) modeling software.

ponents of a water supply system. Storage capacity should equal one day of service. This capacity enhances water security in case of treatment plant stoppage, leads to smaller treatment and pumping facilities, reduces capital and operations costs, and enables optimal pump scheduling and intense treatment during periods of low electricity tariffs (usually nights and off-peak hours). Planners must balance potential savings with the capital costs of constructing storage tanks.

Storm water management
Storm water and rainwater harvesting and flood management: Storm water collection systems may be combined with sewage collection systems. Though this may reduce capital investment, treatment becomes more difficult if wastewater and storm water are reused. For this reason, these systems should normally be separate, complemented by effective and innovative water storage tools. For example, rainwater may be harvested

from the roofs of buildings and used for gardening, flushing toilets, and washing cars.

In general, climate change analysis predicts that East Asia will receive heavier and more intense rains. A significant negative effect is that the sea is expected to rise by about 0.5 meters by 2100 (figures 3.43 and 3.44). Flood management systems must be designed to cope with these expected loads, and systems must be strategically located at nonvulnerable elevations. All new coastal cities in the region should also consider plans to locate infrastructure above the anticipated sea level.

It is often advantageous to bundle the construction of road, water, sewerage, and storm infrastructure. A common underground infrastructure and service corridor is a typical practice in many cities. This may reduce overall costs and ease maintenance.

Nonconventional water resources

Wastewater reuse: Treated wastewater is a potential resource. It may be used to irrigate public parks and landscaped plots and is rich in nutrients for plant life. The strategic use of treated wastewater may cut crop production costs and relieve stresses on the freshwater resources needed to meet domestic, industrial, and environmental demand. Wastewater reuse is practiced, notably, in China, Japan, and Singapore. In situ treatment and the recycling of industrial wastewater may also partially contribute to satisfying industrial water demand.

Water desalination: Most East Asian and Southeast Asian countries have extensive coastlines, and desalination represents a promising means of enhancing water resources. Seawater desalination has undergone significant improvements over the last two decades. Owing to technological advances in membranes and energy recovery devices, the cost of seawater desalination in reverse osmosis plants dropped from around US$3.00 to less than US$0.60 per cubic meter. In addition, energy consumption fell from 8.0 kilowatt-hours per cubic meter to less than 3.0 kilowatt-hours per cubic meter. Thermal desalination processes may be competitive if they are well designed and if they are integrated into power generation plants. Cogeneration plants generally improve the production of both electricity and freshwater.

Reverse osmosis desalination plants are more flexible and do not need to be constructed together with power plants. Because of their significant energy needs, reverse osmosis desalination plants should be equipped with energy recovery devices such as the newly developed pressure and work exchangers that recover nearly all the energy from brine before the brine is discarded in the sea. Alternatively, brine may be processed to produce raw material for the sea minerals industry. In such cases, the plant is

Figure 3.43 Area at Risk If There Were a 0.5-Meter Rise in Sea Level in Asia

Source: World Bank (2007).

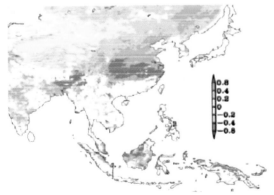

Figure 3.44 Changes in the Annual Mean Daily Precipitation Expected by 2100

Source: World Bank (2007).
Note: The values are expressed in millimeters.

known as a zero disposal plant. The disposal of brine in the sea must be carefully executed to avoid environmental damage. Brine must be dispersed in active water to avoid increased salt concentration or rises in seawater temperature that may harm marine flora and fauna. Special nozzles are available and may be used to limit environmental damage.

Water supply and demand management

Nonrevenue water and leak reductions: Nonrevenue water is the collective commercial and physical losses from a system. Good practice suggests that the physical losses should represent less than 4.4 percent, as in Singapore in 2007 (Lau 2008). Physical losses entail the loss of water as an economic resource, the loss of revenue needed to maintain sector sustainability, and the energy costs involved in producing and transporting the water that is lost. Wasted energy unnecessarily increases greenhouse gas emissions, which are responsible for the global warming and climate change that are already affecting water resources, flood management, and developmental investments.

Leakage negatively affects the environment, the economy, and ecology. Water leaks represent losses of resources and energy and may damage infrastructure installations and ecological facilities. Leaks may be minimized or controlled through pressure control devices (that is, zoning or reconfiguring of online control and distribution systems), district metering, and instruments to detect leaks. Leaks are linearly proportional to pipe water pressure. Thus, pressure should be kept at a level minimally adequate to deliver services. For example, leaks drop by 50 percent if pipe pressure falls from 4 bars to 2 bars. Appropriate pressure modulation may be achieved by introducing variable speed pumps, proper pump scheduling, and pressure control valves. The construction of elevated reservoirs and appropriate configurations of distribution systems also help reduce the risk of leakage. Detection equipment is available and is used by water utilities around the world.

Metering: Metering is crucial to water supply management. Metering enhances equity because consumers pay for only the water they receive. Metering also promotes the management of water demand. Unlike fixed fees, meter-based charges provide incentives to save and conserve water. However, water meters must be frequently checked and calibrated and may need to be replaced each decade. Meters are manufactured in classes (A, B, C, and D) based on levels of accuracy at specified flows. The higher-class meters (that is, C and D) are more accurate in measuring low flows at wider ranges, but class D meters are more expensive. Remote reading technologies are often used to save costs in meter reading.

Conservation and use efficiency interventions: New technological devices have been developed and are available to promote water conservation. Showers, toilets, and laundry facilities represent the most significant sources of water consumption in households. Box 3.10 illustrates experiments in Canada in which water consumption was reduced by 52 percent by introducing modern fixtures, including kitchen faucet aerators, aerated shower heads, dishwashers, and front-loading clothes washers.

In some areas, heating and hot water infrastructure is centralized. This may result in high energy and water losses to flush out cold water in house connections. In most cases, instant heaters with shorter service pipes from gas furnaces or electric heaters are more feasible and energy efficient, and water losses from such devices are negligible.

The practice of washing cars using hoses connected to house taps should be abandoned. Bucket-and-towel washing or specialized commercial car wash terminals should be substituted because these processes consume less water. Special car wash terminals may also recycle water. In addition, street cleaning should take place at night because roads are coolest then, thereby reducing evaporation.

In general, agriculture is the sector that consumes the most water. In urban areas, agri-

BOX 3.10

Canada: Conservation and Domestic Water Consumption

The average domestic water consumption by households in Canada is approximately 350–400 liters per person per day (300 liters per day for indoor use and 100 liters per day for outdoor use). Canadians use considerably more water than do most other nationalities. Consequently, summer water flows from mountain glaciers and annual snowpacks are diminishing. Sustaining access to water is therefore an important conservation goal in Canada.

Domestic water is used in bathrooms (toilets, showers, and faucets), kitchens (dishwashing and food preparation), and laundry facilities. Based on typical values, the charts illustrate the percentages of water used for these activities through conventional and low-flow fixtures.

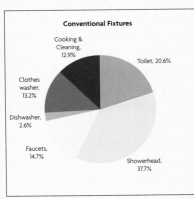

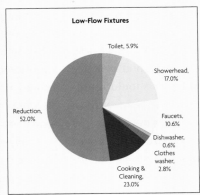

These values are based on a family of four people, as shown in the table:

FIXTURE	FAMILY USE	WATER CONSUMPTION	
		CONVENTIONAL FIXTURE	LOW-FLOW FIXTURE
Showerhead	8 minutes/person/day	15 liters/minute	7 liters/minute
Toilets	5 flushes/person/day	13 liters per flush	6 liters to flush solids
			3 liters to flush liquids
Faucets	5 minutes/person/day	10 liters/minute	7 liters/minute
Kitchen (cooking and cleaning)	15 minutes/day	10 liters/minute	7 liters/minute
Conventional dishwasher	1 use/day	33 liters per use	8 liters per use
Clothes washer (top loader)	7 uses/week	170 liters per use	36 liters per use

The toilet is the most important source of water consumption in the home. Over 70 percent of water use occurs in the bathroom. Toilets and showerheads thus represent the best opportunities for water reduction in the home.

Low-flow fixtures include dual-flush toilets. These toilets have two flush buttons: one provides a 3-liter flush for urine; the other provides a 6-liter flush for solids. Showerheads and faucets may be designed to reduce flows without a noticeable reduction in performance. New appliances such as dishwashers and front-loading clothes washers have significantly lower water needs. Low-flow fixtures may reduce water consumption in the home by over 50 percent, from approximately 1,200 liters per day (interior use) to 600 liters per day. This does not include outdoor use for landscaping.

Source: Adapted from The Living Home (2008).

cultural activities are still common in parks, along streets, and in public and residential gardens. Efficient irrigation in these areas helps improve water use and sector sustainability. Irrigation should be properly timed to avoid peak daily temperatures and reduce evaporation and evapotranspiration losses. Efficient irrigation systems, such as drip irrigation, subsurface irrigation, and sprinkler irrigation should be used.

Wastewater treatment and sludge disposal
Location of the wastewater treatment plant: Wastewater treatment plants are not popular facilities. While wastewater is generated within

residential centers, residents generally pressure governments to locate wastewater treatment plants far from their neighborhoods. In general, plants need to be located as close as possible to sewage generation sources and downwind of communities; they need to discard wastewater downstream of neighborhoods and water works facilities if there is an intermediary river. The plant also needs to be centrally located to minimize the energy consumption of sewage transfer and effluent disposal or reuse. There are often trade-offs among these conflicting interests.

Process of wastewater treatment: There is a wide range of sewage treatment processes. As much as possible, priority should be given to biological treatment processes to avoid the use of hazardous chemicals. It is also important to treat domestic and industrial sewage separately. Moreover, legislation is needed to prohibit non-biodegradable domestic detergents and the disposal of hazardous wastes, such as heavy metals, pesticides, hydrocarbons, and medical wastes, into the city sewerage system. Awareness campaigns and public participation are essential to these efforts. Activated sludge treatment plants are common around the world and lauded for their efficiency and relatively compact size. However, the treatment processes are energy intensive. Treatment plants with extended lagoons consume much less energy and are cheaper to construct, though they need more land.

Sludge management: In addition to treated effluent, wastewater treatment plants produce sludge, which is composed of biomass and settled biological material. If the biological content has been appropriately digested, sludge may be a valuable resource for composting, fertilizing, or generating methane. Generated methane may be captured and used as an energy resource. Commonly, treatment plants are equipped with gas turbines and generators that use methane to produce electricity. The generated electricity may be sufficient to cover most of the electricity demand for treatment, or it may be sold to the distribution grid. Special legislation may encourage plant operators to sell electricity or

subsidize production costs through carbon finance funds. These funds promote technologies that reduce greenhouse gas emissions such as carbon dioxide, which is linked to global warming and climate change.

In traditional plants, sludge is often discarded in the sea or dumped into solid waste landfills. These practices are falling out of favor because they risk harming the marine environment and polluting groundwater.

Energy efficiency

Energy is often the dominant factor that determines the cost of water and wastewater services. The energy needs of these services may vary from less than 1 kilowatt-hour to many kilowatt-hours per cubic meter of treated water. The amount of required energy depends on the following factors:

• Distance and elevation of water sources relative to service areas
• Topography of service areas
• Depth of groundwater aquifers (if applicable)
• Location of wastewater treatment and disposal facilities
• Energy consumption in water production and wastewater treatment facilities
• Energy recovery ratios at wastewater treatment plants via sludge digestion
• Energy recovery ratios at desalination plants
• Levels of technical and commercial water losses
• Configuration and design of the water distribution and wastewater collection systems
• Modes of operation of the water distribution system

There is a strong and direct relationship between water use and energy savings. This link has led to the expression Watergy. Box 3.11 summarizes the scope of Watergy within the water sector, such as in demand management and supply management and the synergy between the two in terms of system design and operation.

Combined Water and Energy Activities in Water Supply Management

	Supply-side efficiency measures	Demand-side efficiency measures *Consumers Residential/Industrial*	Comprehensive demand / supply side approach synergies

Watergy efficiency is cost-effective delivery of water services, while minimizing water and energy use.

Water supply systems offer multiple opportunities to reduce water and energy waste directly, while serving customer needs more effectively.

Reducing demand by helping the consumer use water more efficiently decreases the required water supply, saving both energy and water.

Looking at a water system comprehensively and ensuring thay efficiency projects are designed in tandem create greater efficiency opportunities.

- Leak and loss reduction
- Operations and maintenance
- Pumping systems
- Primary and secondary wastewater treatment

- Water-efficient household appliances
- Low-flow toilets
- Low-flow showerheads
- Industrial water reuse
- Leak and water waste reduction

- Right-sizing pump systems after reducing consumer demand
- Avoiding wastewater treatment by promoting and reducing demand

Source: Alliance to Save Energy (2002).

Box 3.12 illustrates a case study in Brazil. The study reveals that automating the water supply system and providing online control saved 22 gigawatt hours per year, equivalent to US$2.5 million. The control system cost only US$1.1 million.

Stakeholders

The stakeholder dynamic illustrated in figure 3.45 is important to the water sector. This dynamic includes interfaces between customers, service providers, municipalities, regulators, and policy makers. Transparency, accountability, and public participation are needed because the water industry is often highly monopolistic. These factors allow strategic decisions to be made using top-down and bottom-up approach-

es. For instance, tariffs ensure sector sustainability by furnishing revenues to finance sector management expenses. Tariffs are usually set at the municipality level, which represents the local government, though, in some models, tariffs are set by the central government. Tariffs are eventually endorsed by policy makers at the central level. Reviews of prices are carried out when tariffs expire. Thus, there is a constant need to analyze the real costs of service delivery thoroughly, including the rent value of any scarcity in water resources. The regulator is responsible for conducting price reviews. Calculations of the real costs are expected to cover the shadow price of water, the cost of carrying out treatment according to specified standards, and the cost of distribution and delivery. The shadow price of water is governed by the demand of all users, and the policy affect-

The Watergy Case Study in Fortaleza, Brazil

Challenges

The importance of the project in Fortaleza was highlighted during the energy crisis in Brazil in 2000 and 2001. Of the energy generated in the country, 70 percent comes from hydropower. Droughts and energy shortages are thus inextricably linked. During droughts in 2000 and 2001, all consumers were required to reduce energy consumption by 20 percent.

Since 2001, the Alliance to Save Energy has worked with the local utility in northeast Brazil, Companhia de Água e Esgoto do Ceará (CAGECE), to develop and implement measures to promote the efficient use of water and energy. This partnership has been aimed at improving the distribution of water and the access to sanitation services, while reducing operating costs and environmental impacts. The partnership has reduced CAGECE's energy use and served as an example of good practice for other national projects, which is important because the water and sanitation sector accounts for 2.3 percent of the nation's energy consumption.

Background

Designs of water distribution systems are based on population projections derived from statistical and historical data over a 20- or 30-year planning horizon. Because of this method, many systems are overdesigned, particularly in the size of storage, treatment, and distribution facilities. The overdesigning leads to greater energy consumption than needed to provide for adequate demand, especially in booster stations. Design criteria affect not only pumping stations, but also the size of pipes, the capacity of reservoirs, and the construction of treatment facilities and booster stations. Moreover, water systems need to be able to expand to satisfy growths in demand, but without sacrificing the efficient use of energy.

Objectives

The focus of the partnership between the alliance and CAGECE has been the development of a methodology that would provide CAGECE with the tools and expertise to produce initiatives that result in savings and the rational use of energy and distributed water. As the work progressed, it became clear that the model would be useful to other water and sanitation companies exploring ways to increase efficiency.

Approach

An automated water distribution system allows operators to obtain strategic data in real time. The automation of the system in the Fortaleza Metropolitan Region allows for the correction of deficiencies, particularly those linked to overdesign. Along with CAGECE's efforts, alliance actions in 2002 included the following:

- Establishment of a baseline of energy consumed and water distributed for CAGECE
- Implementation of efficiency measures that led to a reduction in operational energy consumption
- Development of a financing proposal, in association with the government's Fight against Electricity Waste Program, to undertake energy efficiency projects with CAGECE operational staff; the technical support provided by the alliance resulted in the development of energy efficiency projects, cost-benefit analyses, and the specifications of the equipment that could be financed
- Arrangement of R$5 million in financing for CAGECE for energy efficiency projects; the projects included the automation of operations, the rewinding and replacement of motors, the maximization of the efficiency of pump systems, and an increase in storage capacity to allow pumps to be shut down during peak hours
- Creation of an operations procedures manual to serve as a reference for daily operations for crews and CAGECE management

Key results

- 88 gigawatt hours of energy saved over four years
- 88,000 households newly connected to water, though water consumption remained constant
- $2.5 million saved per year on an initial investment of $1.1 million
- Standardization of operating procedures and increased reliability of operations data
- Ability to act in real time using system control devices
- 220,000 tons of carbon dioxide emissions avoided per year

Source: Barry (2007).

---

ing resource allocations thus also affects various end users. The cost of treatment is influenced by the quality of the raw water and relevant national standards.

Customers should participate in setting standards. This requires a process of consultation and participation to ensure that the economy can afford to maintain the quality standards and that service providers can achieve the standards. Often, costly capital investment programs are needed, and private sector participation may be required. All costs should be checked against international good practice. The economic regulator needs to set thresholds for operational

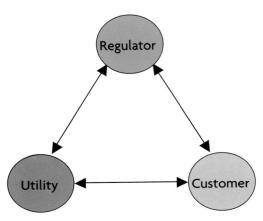

Figure 3.45 The Stakeholder Dynamics and Accountability Triangle

Source: Author compilation (Khairy Al-Jamal).

efficiencies and incorporate these into price setting and the review of prices.

The tariff structure should include incentives to improve services and enhance efficiency, conservation, equity, and social and environmental protection. Progressive block tariffs are appropriate demand management tools to achieve these targets. Poor households with low consumption rates are within the first blocks, which are usually underpriced and subsidized by other consumers who are wealthy and can afford services. The regulators must constantly interact with customers and service providers to ensure that services comply with standards.

Economic and Financial Aspects

In general, the cost of water supply ranges from US$0.20 to US$1.00 per cubic meter, while the cost of wastewater collection and treatment ranges from US$0.50 to US$1.00 per cubic meter. Normally, wastewater services cost twice as much as water supply services. In some cases, the cost of water may reach US$10 per cubic meter if it is sold by vendors. Because of these relatively high rates, leakage or customer abuse leads to real economic costs. Reducing losses from 50 percent to 15 percent and cutting demand in half through water con-

servation practices can produce net water savings for a city of 400,000 inhabitants of around 71 percent (figure 3.46). This is equivalent to about 61.8×10^6 cubic meters of water per year. A new city needs 25.8×10^6 cubic meters of water per year; this means that another city of about 1 million inhabitants may be served at the same cost.

Similarly, if one assumes that an urban water supply and sewerage system consumes energy equal to 2.0 kilowatt-hours per cubic meter of supplied water, annual urban energy consumption will decline from 175 gigawatt hours to 52 gigawatt hours and follow the same trend depicted in figure 3.46. Annual energy savings would total 123 gigawatt hours, which is sufficient to supply a city of more than 120,000 inhabitants with 1,000 kilowatt-hours per capita per year. This electricity savings would also cut carbon dioxide emissions by 307,000 tons per year. These benefits may increase more than threefold if water and wastewater services are more energy intensive. This would be the case if desalination plants are being used or if the demand centers being served are at higher elevations than the water treatment facilities.

As in other infrastructure sectors, attaining economies of scale is important. For instance, the cost of producing freshwater at desalination

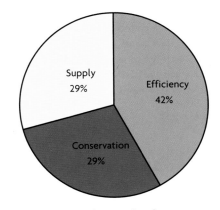

Figure 3.46 Savings in the Supply of Water

Source: Author compilation (Khairy Al-Jamal).
Note: The figure shows the share of savings rising because of the reduction in nonrevenue water from 50 percent to 15 percent and the efficiency gain of 50 percent resulting from conservation efforts.

plants ranges from US$3.00 at smaller capacity plants (1,000 cubic meters per day) to a much lower US$0.60 at larger facilities (30,000 cubic meters per day). The same benefits of scale apply to other water facilities and wastewater treatment plants. Given clear cost savings, cities might consider building joint plants that serve more than one city area.

It is also recommended that construction works be synchronized among sectors and that joint facilities, such as common underground corridors, be used for water, wastewater, storm water, electricity, and telecommunications infrastructure.

Conclusions

To a large degree, all sectors regulate or modify natural resources, serve mostly the same customers, require sources of funding, deliver services managed by the public and private sectors, and face the challenge of higher costs because of urban sprawl and the associated widely distributed infrastructure. A high level of synergy is therefore possible among sectors to boost overall efficiency. Urban policy makers and planners might consider the following options, depending on the situation:

- Establish a common regulator for more than one service, such as water, electricity, and telecommunications.
- Develop similar policy principles for tariff structures, standards, and resource allocation among sectors.
- Establish feasible multifunction utilities.
- Enable investments by supporting incentives, policies, and the enforcement of laws and legal systems to protect investments and improve revenue collection (that is, willingness to pay).
- Increase the efficiency of construction management by coordinating construction activities, developing unified procurement

systems and land use plans, and sharing access roads.
- Develop cross-compatible customer charters delineating the principles of service and share customer complaint centers to minimize costs.
- Consider cross-sector economic and sustainability gains related to cogeneration (electricity and seawater desalination, irrigation, and hydropower generation), performance incentives, conservation, and efficiency.
- Enable a balanced environment for public-private partnerships: all linked sectors need to be feasible to encourage the private sector to invest in the water sector, and those sectors need sufficient tariffs and incentives for infrastructure services.
- Encourage proper land use and the distribution of infrastructure facilities.

References

Alliance to Save Energy. 2002. "Executive Summary: Watergy, Taking Advantage of Untapped Energy and Water Efficiency Opportunities in Municipal Water Systems." Alliance to Save Energy, Washington, DC. http://ase.org/uploaded_files/watergy/watergysummary.pdf.

Barry, Judith A. 2007. "Watergy: Energy and Water Efficiency in Municipal Water Supply and Wastewater Treatment; Cost-Effective Savings of Water and Energy." Handbook, Alliance to Save Energy, Washington, DC. http://www.watergy.net/resources/publications/watergy.pdf.

Lau Yew Hoong. 2008. "Sustainable Water Resource Management in Singapore." Presentation at the United Nations Economic and Social Commission for Asia and the Pacific's "1st Regional Workshop on the Development of Eco-Efficient Water Infrastructure in Asia Pacific," Seoul, November 10–12. http://www.unescap.org/esd/water/projects/eewi/workshop/1st/documents/presentation/Session%204%20National%20Experiences/21.%20Singapore-%20report.pdf.

The Living Home. 2008. "Domestic Water Use," June 26. http://www.thelivinghome.ca/index.php?option=com_content&task=view&id=98&Itemid=132.

Walski, Thomas M., Donald V. Chase, and Dragan A. Savic. 2001. *Water Distribution Modeling,* 1st ed. Waterbury, CT: Haestad Press.

World Bank. 2007. "East Asia Environment Monitor 2007: Adapting to Climate Change." Report 40772, Environmental Monitor Series, World Bank, Washington, DC.

Cities and Transport

Overview

The reasons that cities should care about the transportation sector are myriad and complex. Transportation produces important benefits (principally mobility and accessibility) and enables the economic and social activity that sustains urban life. At the same time, transportation consumes significant shares of land, energy, time, and other resources and generates specific undesired outputs such as pollution and accidents. This chapter addresses urban transportation issues, particularly those in rapidly growing cities that face challenges and various investment options.

Transportation is not an end, but rather a means to one or multiple ends such as access to jobs, markets, and other social or economic opportunities. For this reason, it is difficult to create a single objective recognizing all relevant constraints and trade-offs. An urban transportation system requires many inputs (some more controllable than others) and produces numerous outputs (some desirable and others undesirable) that influence the inputs. These are described and illustrated in figure 3.47.

Independent inputs, which are mostly given or not controlled, include demographic and economic conditions (that is, population, income, and types of industries), geographic constraints (for example, rivers, lakes, coastlines, and mountains), climatic and atmospheric conditions, and social norms and historical practices.

Among other effects, these inputs influence land development patterns (that is, the distribution of land uses and densities), the shape of the city (that is, linear, circular, and semicircular), and spatial and temporal travel patterns (for example, radial, circumferential, and polycentric patterns and trip trends at different daily hours, weekdays, and seasons).

The dependent inputs over which decision makers retain some degree of control (and the focus of this chapter) include (1) policy, legislation, and regulations; (2) institutions; (3) physical systems, technology, and spatial planning; (4) stakeholder dynamics; and (5) economic and financial factors. All these inputs should be considered in analyzing problems and potential interventions. Urban transport interventions, including those aimed at passengers or freight, involve one or more of the following:

- *Land use and travel demand:* interventions that influence travel behavior, including origins and destinations, purposes, modes, frequencies, and trip distances

- *Infrastructure and services:* interventions that enhance the supply or capacity of infrastructure and services such as roads, public transportation, traffic management, and other investments

- *Vehicle fleet and fuel supply:* interventions that alter the number, composition, technologies, or use of vehicles and fuels

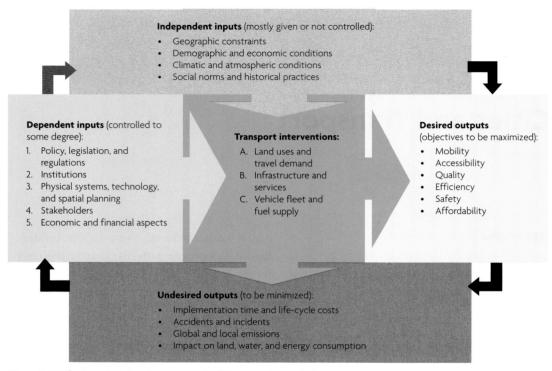

Independent inputs (mostly given or not controlled):
- Geographic constraints
- Demographic and economic conditions
- Climatic and atmospheric conditions
- Social norms and historical practices

Dependent inputs (controlled to some degree):
1. Policy, legislation, and regulations
2. Institutions
3. Physical systems, technology, and spatial planning
4. Stakeholders
5. Economic and financial aspects

Transport interventions:
A. Land uses and travel demand
B. Infrastructure and services
C. Vehicle fleet and fuel supply

Desired outputs (objectives to be maximized):
- Mobility
- Accessibility
- Quality
- Efficiency
- Safety
- Affordability

Undesired outputs (to be minimized):
- Implementation time and life-cycle costs
- Accidents and incidents
- Global and local emissions
- Impact on land, water, and energy consumption

Figure 3.47 The Input-Output Framework of Transportation Interventions

Source: Author compilation (Georges Darido).

The *desired outputs* noted in table 3.17 may be used to define project development objectives and program indicators for monitoring and evaluation.[1] Reducing *undesired outputs* or outcomes is important in the effort to ensure sustainable transportation interventions. For example, international experience suggests that significant shares of private vehicles and low urban densities boost fuel consumption (that is, energy per capita), which increases travel expenditures, infrastructure investment, and the emissions of local pollutants (mono-nitrogen oxides, sulfur oxides, carbon monoxide, and particulate matter) and global pollutants (carbon dioxide [CO_2] and other greenhouse gases). Table 3.18 provides sample transportation sector outcomes in several cities around the world. The spatial, physical, and technological factors contributing to these outcomes are described in subsequent sections.

Sustainable transport interventions should be linked to a continuous and comprehensive planning process that involves incremental implementation or a building block approach. The selection and sequencing of interventions should depend on enabling conditions and the implementation of complementary measures. In other words, maximizing mobility and accessibility may be achieved only if reasonable levels of safety, economic viability, and financial sustainability have been ensured. The World Bank Transport Strategy emphasizes that clean, safe, and affordable infrastructure and services represent the main aims of urban transportation users. This chapter describes sustainable transport interventions under each of the five controllable inputs listed (see figure 3.47), distinguishing between enabling conditions (stage I) and additional measures (stage II) that are substitutes or complementary actions.

Table 3.17 The Typical Objectives or Desired Outputs of Transportation Interventions

OBJECTIVE	DESCRIPTION	EXAMPLES OF INDICATORS
Mobility	Amount and type of travel (both passenger and freight)	Number of trips by mode, passenger, or ton-kilometers; change in travel times or delivery
Accessibility	Connectivity between desired origins and destinations; ability to reach desired goods, services, and activities	Number of jobs within a one-hour radius; retail area within a 10-minute walk of a station
Quality	Quality of travel between origin and destination	Reliability (for example, travel time variability and failures), comfort, convenience, and equity
Efficiency	How resources are used and how the environment is impacted, including: • Local and global emissions • Energy consumption and efficiency • Impact on land and water	• Emissions of NO_x, SO_x, CO, PM, or CO_2 • Quantity of fuel consumed per unit of mobility or economic welfare created • Amount of noise, runoff, sediments, dust, and other impacts on health and welfare
Safety	The safety and security of the transport system	Minimizing intentional and unintentional incidents, fatalities, injuries, property damage
Affordability	Economic and financial sustainability from various perspectives, including • Users (by income group) • Government and general public • Operators and others	• Travel expenditure in relation to income • Implementation time and capital costs • Operating, maintenance, and disposal costs • Other social and economic impacts of the investment (for example, poverty reduction)

Source: Author compilation (Georges Darido).

Note: NO_x, SO_x, CO, PM = mono-nitrogen oxide, sulfur oxide, carbon monoxide, and particulate matter, respectively.

Table 3.18 Urban Transportation Outcomes in Selected Cities

CITY	POPULATION DENSITY PER HECTARE	WALKING, CYCLING, TRANSIT, %	JOURNEY COST, % OF GDP	ANNUAL TRAVEL, KILOMETER PER CAPITA	ENERGY, MEGAJOULE PER CAPITA
Houston, United States	9	5	14.1	25,600	86,000
Melbourne, Australia	14	26	—	13,100	—
Sydney, Australia	19	25	11.0	—	30,000
Paris, France	48	56	6.7	7,250	15,500
Munich, Germany	56	60	5.8	8,850	17,500
London, England	59	51	7.1	—	14,500
Tokyo, Japan	88	68	5.0	9,900	11,500
Singapore	94	48	—	7,850	—
Hong Kong, China	320	82	5.0	5,000	6,500

Source: Mobility in Cities Database (2001).

Note: — = not available.

The Policy, Legislative, and Regulatory Framework

Urban transportation is shaped directly or indirectly by policies, legislation, and regulations at the national, subnational (regional or metropolitan), and local levels. Table 3.19 provides a summary of typical considerations at each policy level. It also distinguishes the enabling conditions for more advanced policy measures that build on them.

Sustainable transportation policies at the national level require institutions, processes, and financial mechanisms that prioritize public transportation and nonmotorized transportation, while discouraging private vehicle use. A lack of any of these elements may undermine a policy's impact. For example, the Chinese government adopted policies to prioritize public transportation and ensure people-oriented projects, but local impacts have been shaped by other factors, including capacity building and financial mechanisms.

Table 3.19 Policies, Legislation, and Regulations Affecting the Transport Sector

LEVEL	STAGE I: ENABLING CONDITIONS	STAGE II: ADDITIONAL MEASURES
National	Vehicle and fuel standards and taxes Roadway design standards Environmental protection and management laws	Sustainable transport policies Energy policies and targets Universal design and participatory rules Capacity building and research
Regional/metropolitan	Urban expansion and land management policies Public transport provision and regulation	Integrated transport improvement and land use plans Financial mechanisms (road pricing) Vehicular restrictions
Local	Zoning and taxation Traffic and parking regulations	Road space allocation Financial mechanisms (value capture)

Source: Author compilation (Georges Darido).

Funding for investments in transportation and, occasionally, operation and maintenance may be financed by vehicle and fuel taxes, bonds, and government-backed loans. The funding of transportation infrastructure typically requires participation by national or subnational governments, while local governments typically provide operations and maintenance with or without private sector participation.

A fuel tax is arguably one of the most important and effective fiscal measures because it directly levies users for consumption, but it is often politically difficult to pass or sustain. Fuel tax revenue is usually collected by national governments and then redistributed to fund roadway and transport investments. Most oil-importing countries impose a tax on transport fuels, but policies vary widely. In the United States, gasoline is taxed roughly US$0.12 per liter, but it is taxed several times more in European countries. The additional revenue collected in Europe has funded an arguably more sustainable mix of high-quality transport infrastructure and services, while encouraging less dependence on automobiles. Opponents argue that a fuel tax is socially and economically regressive because middle- and lower-income groups spend a greater share of their total incomes on fuel. For these and other reasons, national governments commonly adopt other taxes to raise revenue, including vehicle, registration, and licensing fees. A carbon tax is analogous to a fuel tax because greenhouse gas emissions are directly related to fuel consumption.

Many countries mandate road, vehicle, and fuel standards to promote safety, efficiency, and quality. The United States, for example, requires automobile manufacturers to meet fuel efficiency targets under the law on corporate average fuel efficiency passed in 1975, but standards in many Asian and European countries are stricter, as illustrated in figure 3.48. Several countries also require that domestically sourced ethanol be blended into fuels, but the efficiency of the ethanol production process greatly depends on the source of the fuel. Corn is the main fuel source in the United States. However, corn is inferior to sugar cane—a crop used in a successful ethanol program in Brazil—because corn is a major food crop that requires more resources to produce.

Other considerations at the national level include environmental protection laws, energy policies, and participatory regulations. Environmental laws often require a detailed review process to identify and mitigate project impacts on air, soil, water, and the environment (that is, impacts such as noise or visual intrusion). These rules influence transport policies or projects at the subnational level. For example, air quality regulations linked to federal transport funding in the United States have driven cities and states to implement vehicle inspection and maintenance programs that mandate emissions and safety standards for motor vehicles. National policies also include targets for energy efficiency or independence. China, for example, has targeted a 20 percent reduction in energy intensity in all sectors by 2020. In transport, this will require a reduction in consumed energy per unit of GDP (or other measure). Meeting this target will require changes in travel behavior, project design, and implemented technologies. In this context, national research and development efforts should focus not only on evaluating and testing advanced technologies, but also on training and educating transportation professionals. In addition, national governments have also addressed equity concerns through rules on (1) universal designs to accommodate all users, including people with disabilities and special needs (for example, the Americans with Disabilities Act in the United States), and (2) public participation and transparency reviews that give stakeholders ample opportunities to influence planning (such as the National Environmental Policy Act in the United States and the subsequent rules requiring environmental impact statements).

A fundamental challenge to achieving equitable and efficient transportation is the need to charge users the full, long-run marginal costs of travel and parking, including externalities (that is, the undesired outputs in figure 3.47). Many innovative time and price instruments, supported by relevant technologies, have been tested in cities in the last few years. These instruments may raise additional revenue for public transportation and alternative investments. For example, London, Singapore, and Stockholm have implemented congestion and road pricing schemes that require drivers to pay tolls if they enter defined central areas during certain hours of the day. In 2008, Milan, Italy, went a step further and applied a polluter pays principle in the city center by charging vehicles for their expected emissions. Advanced parking management, which includes centralized control and varied rates based on hour of the day, is another example of an innovative revenue tool.

Urban expansion and land management policies are basic considerations at the metropolitan

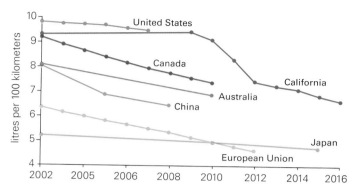

Figure 3.48 Average New Vehicle Fuel Economy Standards
Source: IEA (2007).

level. Regional transportation investment programs that support sprawl or physical decentralization in dense cities may have counterproductive effects, such as encouraging more vehicular traffic and congestion. The integration of transport and land development planning should consider macrolevel urban spatial and land use patterns, site plans, transportation network characteristics, travel patterns, user costs, and environmental impacts. In the United States, integrated transport, air quality, and land use plans for metropolitan and statewide transportation programs are updated at least once every four years. There is a strong consensus that compact cities with a single or only a few large, dominant centers or central business districts are more well suited for traditional fixed-route, fixed-schedule public transportation systems. Lower-density homogeneous cities with many dispersed, weak centers are more well served by individual modes. There is less agreement on the specific instruments that may be used to influence conditions, but they may include (1) road pricing, (2) the incremental taxation of the land and property benefiting from transportation investments, and (3) land development regulations on density requirements (floor area ratios), lot sizes, building setbacks, traffic rules, parking, and zoning.

Metropolitan and local governments are typically responsible for public transportation ser-

vices and planning, which should be linked to demand, available resources, and urban characteristics. Cities typically establish policy-based criteria to define transportation network coverage, the distance between stops, and service frequencies. Network coverage is often defined as the share of the population within walking or bicycling distance of a public transportation stop. In Bogotá, Colombia, for example, the master plan established that the share of residents within 500 meters of a station or stop under the Transmilenio bus rapid transit (BRT) system should reach 50 percent in phase 1 and 80 percent in phase 2. Along with service decisions, many cities also enact policies regulating public transportation fares and subsidies.

The allocation of existing and planned road space among pedestrians, transport (motorized, nonmotorized, and public), and parked vehicles represents one of the most potent, low-cost ways that local governments may use to promote equitable transportation management (World Bank 2008). The goals of allocating street space may be diverse, such as protecting walkers and cyclists, ensuring the safe movement of people, and facilitating public transportation via transit-only lanes. One option for reallocation is constructing dedicated lanes for high-performance BRT on arterial roads (for example, the BRT system presented later in this note in box 3.17). In many large cities, transportation demand may justify reallocating one or more lanes to buses. Yet, local authorities often find this difficult. Unfortunately, because of public pressure, motor vehicle mobility is often emphasized at the expense of nonmotorized transportation and public transportation.

Institutional Context

Efficient and stable institutions are an essential part of an urban transportation system. These institutions may be characterized by their functions and scope, including jurisdictions and modes. The institutional scope may vary from a special city district (for example, the central business district), a major corridor, or a vast multijurisdictional region. Institutional functions encompass planning, including strategic, policy, investment, and financial planning; implementation and service provision, including operations and maintenance by public and private entities; and management and regulation. In table 3.20, good practices are presented for two different institutional scenarios: a single jurisdiction with unintegrated modes and multiple jurisdictions with integrated modes.

Table 3.20 Institutional Functions and Jurisdictions in Transportation

UNINTEGRATED FUNCTION:	STAGE I: (ONE JURISDICTION, EACH MODE SEPARATELY)	STAGE II: (MULTIPLE JURISDICTIONS, INTEGRATED MODES)

Planning and financing	• Roadway investment and maintenance plans
	• Public transportation network planning
	• Pedestrian and bicycle access and facilities
		• Coordinated metropolitan planning and decision making
		  Service and accessibility standards
		  Prioritizing and budgeting
		  Financial mechanisms

Implementation and service provision	• Physical integration (intermodal terminals)
	• Electronic fare systems with prepayment
		• Integrated transport strategy (physical, operations, fare policy, land use, emissions)
		• Private sector participation
		  Joint development
		  Concessions and management contracts

Management and regulation	• Separate management of
	  Road access
	  Regulation of public transportation and taxis
	  Traffic and parking management
	  Freight
		• Centralized control and multimodal optimization with
		  Real-time information systems
		  Signal priority and coordination

Source: Author compilation (Georges Darido).

Difficult institutional issues often include planning, physical and operational integration, public transportation reform, and fare policies (including subsidies). At a basic level, city transportation institutions create forums for discussion and coordination among road planners, economic planners, public transportation operators, traffic management officials, and police. In China, many cities establish leading groups chaired by city officials. At an advanced level, transportation institutions may represent forums for joint decision making and priority setting across multiple jurisdictions and multiple modes. Some good examples of this include London (Transport for London), Madrid (*Consorcio de Madrid*), Paris (STIF, the organizing authority for public transport in Ile-de-France), Singapore (Land Transport Authority), and Vancouver (TransLink). Good examples also exist in emerging countries. Box 3.13 summarizes the essential pillars of sustainable transportation institutions based on international experience in Latin America and other regions.

Ideally, there should be one metropolitan authority overseeing all transportation issues and modes, particularly in regions with multiple jurisdictions. This authority should plan multiple modes, set priorities, and coordinate decisions on investments, taking into account land and environmental plans and the concerns of the public, civil society, and private sectors. The authority should oversee strategic policies and the management of modes, including parking, taxis, public transportation, highways, and arterial roads. The regulation and reform of public transportation are challenging because they must balance the roles of the public and private sectors and respond to local conditions. An excerpt from a recent World Bank operational guide illustrates the challenges:

> Institutional approaches to providing public transport services range from a single publicly owned monopoly operator at one extreme, to numerous weakly regulated or unregulated small-scale, privately owned providers at the other. In some cities a range of approaches co-

---

BOX 3.13

The Four Pillars of Sustainable Urban Transportation Institutions

Before financing major urban transportation projects, decision makers should attempt to put in place the basic elements to ensure the long-term sustainability of the sector. Specifically, policy makers should incorporate a four-point agenda into any urban transportation strategy:

1. Create a regional transportation coordination commission in charge of coordinating policies among federal, state, and municipal governments, giving highest priority to major urban transportation investments in the metropolitan region and promoting modal integration. This will help improve the sector's economic efficiency and long-term sustainability.

2. Adopt a strategy for integrated land use, urban transportation, and air quality that provides a framework for community leaders and decision makers to evaluate future urban transportation investments and policies.

3. Enact into law formal financing mechanisms to ensure that the long-run variable costs of urban transportation systems are covered by operating and nonoperating revenues and by appropriate user charges.

4. Promote private sector participation in the operation, maintenance, and construction of urban transportation systems to lessen the financial burden on government (through, for example, concessions or management contracts).

Source: Adapted from Rebelo (1996).

---

exist. The first extreme tends toward inefficient operations and uneconomic fares both of which map into high subsidies. It may also produce poor services, especially when the subsidy mechanism fails and operators are starved for funds. The other extreme may produce good services at zero public expenditure, but more often provides poor service with high accident and pollution costs. When this regulatory set-up is matched by low fares constrained by regulation or unfettered competition, service levels and quality fall and externalities rise. (World Bank 2008: 8).

Physical Systems, Technology, and Spatial Planning

The design of systems, technologies, and spatial plans should be driven by current or near-term transportation demand and a longer-term credible and transparent transportation plan. The transportation plan is the product of a continuous, comprehensive, and inclusive process. Spatial planning should consider future land uses and existing travel and freight demand. Physical systems and technologies include the supply of infrastructure and services for passengers and freight. Technologies also include the fuels, vehicles, and equipment used to deliver infrastructure and services. Table 3.21 provides a framework for these types of transportation interventions. Table 3.22 summarizes physical, technological, and spatial interventions at basic and advanced levels.

International experience suggests that sustainable transportation investment strategies need to prioritize public transportation and other essential modes, encourage nonmotorized trips, ensure that users of private automobiles internalize the costs they impose, and include urban plans and incentives to support compact cities. The strategies should aim to achieve realistic results at different points in time: (1) in the short term, by improving the fuel efficiency of existing vehicle fleets; (2) in the medium term, by facilitating a shift away from private car use; and (3) continuously, by supporting the development of compact cities built around public

Table 3.21 The Framework of Transportation Interventions

FOCUS	SPATIAL PLANNING	PHYSICAL SYSTEMS	TECHNOLOGIES
Land uses and travel demand	Macro or master planning	Microdesign (for example, transit-oriented development)	Travel demand management
Infrastructure and services	Location-efficient planning	Mobility and freight management (roads, public transportation, nonmotorized transport, traffic management, other facilities)	Intelligent transportation systems
Vehicle fleet and fuel supply	Fleet management and efficiency programs	Standards, inspection, and maintenance programs	Alternative fuels and advanced technologies

Source: Author compilation (Georges Darido).

Table 3.22 Basic and Advanced Transport Interventions

FOCUS	STAGE I: ENABLING CONDITIONS AND MEASURES	STAGE II: ADDITIONAL MEASURES
Land uses and travel demand	• Microdesign: 　Urban densities 　Road patterns and design 　Intersections and crossings 　Basic pedestrian and bicycle facilities 　Parking and access management • Macro plans: 　Origin-destination surveys and calibrated transport model 　City structure and development pattern	• Microdesign: 　Mixed land uses 　Building design and orientation 　Transit-oriented development (Box 3.14). • Macroplans: 　Development along high-quality public transportation corridors 　High-density, mixed-use nodes around public transportation facilities 　Reserving rights-of-way for future corridors 　Energy use and emissions 　Travel demand management (boxes 3.15 and 3.16)
Infrastructure and services	• Road network development • Public transport • Traffic management, including road safety measures	• Integrated public transportation and traffic management network • BRT (box 3.17) • Intelligent transport systems • Freight management
Vehicle fleet and fuel supply	• Cleaner fuels (low-sulfur diesel) • Inspection and maintenance programs	• Alternative fuels • Advanced vehicle technologies

Source: Author compilation (Georges Darido).

transportation corridors that reduce demand. There are four outcomes of sustainable transportation interventions: managed demand, enhanced supplies, shifted modes, and improved performance. The following sections describe specific interventions and expected outcomes.

Land uses and travel demand

The transportation plan should recognize the effects of transportation interventions on future land development and travel demand.[2] The interaction between transportation supplies and land use is a complex, two-way relationship. Existing land uses and developments are served by transportation infrastructure and services, which, in turn, induce certain types of land development and travel patterns.

Spatial planning is one of the most important factors influencing the demand, mode choice, and financing of urban transportation investments. In turn, sustainable development may be the most important goal of transport investments, particularly public transportation. Land development planning should include two complementary approaches: a macro- or top-down approach and a micro- or bottom-up approach. The macroapproach entails viewing the city or region "from 10,000 meters" and with a time horizon of more than a decade. Strategic planning and alternative analysis are initial steps in the macroapproach that allow the selection of appropriate modes and a city's alignment. The microapproach is more focused geographically (that is, on blocks or corridors) and anthropogenically. It also has a shorter time horizon (less than 10 years) and requires more detailed preliminary designs.

At the macrolevel, the major determinants of travel demand include the distribution and character of land use. Transportation investments may positively influence the distribution and character of land use by creating accessible and visible development nodes. For example, properly located public transportation stations may be focal points for development, thus increas-ing public transportation demand, reducing inefficient land consumption (sprawl), and creating nonmotorized transportation opportunities. Other macrolevel planning issues include some of the following.

High-density, mixed uses around public transportation: Public transportation traditionally serves the city center; major activity centers; and, particularly, work-related travel. The locations of activity centers and job sites affect the design and effectiveness of public transportation. Traditionally, jobs are concentrated in the city center, but cheaper land or perverse incentives may attract development and jobs to urban peripheries in which there is little or no infrastructure. Development in the urban periphery reduces economies of scale for public transportation, making operation without substantial subsidies more difficult. Dense development around public transportation is critical.

High-quality public transportation corridors: Focusing development along high-capacity, high-quality public transportation corridors is important, particularly to prevent unplanned development elsewhere. Such a strategy may be pursued in cities in China, but it has not yet been practiced effectively except in Hong Kong, China. In Singapore and Stockholm, urban rail and mass transit have been used effectively to supply high-quality public transportation. BRT systems represent another innovative and cost-effective public transportation approach that has been developed and is now widely applied in Latin America. Road infrastructure also plays a critical role. For example, the ring road model in Chinese cities induces greater private vehicle use and leads to dispersed development that is difficult to serve through public transportation.

City structure and development patterns: Radial city development and structures most effectively facilitate high-capacity rail and bus systems provided major job and activity sites are located in urban centers or along arteries (radial lines from the center). Curitiba, Brazil, is a good example of a radial, transit-oriented city

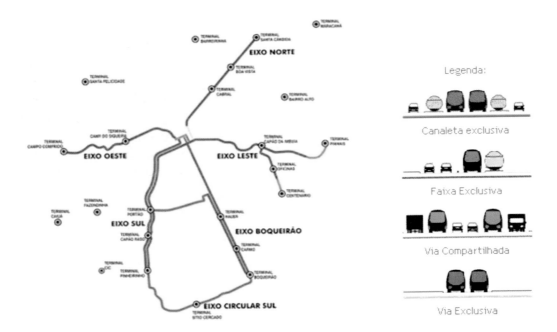

Figure 3.49 The Structure of the Integrated Public Transportation Network in Curitiba, Brazil

Source: Urbanização de Curitiba S.A. http://www.urbs.curitiba.pr.gov.br/PORTAL/rit/index.php?pagina=terminais.

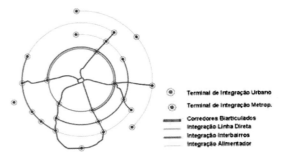

in the developing world with a high-capacity BRT system serving five high-density corridors (figure 3.49). These arteries were planned and rights-of-way were reserved decades before they were fully developed. This degree of urban foresight required a long-term vision and institutions that had sufficient capacity and political independence. Ring road or circumferential development places fewer constraints on land development, but encourages dispersed and inefficient land consumption. Urban development based on satellite cities is also less than ideal. It takes many years for a satellite city to become self-sustaining, and costly new connections to the city center and other poles are needed.

Land management: Setting aside large lots in primary and secondary markets for high-rise developments is important. Migration and urbanization trends in East Asian cities should be directed toward radial corridors rather than ring road development. Cities that offer good examples of effective land management include Hong Kong, China; cities in the Republic of Korea; and Singapore. However, weak land-use policies combined with large lots, significant private vehicle use, and low fuel prices favor urban sprawl and decentralization. In many cities in North America in the second half of the 20th century, these factors decreased the effectiveness of public transportation services and created a vicious cycle that reinforced dependence on automobiles.

Financing mechanisms: It is important to be able to capture and transfer revenue derived from infrastructure efficiently through, for example, a value-based property tax. This approach was institutionalized in Hong Kong, China; Singapore; and Tokyo. In China, a significant share of city revenues comes from land sales or long-term leases, which provide incentives to expand city boundaries and oversupply land, thereby exacerbating sprawl.

Tools and resources for macroplanning include household travel surveys on origins and destinations and transportation models that use survey information for calibration.

Microapproaches are characterized by a narrower geographic and human focus. They also have a shorter time horizon (less than 10 years) than strategic planning and require more detailed preliminary designs. Microdesign is largely analogous to the principles of transit-oriented design or transit-oriented development (box 3.14).

Microdesign may be summarized as follows:

Land use distribution (space and time): All functions in large cities cannot be centralized, and the relative locations of developments and transportation links determine travel demand and mode choice. Mixed land uses are important because they influence how far one must travel to visit a store or reach work or school. Convenient foot and bicycle access (less than 10 minutes) should be provided from residential housing to shops, services, and recreation, complemented by public transportation facilities for work trips.

Urban densities: Population densities and job locations affect transit and land use plans, but are not the only considerations. Lower urban densities tend to boost car use, thereby negatively affecting multimodal sectoral plans. However, high densities without adequate planning and services may impede the quality of the lives of residents. Figure 3.50 shows a sprawled residential development typical in developed countries.

Design and orientation of buildings: Large street offsets, parking lots, fences, and greenbelts around buildings became much more common in the second half of the 20th century, owing to concerns about safety, security, noise, and pollution. However, these elements may discourage walking and bicycling because they impose barriers and make trips more circuitous.

Figure 3.50 A Locality in the State of Colorado

Source: Digital Globe, Google Earth.

Thoughtful design is critical.

Road patterns and design: Traffic volumes on local streets influence the quality of life in residential areas. Road designs influence driving behavior, traffic speeds, and safety. Certain road patterns are characterized by low accessibility and connectivity (for example, dead-end streets and cul-de-sacs). Speed limits should be conservative and enforced rigorously in dense residential and commercial areas using signs, police, cameras, and speed humps.

Intersections and crossings: Local access roads wider than 5 meters, equivalent to approximately two lanes, tend to discourage crossing. Intersections with more than one lane in each direction may require traffic channelization (that is, sidewalks, curbs, pedestrian islands, and markings) and signals. Most congestion on arterials in dense areas is caused by limited throughput at intersections rather than the dimensions of the intermediary road sections. Road width may often be reduced or limited to two or three lanes in each direction and include channelization and modal separation, while minimally affecting travel times.

Pedestrian and bicycle environment: Encouraging walking and bicycling by prioritizing nonmotorized over motorized vehicle access (and parking) is a key objective for the urban environment (figures 3.51 and 3.52). It is also important to provide pathways or greenways in city blocks,

BOX 3.14

Transit-Oriented Development

Transit-oriented development is characterized by

- Proximity to and a functional relationship with transit stations and terminals and service provision by high-quality public transportation (BRT systems, underground trains, and so on)

- Compact, mixed use buildings and infrastructure that, because of their design, encourage walking, cycling, and transit use by residents, employees, shoppers, and visitors

The ingredients of successful transit-oriented development include strategic (macro-) and design (micro-) elements such as

- A strong development climate
- Master plans for multiuse, high-intensity developments supported by implementation plans

The ingredients also include transportation investments that promote the following:

- Easy and direct pedestrian, bicycle, and public transportation access (as pictured in the eco-block example in the figure)
- Good signage and a pleasant environment to attract substantial pedestrian flows
- Significant regional accessibility to major job and activity centers
- Short, direct connections between transportation modes and transit facilities
- Bicycle lanes and parking facilities that feed stations
- Attractive facilities that are well integrated with the surroundings (public spaces, street furniture, and so on)
- Safe and secure designs, including adequate lighting
- Effective parking management around stations
- Environment-friendly technology options, such as shared fleets of alternative (electric) vehicles located in neighborhoods

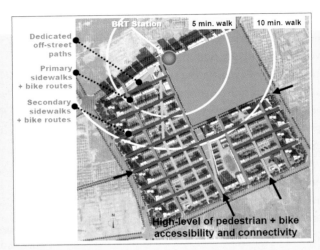

Source: Fraker and Wurster (2009).
Note: The eco-block concept is illustrated using a location in China.

Research shows that the impacts of transit-oriented development are realized in the long term and depend on the quality of related microdesigns and the rate of an area's demographic and economic growth.

- Research by Lund, Cervero, and Willson (2004) on residential and commercial sites in major cities of California shows that factors related to transit-oriented development, particularly proximity to urban and commuter rail stations, increase ridership on rail and buses by as much as a factor of three to four relative to control sites.
- Cervero and Day (2008a, 2008b) have surveyed households relocating to suburban transit-oriented development sites and nontransit-oriented development sites in Beijing and Shanghai, China to assess the impacts on travel behavior.

The latter sites showed significant positive impacts:

- Increased public transportation ridership
- Improved access to regional jobs (as measured by employment locations within a radius equivalent to one hour in travel time)
- Reduced commuter times per household worker

Source: Zimmerman (2008).
Note: The photo on the left shows a high-quality public transportation corridor between Arlington, Virginia, and Washington, D.C., with an underground metropolitan train (the orange line and M) and a feeder bus system. The corridor exhibits many elements of good macrolevel planning and transit-oriented development, including higher densities around high-quality public transportation (the underground) in an otherwise car-oriented environment. After 20 years of mixed use development around stations (such as Clarendon, pictured on the right), the corridor has become a good example of urban form.

major complexes that are safe and shaded, and pervious surfaces that absorb runoff water.

Parking and access management: Surface, underground, and on-street parking in cities is limited because of the high cost of land and construction. Where available, parking facilities should target multiple uses—for example, office parking on weekdays, event parking at night, and fair or market parking on weekends—and be designed to minimize walking distances and impervious surfaces.

Figure 3.51 A Pedestrian-Friendly Street in Curitiba, Brazil

Source: Institute for Research and Urban Planning of Curitiba.

Spatial or land use planning should strive to optimize resources by reducing unnecessary motorized travel and encouraging trips on the most appropriate modes and routes at the most appropriate times. However, land use planning alone has a limited impact on the use of private vehicles (Wright 2004). Travel demand management—preferably linked to public transportation—and other major investments in infrastructure and services are needed. Common, related activities include adequately pricing roads, restricting vehicle use by time or location (for example, car-free Sundays), establishing high-occupancy or priority lanes, encouraging ride-sharing and carpooling, and promoting tele-commuting and flextime. Box 3.15 describes innovative travel demand management and emissions approaches adopted in Milan's Eco-Pass project. Box 3.16 describes Beijing's travel demand management and public transportation approaches linked to its tremendous urban growth and the 2008 Olympics.

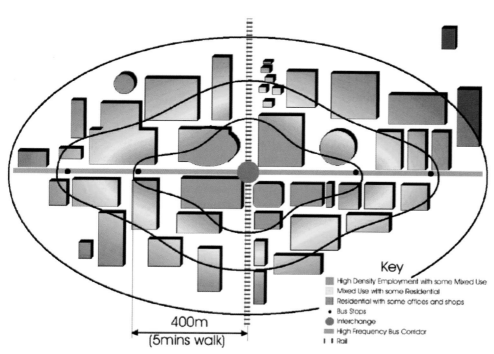

Key
- High Density Employment with some Mixed Use
- Mixed Use with some Residential
- Residential with some offices and shops
- Bus Stops
- Interchange
- High Frequency Bus Corridor
- Rail

400m
(5mins walk)

Figure 3.52 An Example of Microdesign and Walking Isochrones

Source: Colin Buchanan and Partners (2001).

BOX 3.15

Emission-Based Road Pricing in Milan, Italy

In January 2008, Milan introduced EcoPass, a program designed to restrict access to the city center during certain hours of the day by charging fees to drivers of the most heavily polluting vehicles. This is the first urban environmental policy in the region whereby the transportation sector has applied the European Union's polluter pays principle. Significant results have already been achieved through this innovative scheme:

Source: Comune di Milano (2009).

- Traffic has been reduced 19 percent during the enforcement period and 8 percent overall.

- The road speed of public transportation has risen by 11 percent, and public transportation ridership has increased by 10 percent.

- CO_2 emissions have fallen by 12 percent, and the incidence of particulate matter by 19 percent.

BOX 3.16

Beijing: Travel Demand Management and the Legacy of the Olympic Games

Transportation operations during Beijing's 2008 Olympics required not only massive infrastructure investments, but also a new paradigm for travel demand management and unprecedented interagency coordination and public cooperation. According to transportation officials, Beijing has spent over Y 100 billion (approximately US$14 billion) in the past five years on transportation infrastructure and services. Temporary travel demand management measures have included prohibiting half of Beijing's private vehicles from driving on city roads on alternating days based on the last digit of license plates (the 50 percent restriction). One-third of the capital's more than 3 million vehicles were removed from roads, though the city exempted government, emergency, public transportation, taxi, and Olympics-related vehicles from the restriction. The movement of freight vehicles was also restricted, and centers for the distribution of goods and tolls on inbound routes were implemented to reduce traffic in the city center. The government also suspended activities at hundreds of factories and construction sites in and around Beijing. As a result, the city's notoriously poor air quality in August and September 2008 was the best it had been in over 10 years.

Congestion, notoriously severe in Beijing, was significantly reduced despite the Olympics-related traffic and the more than 260 kilometers of lanes on arterials and ring roads reserved for Olympics, press, and government vehicles. According to transportation officials, the share of commuters riding public transportation rose by 35 to 45 percent largely because of restrictions on car use, the expansion of the public transportation network, and a recent reduction in bus and subway fares. In the previous year, the city had opened three subway lines, a light rail line from downtown to the airport, new bus lines, and a new Beijing-Tianjin intercity express railway. Beijing's transit network now includes more than 200 kilometers of rail and 45 kilometers of rapid bus lines. The improvements have diversified the commuting options for millions of residents.

However, more investment is needed. Returning to a pre-Olympics transportation paradigm and ignoring long-term demand will quickly overwhelm Beijing's new infrastructure. Beijing's population of 18 million is growing by roughly a half million a year. Car ownership is not merely a symbol of status, but increasingly a necessity. The number of motor vehicles is growing by over 10 percent per year; the corresponding rate is over 20 percent for private cars. International comparisons are few, but instructive. London, New York, Paris, and Tokyo experienced rapid growth in car use and ownership in the 20th century, but motorization in Beijing appears to be progressing more rapidly than in any time in history. In Tokyo, 20 years (1962 to 1982) elapsed before the number of motor vehicles increased from 1 million to 3 million, but in Beijing, the same increase took place in only 10 years (1997 to 2007). New cars in Beijing have never been more efficient, but travel demand will overwhelm road capacity despite the rapid and ongoing expansion of the network.

Public opinion surveys undertaken since the Olympics have shown that Beijing residents are now more aware of sustainable transportation and air quality issues. While about 70 percent of residents are willing to live with certain car restrictions to sustain reductions in poor air quality and congestion, most car owners oppose the scope of the restrictions on vehicle use. Car owners remain a minority, but this may change because more than 1,000 vehicles are being added to Beijing's roads every day. Moreover, officials have grappled with questions about which measures should be kept and to what degree. The Olympics experiment has represented a unique opportunity to implement change, but it has also heightened public expectations.

After the Olympics, the government announced that private cars would be allowed in the city, which includes the area within Beijing's Fifth Ring Road, on four out of five weekdays (a 20 percent restriction). As before, the last digit of license plates determines the restricted day. This policy was initiated in October 2008 for a six-month trial period. Similar one-day restrictions have been in place for many years in megacities such as São Paulo, Brazil, and Mexico City. Evidence suggests that these restrictions become less effective

BOX 3.16 (continued)

over time because the unused road capacity is eventually taken up by other vehicles. More important, some people find ways to circumvent the rules. For example, some residents buy a cheap, older vehicle that may be legally driven on certain days. For mitigation of such noncompliance, the approaches are often combined with vehicle inspection, maintenance, and scrappage programs.

Looking forward, Beijing transportation officials are supporting a shift from investment scale-up, whereby infrastructure investment consumes a large share of GDP, to optimized operations. New infrastructure and technologies such as intelligent transportation systems will remain important, but effective planning and demand management will be sought to ensure long-term success. Among the most important proposed initiatives are the following:

- Harmonizing land use and transportation, including efficiently designing and locating activity centers and transit-oriented developments, improving accessibility for pedestrians and bicyclists, and adopting other policies and strategies

- Setting appropriate prices and tolls, including road-use fees, vehicle registration, fuel, parking, and public transportation (to encourage improved travel behavior and financial sustainability)

- Efficiently allocating resources among modes and along integrated corridors—in Beijing, as in many of the largest cities, buses will

Source: Photo by Sam Zimmerman.
Note: The photo shows Beijing's Third Ring Road at peak rush hour before the 2008 Olympics.

continue to be workhorse vehicles—and investments in measures to support buses are usually cost-effective

- Adopting the latest, proven technologies for vehicles and fuels

- Improving the institutional system and planning processes

Infrastructure and services

Effective spatial planning considers the location of infrastructure and services relative to demand and other transport supplies in the region. Location efficiency, or transport-efficient development, aims to optimize the location of transportation investments and new major activity centers by maximizing benefits and minimizing environmental externalities such as energy use (Zegras, Chen, and Grutter 2009). Information gathering and modeling are crucial to transport-efficient development. The type of land development is also important, because the type of development affects the economies of scale and determines the physical constraints that make certain systems, technologies, or spatial plans more or less viable. Table 3.23 delineates the characteristics, opportunities, and challenges of typical types of land development.

Though transportation interventions typically influence more limited spatial dimensions (for ex- ample, radial or circumferential corridors that may form a network or a grid), demand normally has a wider scope (for example, block, district, city, and region, as well as time). The time dimension is also important because some interventions require many years and may be implemented incrementally. Urban planners normally strive to achieve a spatially balanced flow of passengers (encouraging trips in both directions on a road or rapid transit corridor) and reduce peak demand, which is the most costly to satisfy.

Mobility management includes a range of transportation interventions that enhance supply and induce modal shifts. Road development is perhaps the most common type of mobility management intervention, but it is most effectively practiced to balance transportation projects with other types of investment such as public transportation, nonmotorized transportation, freight management, traffic management, and road safety. Table 3.24 provides a

Table 3.23 Type of Development and the Implications for Transportation

DEVELOPMENT	CHARACTERISTICS	OPPORTUNITIES	CHALLENGES
Greenfield (urban expansion)	Site of former agricultural or other nonurban land, usually on the edge of cities	• Application of best land use practices • Reserving rights-of-way (ROW) for future corridors • Smaller resettlement requirement	• No existing demand or services • Requires costly new infrastructure
Greyfield (redevelopment)	Site of existing or former residential, institutional, commercial, or industrial establishments	• May capitalize on existing demand and services • Benefit of upgrading obsolete facilities or land uses (parking lots) • Easier to get public support	• Redesigning existing facilities and services to serve new demand • Resettlement and appropriation • Potential contamination site (brownfield)
Infill (development)	Open site available within already developed area or next to other built sites and existing services	• May capitalize on existing demand and services	• Rarely available or more costly • Reduction of open/green spaces

Source: Author compilation (Georges Darido).

Table 3.24 Mobility Infrastructure Hierarchy

INFRASTRUCTURE	GENERAL CHARACTERISTICS	FUNCTIONS
Good practice: A balanced network		
Urban highways and ring roads	• Highest speeds • Controlled access, with grade separation and interchanges • Highest cost and lowest network density (<0.2 km/km^2)	• Long-distance trips • Divert through traffic, particularly trucks • Evacuation routes • Encourage dispersed land development patterns
Primary roads or arterials	• Medium-high speed • Sidewalks and signalized crossings at every intersection • High cost and medium network density	• Major thoroughfare and interdistrict trips • Access to highway networks and major activity centers
Secondary roads	• Limited on-street parking • Sidewalks and pedestrian crossings at major intersections • Medium cost and medium network density	• Intradistrict trips • Access to primary roads; high-density commercial, office residential, and institutional developments
Local roads or collectors, distributors	• Low speeds and unsignalized intersections • Limited on-street parking • Sidewalks • Low cost and high network density	• Access to major roads • Motorized and nonmotorized access to commercial developments and residential areas
Additional elements: Best practice		
Traffic management and road safety	• Centralized and coordinated signals and cameras • Channelization (islands) and pedestrian signals at intersections • Analysis of accidents and incidents	• Adapt to current conditions and give priority to special vehicles • Incident management and enforcement • Target road safety investments
Public transportation facilities	• Dedicated or exclusive lanes • Intersection priority • Stations and terminals, park-and-ride facilities	• Give priority movement to public transportation • Maximize the coverage of the network, while minimizing the burden of transfers
Bicycle paths	• Crossings at major intersections • Amenities (secure bicycle parking, shade, and so on)	• Recreation (parkway) • Feeders to public transportation facilities • Commuter routes
Pedestrian streets or areas	• Downtown or shopping areas with high pedestrian traffic • Amenities (shade, benches, and so on) • No surface parking	• No vehicular access to buildings • Feeders to public transportation facilities • Public spaces for events
Freight facilities	• Multimodal terminals • Designated loading and parking	• Optimize operations by more closely matching vehicles and freight

Source: Author compilation (Georges Darido).

summary of the most common types of mobility infrastructure by function and characteristics.

Balancing transportation investments is important because it influences current and future mode shares and the sustainability of the system. For example, providing free or ample parking for private vehicles may severely reduce the viability of public transportation. Moreover, maintaining or increasing the mode share of public transportation and nonmotorized modes becomes more difficult as the fleet and the infrastructure for private vehicles grows. In addition, new infrastructure, particularly for private vehicles, may induce a rebound effect (that is, new demand because of greater capacity).

Public transportation is a particularly important type of mobility intervention in the urban context. Table 3.25 summarizes the main elements of a public transportation network by function, capacity, and characteristics. At a basic level, most elements are applied, but not in a coordinated or balanced way. Best practice recommends creating an integrated network that is scaled to demand. Because the full cost of private vehicle travel is not internalized in most cities, individual decisions to use a private vehicle instead of public transportation may be based on inaccurate costs. To be successful, public transportation systems require specific land development patterns. Some corridors are more well suited for rapid transit, including rail, metro, or BRT systems. At the least, cities need sufficient available or redevelopable land around potential stations and good pedestrian and bus access.

Intelligent transportation systems aim to increase the capacity or efficiency of infrastructure by harnessing appropriate technology. These systems have the potential to improve highway operations by mitigating congestion, managing speeds, and smoothing traffic flows. Specific

Table 3.25 Elements of a Public Transportation Network

SERVICE TYPES	FUNCTION	CAPACITY	CONDITIONS	REQUIREMENTS
Feeder or circulators (collectors and distributors)	Shortest trips (usually 1–3 km) within district or neighborhood	Low: small buses (7–20 m in length, 20–40 passengers)	Lowest population density, but with defined nodes	Local streets, low costs
Local (bus)	Medium trips (3–8 km) from district to city or district to district	Intermediate schedules: if more than 1 minute headways	Medium-density nodes or corridors	Arterial roads, bus stops, and other facilities
Commuter express bus or suburban rail	Long trips (>20 km) from regional suburbs to city center or district	Intermediate	Few origins, limited destinations	Highways or arterials, bus stops, and other facilities
Surface mass transit— BRT or light rail transit	All trips from district to city (usually 5–20 km)	Intermediate to high: ridership of 100,000– 300,000 daily, 10,000– 30,000 in peak hour	High population density: 5,000–10,000 persons per km^2	Exclusive lanes on major arterials; 10–20 m of right-of-way; stations and terminals; intermediate investment, typically US$1 million to US$10 million per km, depending on infrastructure
Grade-separated mass transit (elevated or underground)	All trips from district to city (usually 5–20 km)	High: ridership of 200,000–500,000 daily, 20,000–50,000 in peak hour	Highest population density: >15,000 persons per km^2	Underground or elevated stations and terminals; highest investment, typically US$50 million– US$200 million, depending on infrastructure
Intercity (bus or rail)	Longest trips from region to region	Medium to high	Limited origins and destinations	Intermodal stations and terminals

Source: Author compilation and estimates (Georges Darido); adapted from PPIAF and World Bank (2008).
Note: km = kilometer; km^2 = square kilometer; m = meter.

measures for each of these strategies are described in the bubbles in figure 3.53. In California, each strategy has reduced fuel consumption and on-road CO_2 emissions by 5–10 percent. Mitigating congestion and smoothing traffic boost average travel speeds closer to the optimal 35 miles per hour (roughly 55 kilometers per hour). Managing speed entails increasing vehicular speed to more efficient levels (below 65 miles per hour or 100 kilometers per hour). This scenario, however, assumes a congested environment and no rebound effect. These strategies correspond to the market packages of intelligent transportation systems pictured in figure 3.54, namely, freeway management and emergency management. (For a complete data-

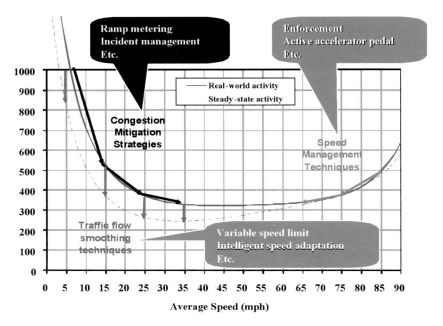

Figure 3.53 The Benefits under Speed Conditions of Select Highway Applications of Intelligent Transportation Systems

Source: Matthew Barth, Center for Environmental Research and Technology, University of California–Riverside, Riverside, CA, http://www.cert.ucr.edu/research/tsr/.

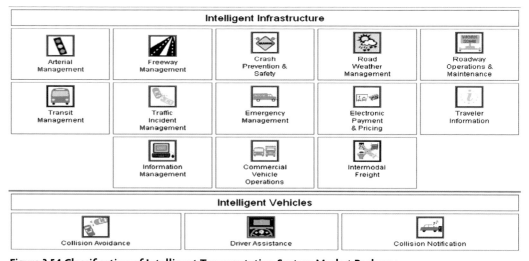

Figure 3.54 Classification of Intelligent Transportation System Market Packages

Source: U.S. Department of Transportation (2009).

base on the costs, benefits, and lessons learned in intelligent transportation systems, visit http://www.benefitcost.its.dot.gov/.)

Vehicle fleet and fuel supply

Technological interventions targeting vehicle and fuel savings should normally be implemented together to ensure performance and cost-effectiveness. There is usually little point in investing in advanced vehicle technologies without complementary fuel measures (and vice versa). For example, filters that capture dangerous particulate matter in exhaust will not work effectively without ultralow sulfur diesel fuel, and these filters may be counterproductive if they are not properly installed. It is also important that basic steps be implemented before advanced technologies are pursued. These steps might include adopting minimum fuel and economy standards for new vehicles and banning the most polluting vehicles (such as vehicles with highly inefficient two-stroke engines). In many countries, periodic inspection and maintenance schemes are common to compel the repair of the most polluting vehicles or their removal from roads. Table 3.26 provides a partial overview of common vehicle and fuel technologies and practices, along with illustrative examples.[3] However, only a few advanced interventions are suggested because they are highly dependent on the local context.

Good management and operational practices are also essential. For example, a bus with a highly sophisticated hydrogen fuel cell would provide little benefit if the bus carries few passengers or the primary fuel source used to produce the clean hydrogen was something more polluting than standard diesel. Table 3.27 provides estimates of CO_2 emissions from various vehicle types based on realistic assumptions of occupancy, maintenance, and primary energy sources.

Stakeholder Dynamics

Incentives for the public sector, private sector, and citizen stakeholders need to be aligned and coordinated. Table 3.28 outlines the various stakeholders and their interests.

The main interests of the stakeholders are as follows:

- *Decision makers:* Elected and appointed officials (typically with four- to five-year terms) gauge the political and economic feasibility of innovative urban projects. Innovative projects are often linked to city goals, efforts to boost the quality of life, and major special events that may catalyze transformative change (such as the Olympic Games in Bei-

Table 3.26 Summary of Select Vehicle and Fuel Interventions

STAGE	VEHICLES	FUELS
Stage I: *Enabling conditions*	• Minimum safety and efficiency standards for new vehicles • Ban existing two-stroke engines	• Minimum fuel-quality standards (unleaded, low-sulfur, reformulated fuel, and so on)
Stage I: *Additional measures*	• Emissions control equipment Catalytic converters for gasoline Particle traps for diesel • Inspection, maintenance, and scrappage requirements	• Alternative and bio-fuels, where appropriate and cost-effective Natural gas Ethanol • Reduce leakages and inefficiencies in the vehicle and distribution system
Stage II: *Additional measures*	• Advanced technologies for managed or shared fleets Eco-driving options and idle reduction Hybrid-electric Plug-in electric	• Renewable alternative fuels with distribution network, where appropriate Solar Wind

Source: Author compilation (Georges Darido).

Table 3.27 CO₂ Emissions from a Range of Vehicle Types

VEHICLE TYPE	LOAD FACTOR (AVERAGE OCCUPANCY)	CO₂ EQUIVALENT EMISSIONS PER PASSENGER KM (FULL ENERGY CYCLE)
Car (gasoline)	2.5	130–170
Car (diesel)	2.5	85–120
Car (natural gas)	2.5	100–135
Car (electric)[a]	2.0	30–100
Scooter (two-stroke)	1.5	69–90
Scooter (four-stroke)	1.5	40–60
Minibus (gasoline)	12.0	50–70
Minibus (diesel)	12.0	40–60
Bus (diesel)	40.0	20–30
Bus (natural gas)	40.0	25–35
Rail transit[b]	75% full	20–50

Source: Sperling and Salon (2002).

Note: All numbers in this table are estimates and approximations and should be treated as illustrative.

a. The ranges have arisen largely because of the varying mix of carbon and noncarbon energy sources (consisting of 20–80 percent coal) and the assumption that the electric battery vehicle will tend to be somewhat smaller than conventional vehicles.

b. This category assumes the application of heavy urban rail technology (metro) powered by electricity generated from a mix of coal, natural gas, and hydropower and high passenger usage rates (75 percent of seats filled, on average).

Table 3.28 Basic and Advanced Stakeholder Interests

STAKEHOLDER	STAGE I: ENABLING CONDITIONS OR MEASURES	STAGE II: ADDITIONAL MEASURES
Decision makers	• Windows of (political) opportunity • Economic feasibility	• Special events and larger vision • Quality of life issues
Users	• User surveys (on-board, intercept)	• Participatory process • Universal accessibility
Public	• Public opinion surveys	• Transparent process
Operators	• Financial sustainability	• Social sustainability

Source: Author compilation (Georges Darido).

jing; see box 3.16). Key relevant issues for decision makers: What will be the impacts of the projects on key interests? Are the impacts reversible if something goes wrong? May projects be implemented if there is a window of opportunity? Are they sustainable beyond the current term?

• *Users or the riding public:* The interests of this group include maximizing personal mobility and the accessibility, quality, safety, and affordability of services and mini-mizing personal expenses. These interests may be measured through user surveys. International experience shows that users value participatory processes and the principles of universal accessibility.

• *The public:* The public, including nonusers, is generally interested in the performance and cost-effectiveness of investments. Public opinion surveys, including household surveys, may form part of a transparent vetting process that also publishes public information on Web sites and through other media and channels of communication.

• *Transport planners and operators:* Transportation officials usually strive to ensure the financial and social sustainability of infrastructure and services. Subsidies may be required for some services or in areas in which affordability is a main concern.

• *The business community:* Transportation infrastructure lubricates a city's economic engine, and business leaders are often keenly interested in the development of key projects.

Economic and Financial Aspects

Best practice dictates that planning and feasibility studies on transportation projects should thoroughly analyze viable alternatives. Analyses should consider the life-cycle costs and time horizons of investments. Studies of public transportation corridors often assess BRT and urban rail alternatives. BRT systems have a slightly lower capacity and a shorter life cycle because buses and busways are not as durable as rail cars and tracks. However, a BRT system is considerably more rapid and cheaper to build than rail if rights-of-way are available (box 3.17). A BRT system is also more flexible, may be implemented incrementally, and is more easily altered.

Table 3.29 summarizes key aspects in economic and financial assessments of transportation projects.

Economic analyses of transport alternatives typically rely on cost-benefit analyses and the calculation of rates of return.[4] For World Bank–financed projects, the economic and financial indicators are usually linked to project development objectives and the monitoring and evaluation framework. These indicators include calculations of net present value and an internal economic rate of return, which are typically estimated over the life of the project and encompass the following elements:

- Costs:
  - Capital (fixed or up-front investment costs)
  - Operating (variable or operating costs, maintenance, disposal costs)

- Benefits (listed from primary to secondary benefits):
  - Travel time savings: quantified using transportation models, including demand forecasts and mode choices; this value theoretically captures most of the potential gains in land values from improved transportation services and accessibility

  - Vehicle operating cost savings: quantified on the basis of the wear and tear on vehicles and fuel savings
  - Road safety benefits: sometimes quantified in terms of avoided injuries, fatalities, and property damage using statistical values from local data sources
  - Air quality benefits: sometimes quantified on the basis of estimates of the economic and health impacts of reduced emissions of local pollutants
  - Greenhouse gas emissions: directly related to fuel consumption and sometimes quantified, particularly to evaluate the possibility of selling carbon credits (see the section on innovative financing below); greenhouse gas emissions are usually normalized by person or unit of economic welfare (such as GDP), and issues related to estimating transport-derived greenhouse gases are discussed
  - Other impacts of infrastructure on employment and poverty: sometimes considered; however, the broader, long-term economic impacts of integrated transport systems and technologies (for example, impacts on the small business market and on technology exports) are rarely quantified

- Sensitivity analyses to assess the viability of investments under different scenarios, based on changes in at least three variables:
  - Costs (that is, increases in the capital or operating costs)

Table 3.29 Economic and Financial Aspects

ASPECT	STAGE I: ENABLING MEASURES	STAGE II: ADDITIONAL MEASURES
Economic	• Feasibility or planning study • Cost-benefit analysis (primary benefits)	• Alternatives analysis • Evaluation of primary and secondary benefits
Financial	• Financial analysis	• Innovative financing options

Source: Author compilation (Georges Darido).

Bus Rapid Transit

BRT is an integrated system of high-quality bus interventions that may be implemented incrementally and catalyze more substantive reform. Among the key elements of a BRT system are the following:

- Exclusive or segregated busways
- Stations with boarding at level and the prepayment of fares
- Large vehicles with multiple doors
- Advanced service and operations plans, including plans for trucks and feeders
- Electronic and integrated fare collection systems
- Intelligent transportation systems, including the centralized control and effective management of passenger information
- Marketing and branding to reinforce a distinct image

BRT systems have been shown to improve service by reducing waiting, boarding, and travel times and by offering modern, comfortable, and convenient services more cost effectively than do rail investments. Public BRT services have increased viability relative to other, more polluting energy-intensive modes.

Milestones in the evolution of BRT systems include the following:

- Since the 1970s, Curitiba, Brazil, has been a pioneer in developing BRT systems as part of a long-term vision and implementation strategy, including reserving rights-of-way for structural axes (major city public transportation corridors) and building institutions with significant technical capacity that have endured political changes.

- The Transmilenio BRT in Bogotá, Colombia has achieved myriad milestones: (1) a high capacity of up to 35,000 passengers per hour per direction, (2) a shorter implementation period as part of an urban redevelopment plan, and (3) recognition as the first public transportation system approved under the Clean Development Mechanism for the sale of carbon credits.

- The Colombian National Urban Transport Program is a framework for technical collaboration and financing for BRT systems in seven participating cities to replicate and scale up the success of Transmilenio. The national government has financed most infrastructure investments, while the cities oversee the operations of pri-

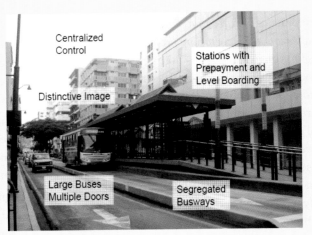

Source: Hidalgo, Custodio, and Graftieaux (2007).

vate concessionaires that earn profits based on system efficiency. Pereira (see photo) was the first city to implement a scaled-down version of the Transmilenio system. The system featured one-way streets in a narrow downtown area and improved solutions for feeders, including an electronic fare system.

- Santiago and Seoul have both chosen to implement ambitious public transportation reforms and investments that have included BRT-type corridors, the integration of bus networks and express trunk feeders, integrated smartcard systems, and centralized controls. Among the main lessons drawn from the Transmilenio experience has been the realization that a realistic or incremental implementation plan (with pilot projects) is critical.

- BRT systems may complement other public transportation investments. Several Asian cities have taken steps to discourage private car use and strengthen public transportation by improving bus systems and building or expanding urban rail systems. These cities include Hong Kong, China; Seoul; Singapore; and Tokyo (Wright 2004).

For more information about relevant costs, benefits, and lessons learned, see Wright and Hook (2007), Levinson and others (2006), and U.S. Federal Transit Administration (2004).

- Start state (that is, to account for the cost of delays)
- Revenue or demand variation (that is, less than expected traffic volumes or ridership)

Forecasts of demand and mode choice are critical elements of an economic analysis. Theoretically, an individual chooses an available mode or service that maximizes utility or minimizes total cost. Utility comprises several variables, including total travel time and total travel cost (figure 3.55). Travel time may be monetized into a value of time among different users, and travel purposes may be measured through surveys. Monetary estimates of time are valid only to the extent that there is a real trade-off between time and money. Another limitation of mode

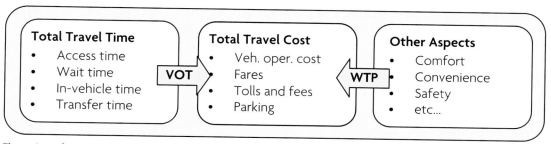

Figure 3.55 Elements of Utility in Models for Choosing a Transportation Mode

Source: Author compilation (Georges Darido).
Note: VOT = value of time. WTP = willingness to pay.

choice models is the valuation of other qualitative aspects of travel, such as comfort, safety, and convenience. Some of these factors may be monetized using surveying techniques to estimate willingness to pay.

Any large sustainable transportation investment is a long-term financial commitment, and its management requires long-term fiscal discipline and institutional capacity. Examples include urban rail or metro investments that may cost hundreds of millions of dollars to construct and millions per year to operate. The government finances public transportation infrastructure in many cities, and services are often subsidized even if they are operated by the private sector. This is because local conditions often do not allow profitable operations, while fulfilling social objectives. The financial indicators for transportation projects depend on the type of project (that is, investment, reform, or public-private partnership) and should measure liquidity (working capital ratio), performance (operating ratio for public transportation companies), and financial sustainability (debt service coverage ratio).

There are many innovative financing options that may be considered:

• *Public-private partnerships* are frameworks for joint investment planning (and asset ownership or revenue sharing) by the public and private sectors. They are usually selected because of fiscal constraints on public funds or the rigor and efficiency contributed by the private sector. One example is the use of public land as a direct payment to private partners in return for the construction of public transportation facilities and stations. The partnerships are not always the lowest cost option; therefore, an appropriate allocation of risks (for example, construction, economic, and traffic risks) and accurate estimations of value are important. Good examples include public transportation terminals financed by the private sector, such as cases in Brazil and Japan. Terminals also represent good locations for leased retail space and other public services (figure 3.56).

• *Land development and value capture* cover techniques whereby the public entity sells surplus land to developers (joint developers) or develops land around transportation investments. A good example is the public transportation system in Hong Kong, China.

Figure 3.56 Curitiba: Terminal Carmo, Adjacent Shops, and Citizenship Street
Source: Institute for Research and Urban Planning of Curitiba.

- *Tax increment financing* provides a dedicated stream of land taxes to finance projects.

- *Carbon financing* involves the sale of greenhouse gas emission credits to finance capital or operating costs. One example is the methodology that was approved through the Clean Development Mechanism and used in Bogotá's Transmilenio.

- *Other tolls and fees* may be imposed to fund specific projects. For example, impact fees have been used in some parts of the United States to tax developers for the expected impacts of the development of transportation networks.

Integration Opportunities

There are a number of analytical resources and modeling tools used in transportation that are potentially applicable to other sectors. These are summarized as follows:

- *Macroplanning tools* include household and user surveys of travel origins and destinations. These surveys should be executed at least once every 7 to 10 years and survey all modes (walking, bicycling, private vehicle use, public transportation modes of various types, taxis, and trucks) aggregated by the travel purpose and income level of the users.

- *Transportation model.* The data on origins and destinations may be applied to develop and calibrate a four-step metropolitan transportation model, which encompasses trip generation, trip distribution, mode choice, and network assignment. Officials may use this model to support decisions on major transportation policies or investments in a metropolitan transportation plan. Carbon footprint and greenhouse gas analyses also require recent survey data on origins and destinations (preferably less than 7 years old) and a calibrated transportation model.

A common pitfall in the development of such a model is the tendency to underestimate or ignore shorter, nonmotorized trips within zones and overestimate longer trips on major corridors. Freight modes must also be accurately considered, especially in large and rapidly growing cities in which truck transport has significant impacts.

- *Emissions inventories.* There are at least two ways to estimate urban transportation emissions. A bottom-up approach entails collecting data on vehicle fleets, including number, type, average fuel efficiency, and annual vehicle kilometers traveled. This information may be replaced or supplemented by motorized trip tables and average distances taken from origins and destinations studies. A top-down approach analyzes the amount of fuel sold or consumed in an area. The latter approach is often used to check the data in the former, but the two approaches are normally difficult to reconcile in drawing conclusions on urban emissions.

- *Important microdesign tools* include site plans, station area plans, and zoning ordinances.

There are also important bottlenecks in sustainable transportation planning and the implementation of potential solutions. These include the following:

- *Road space allocation.* Traffic engineers tend to focus on vehicle volumes on roadway segments and intersections. Investments often thus optimize networks for moving vehicles rather than people or goods. By considering vehicle occupancy and appropriately valuing passengers in high-capacity vehicles such as buses, one may easily argue for an increased priority for the use of high-occupancy vehicles on streets and intersections. These concepts are illustrated in figure 3.57.

- *Energy consumption* may be incorporated in a city's master planning process if modal

Figure 3.57 The Amount of Roadway Used by the Same Passengers Traveling by Car, Bicycle, or Bus

Source: Petersen and WI (2004).
Note: The photo shows a street in Münster, Germany.

origins and destinations data and fuel consumption data are systematically captured. These data may be harnessed to develop multisector spatial plans and policies for energy use and climate change mitigation and adaptation (that is, infrastructure standards and codes and emergency procedures).

- *Transport regulations and city finances.* Public transportation provisions, regulations, fares, subsidies, and service levels directly affect a city's finances. In public-private partnerships, the involvement of private companies and their associated exposure to risks typically progress from management and operations to the ownership of fleets, facilities, and infrastructure. Multiple options are available to allocate risks among public and private partners; the choice depends partly on the extent of development of legal systems and market institutions in a city and country. Regulatory changes in urban transportation are often linked and aligned to complementary reforms in the water and energy sectors, including in ur-

ban utilities and national enterprises.

In sum, there are numerous approaches that integrate transportation and other sectors and may improve the urban ecological and economic environment. Table 3.30 summarizes cross-sector integration opportunities.

Notes

1. It is good practice and the policy in World Bank–financed projects to require ex ante identification of the development objectives of projects and, within a results framework, ongoing supervision over monitoring and evaluation indicators in light of targets.

2. The discussions in this section on public transportation, microdesign, and macrolevel approaches are based on presentations at the World Bank Urban Rail Workshop in Beijing in 2008 and discussions with World Bank experts Shomik Mehndiratta and Sam Zimmerman.

3. A more comprehensive review is available in World Bank (2001). A review of economic instruments such as fuel taxes and efficiency incentives may be found in Timilsina and Dulal (2008).

4. For more detailed information, see the World

Table 3.30 Summary of Cross-Sector Integration Opportunities

DIMENSIONS	URBAN	ENERGY	WATER	SOLID WASTE
Policy, legislation and regulations	Land use zoning for residential, commercial, and institutional properties	Energy and emissions inventories and targets Fuel security Air quality standards	Roadway design standards for drainage	Litter prevention programs Materials recycling programs
Institutional context	Metropolitan coordinating institutions	Reform and regulation Conservation programs	Reform and regulation Conservation programs	Reform and regulation Conservation programs Enforcement programs
Physical systems, technology, and spatial planning	Master planning of non-motorized transport facilities, accessibility, amenities, and urban furniture Public spaces	Fuel and vehicle standards Inspection and maintenance Location efficiency Fleet management and efficiency programs	Storm water runoff Production and disposal of infrastructure, vehicles, and systems	Production and disposal of infrastructure, vehicles, and systems Location of collection and disposal facilities
Stakeholder dynamics	Special events to change behaviors and spur investment			
Economic and financial aspects	Savings through coordination	Savings through coordinated utility development (power lines, lighting)	Savings from coordinated road construction	Savings through coordination

Source: Author compilation (Georges Darido).

Bank's guidance available through "TRN-5 to TRN-26: Economic Evaluation Notes," at http://go.worldbank.org/ME49C4XOH0. TRN-5 provides the context within which we use economic evaluation in the transport sector; TRN-6 to TRN-10 provide criteria for the selection of a particular evaluation technique or approach; TRN-11 to TRN-17 address the selection of the values of various inputs to the evaluation; and TRN-18 to TRN-26 deal with specific problematic issues in economic evaluation.

References

Cervero, Robert, and Jennifer Day. 2008a. "Residential Relocation and Commuting Behavior in Shanghai, China: The Case for Transit-Oriented Development." Working Paper UCB-ITS-VWP-2008–4, UC Berkeley Center for Future Urban Transport, University of California–Berkeley, Berkeley, CA. http://escholarship.org/uc/item/0dk1s0q5.

———. 2008b. "Suburbanization and Transit-Oriented Development in China." *Transport Policy* 15 (5): 315–23.

Colin Buchanan and Partners. 2001. *Key Sites Appraisal Methodology for Development Planning: Final Report.* Edinburgh: Scottish Executive.

Comune di Milano. 2009. "EcoPass: Meno traffico, più aria pulita." Comune di Milano, Milan. http://www.comune.milano.it/dseserver/ecopass/index.html.

Dalkmann, Holger, and Charlotte Brannigan. 2007. "Transport and Climate Change." Module 5e of *Sustainable Transport: A Sourcebook for Policy-Makers in Developing Cities,* rev. ed. Sustainable Urban Transport Project. Eschborn, Germany: Transport Policy Advisory Service, Division 44, Environment and Infrastructure, German Agency for Technical Cooperation.

Fraker, Harrison S., Jr., and William Wurster. 2009. "Sustainable Neighborhood 'Eco-blocks' in China: Qingdao Sustainable Neighborhood Demonstration Project." Urban Sustainability Initiative, Berkeley Institute of the Environment, University of California–Berkeley, Berkeley, CA. http://bie.berkeley.edu/ecoblocks.

Hidalgo, Dario, Paulo Custodio, and Pierre Graftieaux. 2007. "A Critical Look at Major Bus Improvements in Latin America and Asia: Case Studies of Hitches, Hic-Ups and Areas for Improvement; Synthesis of Lessons Learned." Presentation, World Bank, Washington, DC. http://go.worldbank.org/W8FO3NQ680.

IEA (International Energy Agency). 2007. *World Energy Outlook 2007: China and India Insights.* Paris: IEA.

Levinson, Herbert S., Samuel Zimmerman, Jennifer Clinger, James Gast, Scott Rutherford, and Eric

Bruhn. 2006. *Implementation Guidelines.* Vol. 2 of *Bus Rapid Transit,* TCRP Report 90, Transit Cooperative Research Program. Washington, DC: Transportation Research Board of the National Academies.

Lund, Hollie M., Robert Cervero, and Richard W. Willson. 2004. "Travel Characteristics of Transit-Oriented Development in California." Caltrans Statewide Planning Studies, FTA Section 5313 (b), California Department of Transportation, Sacramento, CA. http://www.csupomona.edu/~rwwillson/tod/Pictures/TOD2.pdf.

Mobility in Cities Database. 2001. International Association of Public Transport, Brussels. http://www.uitp.org/publications/Mobility-in-Cities-Database.cfm. Cited in Petersen and WI 2004, 8.

Petersen, Rudolf, and WI (Wuppertal Institute for Climate, Environment, and Energy). 2004. "Land Use Planning and Urban Transport." Module 2a of *Sustainable Transport: A Sourcebook for Policy-Makers in Developing Cities,* rev. ed. Sustainable Urban Transport Project. Eschborn, Germany: Transport Policy Advisory Service, Division 44, Environment and Infrastructure, German Agency for Technical Cooperation.

PPIAF (Public-Private Infrastructure Advisory Facility) and World Bank. 2008. "Introduction to Public Transport Service and Operations Planning." Training materials presented at "Introduction to Public Transport Planning and Industry Reform," Asian Development Bank, Manila, February 4–5.

Rebelo, Jorge M. 1996. "Essentials for Sustainable Urban Transport in Brazil's Large Metropolitan Areas." Policy Research Working Paper 1633, World Bank, Washington, DC.

Sperling, Daniel, and Deborah Salon. 2002. "Transportation in Developing Countries: An Overview of Greenhouse Gas Reduction Strategies." Reports on Global Climate Change, Pew Center on Global Climate Change, Arlington, VA, May.

Timilsina, Govinda R., and Hari B. Dulal. 2008. "Fiscal Policy Instruments for Reducing Congestion and Atmospheric Emissions in the Transport Sector: A Review." Policy Research Working Paper 4652, World Bank, Washington, DC.

U.S. Department of Transportation. 2009. "National ITS Architecture." Version 6.1. U.S. Department of Transportation, Washington, DC. http://www.iteris.com/itsarch/ (last modified January 7).

U.S. Federal Transit Administration. 2004. *Characteristics of Bus Rapid Transit for Decision-Making.* Washington, DC: Office of Research, Demonstration and Innovation, Federal Transit Administration, U.S. Department of Transportation.

World Bank. 2001. "Vehicular Air Pollution: Setting Priorities." South Asia Urban Air Quality Management Briefing Note 1, Energy Sector Management Assistance Programme, World Bank, Washington, DC.

———. 2008. "A Framework for Urban Transport Projects: Operational Guidance for World Bank Staff." Transport Papers TP-15, Transport Sector Board, World Bank, Washington, DC.

Wright, Lloyd. 2004. "Bus Rapid Transit." Module 3b of *Sustainable Transport: A Sourcebook for Policy-Makers in Developing Cities,* rev. ed. Sustainable Urban Transport Project. Eschborn, Germany: Transport Policy Advisory Service, Division 44, Environment and Infrastructure, German Agency for Technical Cooperation.

Wright, Lloyd, and Walter Hook, eds. 2007. *Bus Rapid Transit Planning Guide.* New York: Institute for Transportation and Development Policy.

Zegras, Christopher, Yang Chen, and Jurg Grutter. 2009. "Potentials and Challenges for Using the Clean Development Mechanism for Transport-Efficient Development: Case Study of Nanchang, China." Paper 09–2864, Transportation Research Board 88th Annual Meeting, Transportation Research Board of the National Academies, Washington, DC, January 11–15.

Zimmerman, Sam. 2008. "Land Use and Metros." Presentation at the World Bank workshop, "Urban Rail Development," Beijing, June 27.

Cities and Solid Waste

Overview

Waste management is often viewed as an end stage in a product's life cycle. However, it also provides opportunities to renew the useful life of materials by recycling, composting, and recovering energy through thermal processes such as incineration or methane capture at landfills. The energy released through thermal treatment or methane combustion may then be used to generate electricity or other power, thereby creating a synergistic loop.

The main desired aims of waste management systems may be summarized as the protection of public health by preventing the accumulation of food for rodents, insects, and other disease vectors and the protection of the environment primarily by controlling and eliminating air and water pollution. The conservation of materials and energy resources is another important desired aim of well-designed waste management systems.

Figure 3.58 delineates the main inputs, interventions, and outputs of a waste management system. Inputs may be independent or dependent based on the degree of controllability. For example, geographical location normally represents an independent input over which authorities have little if any control, while policy frameworks are dependent because cities are able to influence legislation. Outputs may work toward set aims, or they may be undesirable; thus, they may include air and water pollution levels that are greater than expected.

Why is waste management important?

Effective waste management is critical to the ecological and economic vitality of urban areas. A properly designed and operated waste management system provides the following:

- *Protection of public health:* Inadequately collected waste and poor disposal are breeding grounds for disease-carrying vermin such as rodents and insects. In addition, bacteria such as salmonella and shigella thrive in food waste, which accounts for over half of the municipal waste in developing countries (Pablo 2007).

- *Ecological protection:* An effective system will regulate or prohibit harmful practices, such as open burning and improper waste disposal, to protect local and regional ecologies and mitigate the negative impacts on air, land, and water.

- *Effective budget management:* Efficient solid waste management is important to the fiscal health of a city because solid waste management in medium-size cities can reach 50 percent of total municipal budgets (Pagiola and others 2002).

- *Employment:* Waste management provides significant formal employment in areas ranging from collection to disposal. Informal employment may also be important, such as in the case of waste collectors who recover materials for recycling prior to collection at disposal sites.

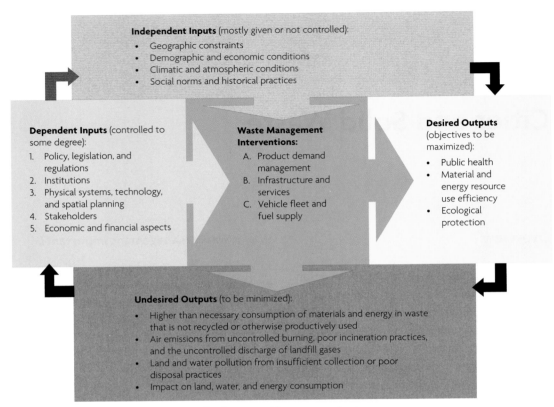

Figure 3.58 The Input-Output Framework of a Waste Management System

Source: Author compilation (Charles Peterson).

- *Aesthetics:* Effective systems protect the visual and sensory appeal of cities by ensuring that waste is effectively managed and that practices do not unduly affect residents and visitors.

What does solid waste comprise?

Most people rightly attribute significant portions of municipal solid waste to residents and commercial operations. However, municipal solid waste includes other sources, some of which require special waste management approaches:

- *Medical facilities:* Hospitals and clinics generate solid waste. Medical facilities also generate infectious waste and, occasionally, radioactive waste. In general, infectious and radioactive waste material must be managed separately from other medical waste, although infectious waste that has been rendered noninfectious through incineration and autoclave or microwave treatment may be discarded in municipal landfills.

- *Industry:* Municipal landfills may receive industrial waste, but hazardous waste must be managed separately.

- *Construction and demolition debris:* This waste comprises discards from new building construction and the renovation of older facilities and residue from structures that have been torn down. Much of this debris may be recovered for recycling, which could minimally include clean fill if problem materials such as wood mixed with lead paint or asbestos are extracted.

- *Slaughterhouses:* These facilities produce animal waste and various excreta that must be properly managed.

- *Sewage treatment plants:* Sludge, the residue of sewage treatment, may be discarded in a landfill in limited quantities to maintain the stability of disposal sites if it has been mechanically dewatered. This is because sludge typically has a high moisture content (70–80 percent). Alternatively, dewatered sludge may be applied to land as a soil conditioner. Dried sludge, which is more suitable for storage and longer-distance transport to agricultural areas, may also be applied to land. In addition, sludge may be used as compost. Such uses require that sludge meet regulatory standards on pollutants such as metals. The biological digestion of sludge at treatment plants may reduce volatile organics by more than 50 percent. Such digestion reduces the amount of sludge that must be treated or discarded. The digestion process, which is anaerobic, also produces gas that may be up to 60 percent methane in composition. The gas may be used as an energy source.

- *Combustion residue:* This comprises ash from waste incineration or the combustion of solid fuels in central facilities or in households (that is, ash generated from cooking and heating).

What are the characteristics of waste?

The options for urban solid waste management depend in part on the quantity and composition of the discarded waste.

Generation: The total quantity of discarded waste depends on the per capita generation rate, which is highly correlated to residential income. Cities with higher incomes tend to generate more waste than cities of similar size, but with lower incomes. People with higher incomes purchase more goods and services, which results in more waste. Table 3.31 illustrates the relationships between income level, per capita waste generation, and total urban-generated waste. If one assumes a consistent population of one million residents, a low-income city would generate 500 tons of waste per day, but this would more than triple, to 1,600 tons per day, in a high-income city.

Composition: The composition of waste also varies according to the income level of the people producing the waste (table 3.32). The amount of food waste tends to be greatest among lower-income earners. As income increases, food waste generally falls because consumers purchase greater amounts of prepared food relative to fresh food. Fresh food results in more waste from peels, pits, and other residue.

Waste composition helps determine the appropriate approaches to waste management. A city with a high level of food waste, for example, should provide more frequent collection to minimize the potential to attract vermin, which may transmit diseases to residents. Aerobic composting is also suggested for areas with significant food waste, which decays rapidly in compost operations given its high moisture content. Conversely, areas with significant food waste are not good candidates for incineration systems. Waste only autocombusts if the mois-

Table 3.31 Waste Generation Rates

INCOME LEVEL	GENERATION RATE (KILOGRAM PER CAPITA PER DAY)	WASTE QUANTITY[a] (TONS PER DAY)
Low	0.5	500
Middle	0.7	700
High	1.6	1,600

Source: Author compilation (Charles Peterson).
a. The size of the population is assumed to be 1 million.

Table 3.32 The Composition of Waste by the Waste Producers' Income

percent

| MATERIAL | INCOME LEVEL | | |
	LOW	MIDDLE	HIGH
Food	40–85	20–65	20–50
Paper	1–10	15–40	15–40
Recyclables	4–25	5–26	11–43
Miscellaneous	15–50	15–50	5–20
Moisture	40–80	40–60	20–30

Source: Author compilation (Charles Peterson).

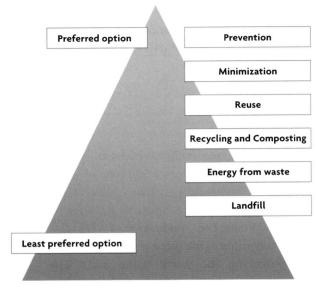

Figure 3.59 Waste Hierarchy

Source: Author elaboration (Charles Peterson).
Note: The most favored options are at the top of the pyramid, declining to the least favored at the bottom.

ture content is less than 50 percent, which means that a supplemental fuel is needed to burn food waste because of the higher moisture content of the waste. The incidence of this factor is greater during rainy seasons.

What are the common approaches to waste management?

The options for managing waste have been ranked in a number of forums. A universally accepted pyramid of choice is provided in figure 3.59. Under this schema, waste prevention is the preferred option at the top of the pyramid, while waste disposal is the least favored option at the pyramid's foundation. Waste prevention and minimization are equivalent to waste reduction and involve the prevention of waste at the source by redesigning products or changing patterns of production and consumption (Peterson 1989).

Waste reduction (that is, prevention and minimization) comprises demand management practices such as the creation of durable products with longer lives. Reuse describes products that may be used more than once, unlike dispos-

able or single-use products such as diapers and soft drink containers. The recycling option includes composting. Energy recovery includes technologies, such as methane capture, that harness waste or by-products to generate usable energy.

Policy, Legislative, and Regulatory Dimensions

The policy, legislative, and regulatory framework for solid waste management involves the following dimensions.

The policy dimension: National and local governments must be committed to improving urban environments to protect the health of residents. Additionally, especially at the national level, governments should emphasize demand management to control the quantity of generated waste. Demand management includes programs to reduce waste, encourage reuse, and recycle.

In Japan, for example, Yokohama took action to reduce the demand for waste treatment (incineration) and disposal. In 2003, the city launched the G30 Action Plan (G = garbage; 30 = 30 percent reduction in waste generation by fiscal year 2010). The program targets residents, businesses, and government and focuses on the 3Rs (reduce, reuse, and recycle). The program has exceeded its goals. By fiscal year 2007, waste had been reduced by 38.7 percent, or 0.1 million tons per year. The reduced waste enabled officials to close two incinerators, yielding savings in operating expenditures of US$30 million per year. (We base the calculations here and below on an exchange rate of approximately ¥100 = US$1.) After subtracting the US$24 million in operating expenditures for the additionally required operations, such as separating waste and subcontracting recycling, the city realized net savings of US$6 million per year. The closing of the two incinerators also obviated the need for US$1.1 billion in capital expenditures that would have been required to renovate the incinerators.

At the local level, effective wastewater management and air pollution control are needed to address risks. In cities with open drains, for example, uncollected waste dumped into drains may block or hinder water flow, which may cause severe flooding during periods of significant rainfall. Because waste disposal sites are often in low-lying areas near waterways or wetlands, inappropriately designed landfills or waste disposal may also contaminate surface and groundwater, with concomitant risks for drinking water and residents who live downstream.

Even if disposal sites are not near water bodies, groundwater may still become contaminated. Leachate, which includes contaminated rainwater and waste with a high moisture content (such as discarded food), may percolate through the soil under an inadequately designed disposal site and pollute groundwater. Leachate may be controlled by lining the bottom of a disposal site with clay or a synthetic material such as high-density polyethylene. Such materials inhibit the flow of leachate, which may then be collected through pipes on top of the liner and treated.

The combustion of solid waste, whether in an incinerator or by fire at an open dump, may add particulates (smoke) and other toxic and nontoxic air pollutants to the atmosphere. Emissions from incineration may be controlled, but a comprehensive emissions control system may be nearly as expensive as a combustion system.

Government policies also need to address the disposal of other types of wastes, such as infectious medical waste, hazardous waste, sewage sludge, residue from livestock slaughterhouses in urban areas, and residue from energy production facilities (especially those using solid fuel such as coal or biomass). These wastes should be managed separately from municipal waste and may require treatment prior to disposal (for example, to render medical waste noninfectious).

Whenever possible, policies should be benchmarked against policies in other cities in the region and elsewhere in developing and developed countries. Benchmarking helps evaluate the expected effects of policies through the analysis of past practice and lessons learned.

The legislative dimension: Appropriate laws at all levels of government are important to the development and operation of the waste management sector. Legislation supporting a viable waste management framework should be instituted if laws are not already in place.

Environmental protection and solid waste management measures are often complementary. For example, improved waste management is frequently a component of environmental legislation to protect water sources and reduce air pollution. The appropriate siting and operation of landfills strengthen efforts to protect water sources. In addition, air quality may be improved by reducing the open burning of waste, installing air pollution control equipment at incinerators, and eliminating poor operational practices.

The regulatory dimension: Enactment of suitable national and local legislation should be followed by passage of appropriate regulations and enforcement measures for waste management laws. Multiple regulations should be considered and approved by governments based on waste management requirements and the existing environmental and waste management policy framework. The following measures should be considered:

- *Waste classification and standards for collection, treatment, and disposal:* Regulatory definitions are important tools in setting standards for waste management practices. For example, management programs should be much more rigorous for infectious medical waste than for noninfectious medical waste. Treatment standards for the various types of infectious medical waste, such as medical sharps and body parts, differ in terms of sterilization. The type of waste also influences the way waste is disposed. Infectious waste, for example, may have to undergo incineration, autoclave, or microwave treatment and may have to be buried in disposal sites with secure perimeters that restrict access.

- *The provision of cost recovery mechanisms, such as user fees:* Those measures should be considered and tied to the amount of disposed waste. For example, if waste is collected from households or taken to a staffed collection point, the generators of waste might be charged per container or bag. To encourage reductions in waste, governments might levy lower fees for recyclable and compostable materials. (See elsewhere below for discussions on other mechanisms to cover the cost of waste management services, such as product charges, packaging fees, and carbon finance to reduce greenhouse gas [GHG] emissions.)

- *Systematic programs to monitor and measure solid waste services:* Such programs may be valuable tools in evaluations of the performance of public or private waste management systems. Such programs should be developed on the basis of appropriate metrics and benchmarks that provide a basis for analyzing performance trends. A performance database might be considered that would include historic data. Benchmark information on other cities in the metropolitan area, region, and country might be collected for such a database. Evaluations of trends might reveal positive or negative performance or progress that is slower than anticipated by local municipal officials. By evaluating trends, the programs would be able to respond effectively to shortfalls or poor performance.

- *Local and regional planning tools for waste management:* Regional collaboration should be actively pursued to promote the most comprehensive solutions to waste management and to achieve economies of scale. Though a city might be capable of reaching suitable economies of scale without seeking regional solutions, a regional approach may enhance the benefits of a project by expanding the scope beyond the urban area.

An important, but often overlooked aspect of regulation is enforcement by independent government agencies. Enforcement is critical to implementing regulations and ensuring ongoing regulatory compliance. An effective enforcement program requires that enforcement agencies have the tools to address regulatory deviations. An enforcement agency or program should have the capacity to levy fines and apply other mechanisms to encourage compliance by punishing noncompliance.

Institutional Context

An effective policy, legislative, and regulatory framework should be combined with institutional measures to eliminate gaps and overlaps in the operational structure of the waste management sector.

Government: An effective program should involve ongoing senior-level coordination among municipal and national government officials and agencies. Local government programs should involve more than merely those people and agencies directly active in the city, but, to encourage regional cooperation, also people and agencies in the wider metropolitan area. In the waste management sector, regional cooperation could achieve efficiencies and economies of scale in treatment and disposal programs. Efforts should also be undertaken to strengthen regional coordination in strategic planning.

Operational structure: In many cities, a public agency provides solid waste services, which may be an efficient means of achieving community public health and environmental goals. However, promoting competition in delivering waste management services may improve efficiency and services. Promoting private sector participation is an option for encouraging competition in the waste management marketplace. Private sector participation may make particular sense in certain areas, such as waste collection and disposal (for example, landfill construction and operations), but it may be considered in all services. Private sector participation may also be useful in ancillary activities. Vehicle maintenance, for example, might be contracted to a private sector provider, possibly with the contractual participation of vehicle fleets in other sectors. An effective private sector participation program would leverage a competitive environment in which private companies provide sound pricing and high-quality services. For assistance in assessing the options for private sector participation, relevant metric and benchmark information should be collected on service levels and market trends within the city and in other cities with a mix of public and private service providers.

However, waste management service decisions should be based on a consideration of ongoing and planned efforts to reduce waste. This minimizes potential conflicts, such as the problems raised if a contract has been awarded for the operation of an incinerator or landfill, but the contract is no longer needed because waste has been sufficiently reduced.

A combination of metrics and benchmarks may be used to monitor system performance as the sector becomes modified (box 3.18). For example, collection metrics might address the issue of uncollected waste in designated areas, the quantities of waste gathered per collection vehicle (requires scales at disposal sites or other locations), and the average number of daily trips per collection vehicle. It is advisable to consider the development of systemwide benchmarks rather than benchmarks for individual vehicles because aggregate benchmarks strengthen comparisons with other cities and are more useful in evaluating sectorwide interventions.

To strengthen waste management, small- and medium-size enterprises might be tapped to provide targeted waste collection and recycling services. These enterprises often supply better services in areas inaccessible to collection trucks. Additionally, recycling may promote informal and creative employment opportunities, such as waste picking. Recovered materials are also valuable for primary industries, which use them to produce new products, thereby stimulating the local economy and providing income to additional workers.

It is important that any private sector participation alternatives pursued by local govern-

---

BOX 3.18

Performance Metrics

Specific waste management metrics in cities vary depending on the services provided. A waste collection operation might include the factors noted below. Waste collection is a good sample subsector because it is an integral part of a waste management system.

- Population served
- Annual quantity of waste collected
- Annual quantity of recyclables collected
- Annual total waste and recyclables collected
- Annual collection cost
- Collection cost per ton
- Collection routes per week
- Total stops serviced per week (the number of times a truck stops to collect waste)
- Average stops per route

- Average tons collected per stop
- Cost per stop
- Population served per stop
- Cost per resident

Common indicators for GHG projects might include the following:

- Average system GHG emissions per ton of collected waste transported to a transfer station, treatment facility, or disposal site
- Average system GHG emissions per ton of collected waste recycled, incinerated, or landfilled with or without gas recovery

These indicators may be used to calculate a collection cost benchmark that might then be used to track performance or assess economic efficiency relative to other programs.

ments be supported by proven procurement methods, particularly those successfully tested in other cities and regions. Procurement methods should address tendering (including the transparency of the procurement process), contract design, contract management (including accountability), and other issues.

Physical Systems, Technology, and Spatial Concerns

The physical, technological, and spatial aspects of waste programs depend on factors unique to each city. An initial determination of the physical parameters of the existing waste management system, the technologies used, and the spatial requirements is crucial. A second step is to assess alternative options that fit the guidelines of the program.

Spatial characteristics

Understanding the physical nature of a city is critical to the development of a clear definition of existing and potential parameters. Baseline factors that need to be defined include the following:

- Population
- Surface area (population density)
- Terrain and topography of a city and its surroundings
- Per capita income
- Mix and location of economic activities, including industrial (types), transport (freight consolidation and warehousing), commercial (types), special institutions (types, such a medical facilities that generate infectious wastes), residential (multifamily and single family)

Waste characteristics

After establishing the factors that affect waste generation and waste management operations, one should seek to understand the characteristics of the waste generated in a target city. The two critical characteristics are waste quantity and waste composition.

Quantity: Quantity is crucial in determining the size of a waste management system. As the population or economy of a city grows, per capita waste and total waste tend to increase (see table 3.31). Reliable data on the quantity of waste being generated and managed in various urban areas are important in determining the size of the system, planning equipment purchases, and identifying the services needed for managing a city's waste.

Composition: Low-income cities tend to produce substantial organic discards, mainly food waste. Food waste has a high moisture content, which can affect a waste stream's suitability for alternative treatment processes. Systems that must treat greater shares of food waste lend themselves to composting rather than incineration because food waste will not generally autocombust.

Waste system components

The processes used to manage solid waste might include the following:

- *Waste storage at households or commercial establishments.* Collection may also occur at selected points that serve a number of these groups.

- *Vehicle collection:* This entails defining pickup frequency, equipment, and crew size. Besides truck collection, *underground pneumatic collection* might offer advantages in urban areas with narrow streets that impede truck collection. Under this scenario, waste is drawn by vacuum pressure to a collection point where it is deposited into a container. A pneumatic system includes trash receiving stations, an underground piping system, a vacuum blower that pulls waste through pipes, air filters for particulates and odor, and a facility where collected material may be stored until it is hauled to a treatment or disposal facility. Pneumatic systems have been used in smaller urban areas in Japan

BOX 3.19

An Innovative Waste Collection Approach

In Curitiba, Brazil, the Green Exchange Program has been undertaken in slum areas inaccessible to waste collection vehicles. To encourage the poor and slum dwellers to clean up the areas and improve public health, the city started to offer bus tickets and vegetables to people who brought garbage to neighborhood centers. In addition, children were allowed to exchange recyclables for school supplies, chocolate, toys, and show tickets. The city purchases vegetables at discounted prices from farmers who have trouble selling abundant products. Through this program, the city saves the costs of arranging waste collection in slum areas, which often have inadequate roads, and helps farmers unload excess produce. The program also helps improve nutrition, transportation accessibility, and entertainment opportunities among the poor. Most important, slums have become cleaner, and there is less litter, less disease, and less garbage dumped in sensitive areas such as rivers.

Citizens bringing their garbage for collection.

Source: Institute for Research and Urban Planning of Curitiba.

BOX 3.20

A Recycling Program Involving Citizens

Curitiba's Garbage That Is Not Garbage Program encourages people to separate discards into recyclable and nonrecyclable waste. To raise awareness of the program, the city is educating children on the importance of waste separation and environmental protection. Campaign mascots have been created, and school activities have been launched. One to three times a week, trucks collect paper, cardboard, metal, plastic, and glass that have been sorted by households. This recycling saves the equivalent of 1,200 trees a day, and local parks contain displays on the numbers of trees saved (Rabinovitch and Leitmann 1993; Hattori 2004). The money raised from selling recyclables supports social programs, and the city employs the homeless and people in alcohol rehabilitation programs in the garbage separation plant. Recycling leads to other benefits. For instance, recycled fiber is used to produce asphalt for road construction. Tire recycling has removed piles of discarded tires, which can attract mosquitoes that transmit dengue disease. Proper tire collection has decreased dengue disease by 99.7 percent (Vaz Del Bello and Vaz 2007). Nearly 70 percent of the residents participate in Curitiba's recycling program. Around 13 percent of Curiti-

The Garbage That Is Not Garbage Program.

Source: Institute for Research and Urban Planning of Curitiba.

ba's waste is recycled, which greatly exceeds the 5 percent and 1 percent recycling rates in Porto Alegre and São Paulo, respectively, where education on waste dissemination has not translated into significant impacts (Hattori 2004).

and Sweden and in large complexes such as airports, shopping centers, and hospitals. In many developing countries, particularly in slums, waste collection in urban areas with narrow streets may be accomplished through innovative solutions. For instance, local residents and the poor may become directly involved in waste collection (box 3.19).

- *Formal and informal (that is, waste picking) recycling:* Recycling may take place at collection points, treatment facilities, or disposal sites. In Curitiba, the city generated employment and formalized informal activities through an innovative recycling program (box 3.20).

- *Collection at transfer stations:* This collection consolidates waste for transport to treatment and long-term disposal facilities.

- *Treatment facilities, including organic waste composting (figure 3.60) or incineration with or without energy recovery:* The heat released during combustion or other thermal processes may be used to generate both electricity and heat through cogeneration. However, given the high moisture content in much municipal solid waste in Asia, officials there should be cautious in pursuing thermal processes that depend on autocombustion because waste needs to have a moisture content of less than 50 percent to autocombust. Though a supplemental fuel such as coal may be used to heat waste so to evaporate water and compel combustion, this fuel increases energy needs, costs, and pollution. Municipal officials should also avoid treatment processes that have not been commercially tested.

- *Disposal sites for direct and residual wastes from households, commercial establishments, and treatment facilities:* Disposal sites range from open and modified dumps (at the lower end of environmental and public health protection) to safer engineered landfills (figure 3.61). Disposal sites, especially engineered landfills, create anaerobic conditions as organic waste decomposes.

A by-product of anaerobic decomposition is landfill gas (LFG). Typically about 50 percent methane (a GHG), LFG may be recovered and used for energy or burned (flared). (Natural gas is composed of about 99 percent methane.) The recovery and combustion of LFG are potentially eligible for carbon finance if a project meets the conditions of the Clean Development Mechanism (CDM), which is a provision of the Kyoto Protocol.

A recovery system that uses LFG to generate electricity is shown in figure 3.62. The complex

Figure 3.60 Waste Sorting Plant and Windrow Composting Operation, Cairo

Source: Photos by Charles Peterson.

Figure 3.61 A Compactor Operating on a Landfill

Source: Photo by Charles Peterson.

Figure 3.62 Central Electricity Generation Facility and Flare for Landfill Gas, Tianjin, China

Source: Peterson and others (2009).

includes blowers (vacuum pumps) to extract gas from the landfill, combustion generators for power production, and a flare to burn excess LFG.

Stakeholder Dynamics

Urban residents, workers, and visitors generate waste in their daily lives by preparing meals, transacting business, and participating in activities that use the goods and services of cities. Successful waste management interventions depend on the cooperation and participation of these stakeholders in the waste management system.

The involvement of stakeholders in a waste management program should be ongoing. It is important that stakeholders be interested in the program and that their suggestions and ideas be considered as the program evolves.

A stakeholder participation program should provide a thorough plan for soliciting and tracking comments and reporting on relevant actions. Periodic meetings should be organized during a program's lifespan to engage stakeholders. In addition, an independent call center should be considered to enable stakeholders to register comments on the waste management system (questions, complaints, and praise). Public meetings, call centers, and other communication mechanisms should track the nature of contacts, and efforts should be made to monitor actions to resolve issues, questions, and complaints. The stakeholder program should monitor the time needed to resolve issues. Summary information on comments, issues, actions, and performance should be presented to stakeholders on a regular basis (often quarterly). Yokohama, Japan, for example, has been successful in involving stakeholders in waste reduction activities through

---

BOX 3.21

Waste Reduction through Stakeholder Engagement, Yokohama

Yokohama's G30 Action Plan identifies the responsibilities of stakeholders, including households, businesses, and the city government, in achieving waste reduction through the 3Rs (reduce, reuse, and recycle), the principle that the polluter pays, and extended producer responsibility (City of Yokohama 2003). The plan provides mechanisms and detailed action programs for an integrated approach to reduce waste. For example, citizens must separate their waste into 15 categories and dispose of it at designated places and times based on the relevant waste category. Businesses are asked to provide products and services that create less waste and to implement the 3Rs. As one of the largest generators of waste, the city is committed to reducing waste and working together with citizens and business.

To raise awareness of the G30 approach, the city has conducted environmental education and promotional activities and requested public action to achieve the G30 goal. To promote waste separation, it has held more than 11,000 seminars among neighborhood community associations and the public to explain waste reduction methods, including ways to separate waste. (In Yokohama, 80 percent of the population participates in neighborhood community associations; see City of Yokohama 2008.) The city has also sponsored 470 campaigns at all railway stations and about 2,200 awareness campaigns at local waste disposal points and other places (City of Yokohama 2006). Campaign activities have also been carried out at supermarkets, on local shopping streets, and at various events. The logo for G30 has been printed in city publications and displayed on vehicles and at city events.

As a result of these efforts, by fiscal year 2005, Yokohama achieved the 30 percent waste reduction target that had been set for fiscal years 2001 to 2010. By fiscal year 2007, the city had reduced waste by 38.7 percent despite an increase in the city's population of 165,875 since 2001. The benefits are outlined as follows:

BENEFIT	AMOUNT
Total waste reduction, fiscal years 2001 to 2007	623,000 tons (−38.7 percent)
Economic benefit	US$1.1 billion capital costs saved because of incinerator closure US$6 million operating costs saved because of incinerator closure Life of landfill sites extended
CO_2 reduction, fiscal years 2001 to 2007	840,000 tons

Note: The calculations are based on an exchange rate of ¥100 = US$1.

comprehensive public campaigns and other efforts to raise awareness (box 3.21).

Economic and Financial Aspects

Three economic and financial factors influencing the waste management sector are institutional capacity, financial sustainability, and the cost efficiency of service delivery.

Institutional capacity
The important element is the capacity of the municipal financial management team, the success of which depends on effectively tracking and managing cash inflows from cost recovery instruments (see below). The financial team also needs to manage cash outflows successfully. Cash outflows are incurred for capital and operating expenditures.

As a complement to cost recovery instruments, budget allocations are normally needed to support capital and operating expenditures. Budget allocations are usually supplied by local governments, but also provincial and national agencies. International donor assistance may be a source of revenue, especially for capital investments.

Financial sustainability
The financial sustainability of a program depends on the program's ability to generate sufficient cash flows to cover program expenditures through various measures.

User charges
A common way to generate cost recovery is to charge users for received services. From a fairness perspective, users should be charged on the basis of quantities of discarded waste. However, this may be administratively challenging even if households and commercial operations receive services directly. Because waste pickup in many cities targets multiple households and commercial operations, charges may be based on other factors, such as floor space or percentages of electrical bills. Such approaches do not provide economic incentives for the generators of waste to increase recycling or reduce waste. Nonetheless, such imperfect proxies at least provide a basis for the recovery of all or a share of system costs.

Product charges
A second option for recovering costs that might stimulate recycling or waste reduction is the application of a charge on products in the waste stream. This polluter pays principle has been applied in some European countries through a packaging fee, which is used to provide revenue to help pay for waste management. This concept has yet to be applied in East Asia, but it might be a workable solution and a step forward.

Carbon finance
Carbon finance, a program designed to support GHG emission reductions through waste management and other technologies, may generate revenue to cover program costs. Carbon finance procedures are detailed in the Kyoto Protocol's CDM provision (box 3.22). The mechanism enables industrial countries that have ratified the Kyoto Protocol to purchase emission reductions to meet their Kyoto Protocol targets through projects in developing countries.

In the waste management sector, emissions may be reduced by capturing and using LFG, composting organic waste, and incinerating waste. Box 3.23 details an example of an LFG capture project in Tianjin, China. As the organic fraction of solid waste decomposes in a landfill, anaerobic conditions lead to the production of methane, a combustible GHG that has 21 times the global warming potential of carbon dioxide.

Efforts are under way to devise a replacement accord for the Kyoto Protocol, which expires at the end of 2012. The current situation presents short-term uncertainty about the future form of carbon finance. However, many projects with expected preparation times of up to two years or longer are still being developed.

BOX 3.22

The Clean Development Mechanism and Waste Management

The Kyoto Protocol. One of the early efforts to address global warming was the formation of the Intergovernmental Panel on Climate Change by the World Meteorological Organization and the United Nations Environment Programme in 1988. The panel compiles published scientific and technical literature on global warming, its potential impacts, and the options for adaptation and mitigation.

Another effort to stabilize GHG emissions was the establishment, in 1994, of a voluntary program under the United Nations Framework Convention on Climate Change (http://unfccc.int/2860.php) following the Earth Summit in Rio de Janeiro. The failure of the voluntary program to achieve the desired results led to the legally binding Kyoto Protocol. The Kyoto Protocol entered into force in February 2005.

The Clean Development Mechanism (http://cdm.unfccc.int/index.html), a provision of the Kyoto Protocol, allows industrialized countries listed in annex 1 of the protocol to purchase emission reduction credits from developing countries to meet their established emission reduction targets. More specifically, annex 1 countries generate credits by supporting projects that reduce emissions in developing countries, sometimes through new technologies in waste management. Emission reductions under CDM programs may be traded among buyers in annex 1 countries. Because emission reductions are tied to performance, carbon finance programs are a source of operating revenue rather than capital investment.

A range of methodologies has been approved by the CDM Executive Board to help determine eligible projects, including those in waste management. A common feature of all methodologies is the requirement to establish the baseline emissions that would have occurred in the absence of a CDM project. In waste management, a common baseline assumes waste disposal exclusively in landfills and in calculations of the associated GHG emissions based on this assumption.

A CDM intervention must also be additional (that is, it must create additional emission reductions that would not have been achieved without the intervention). The assessment of project additionality is based on any of the following conditions: Investment analysis may demonstrate that the internal rate of return to a project without carbon finance revenue would be insufficient to justify the implementation and maintenance of the project. Additionality may also be demonstrated by showing that the applied technology is not used in the country where the project is located or that the proposed CDM project is not a common practice in the host country.

Methane. The most frequent GHG associated with waste management is methane, which is generated amid anaerobic conditions in waste disposal sites. Methane has a global warming potential that is 21 times greater than the corresponding potential of carbon dioxide.

Waste management programs offer several options for generating emission reductions. Two common options involve capturing and using methane in landfills and avoiding methane by composting organic waste. Emission reduction credits may be earned by incinerating waste with or without energy recovery. The World Bank is also developing a methodology for earning emission reduction credits through recycling.

Waste management methodologies. Because landfill disposal is the baseline for assessing emission reductions, potential emission reductions may be estimated using a first order decay model that relies on multiple variables and default values. Critical variables include the composition rates of organics in a waste stream (food, other putrescibles, paper and textiles, and wood) and average annual precipitation and temperature. Areas with higher precipitation and temperature normally show higher decomposition rates based on data in the recent Intergovernmental Panel on Climate Change guidelines (see http://www.ipcc-nggip.iges.or.jp/public/2006gl/vol5.html).

For liquefied petroleum gas projects, estimates may provide reasonable forecasts of emission reductions, but monitoring equipment is needed to track actual gas flow and composition, which permits calculations of combustion efficiency and the methane captured and destroyed. The combustion of captured methane in a flare or in the production of energy (commonly, electricity) releases carbon dioxide. However, because the carbon dioxide is associated with biomass, it is considered carbon neutral. Methane avoidance projects such as aerobic composting depend entirely on calculated emission reductions because there is no way to measure methane that is not produced.

There are two common types of methodologies in waste management projects (liquefied petroleum gas and methane avoidance). The specific methodologies for both large- and small-scale projects (less than 60,000 tons of equivalent carbon dioxide per year) are described in box 3.23.

Emission reductions may be gained by displacing conventional (fossil fuel) power generation sources on the electricity grid. Methane recovered from gas capture programs is used to generate electricity. The methodology described in box 3.23 has been applied in the context of a small-scale power generation project (less than 15 megawatts).

About 15 percent of GHG consists of methane attributable to anthropogenic sources. More than 23 percent of GHG is methane and other non-carbon dioxide (non-CO_2) gases from anthropogenic sources. Methane emissions from municipal waste disposal sites account for 12 percent of global methane emissions, or an estimated 730 million tons of equivalent carbon dioxide. Globally, municipal waste disposal sites are the fourth-largest contributor of non-CO_2 GHGs. Although proportionately small, non-CO_2 GHGs have a much greater effect on global warming relative to carbon dioxide.

In November 2008, there were 1,587 registered CDM projects. Registration is the final step in the development of a mechanism project prior to the start-up of operations. An independent validator must annually verify achieved emission reductions. Registered projects cover a range of activities designed to reduce GHG emissions, including LFG capture. LFG projects account for about 5 percent of registered mechanism projects, most of which use captured methane for energy recovery (typically, electricity). Projects without suitable access to energy markets or modest gas flow rates flare the recovered methane. In China, LFG projects recover energy and flare only as a backup because the Designated National Authority (the entity responsible for CDM oversight in a country) encourages electricity generation.

Cost-efficient service

Another economic factor in waste management is promoting the cost-efficient delivery of services through appropriate capital investments and operations.

Landfill Gas Capture and Use in Tianjin, China

Tianjin's Shuangkou landfill, the first modern landfill built in Tianjin, conforms to China's standards, which mandate a bottom liner and a system for leachate collection and treatment. The design and construction of the landfill began in 1999, and the site started receiving waste in 2001. The landfill was financed by the World Bank as part of a broader loan program for Tianjin.

A daily average of about 1,300 tons of household waste is delivered to the landfill. The 60-hectare landfill has a capacity of 8.5 million cubic meters, equivalent to 7.4 million tons of waste and about 15 years of life at the current rate of infill. At closure, the depth of waste will be about 34 meters.

The decomposition of waste in the anaerobic conditions of a landfill generates methane. The methane is collected in pipes from a series of wells that have been drilled where the waste is deposited. Additional wells will be drilled as waste is deposited in new areas. The captured gas is transported in pipes to a central facility where it is burned to produce electricity. The electricity is sold to the North China Power Grid. A flare is used if there is excess methane or if the generators are out of service, such as during maintenance. The recovery system began operations in June 2008.

The LFG is about 50 percent methane; the balance is composed of carbon dioxide and other gases. The combustion of the methane during power generation or flaring destroys the methane. Under the agreement signed with the World Bank, the Tianjin Clean Environment and Environmental Engineering Company Ltd will sell 635,000 tons of equivalent carbon dioxide in GHG emission reductions, which is 70 percent of the expected reductions during the first seven years of operation. The World Bank has an option to purchase an additional 470,000 tons of equivalent carbon dioxide.

Tianjin Clean Environment and Environmental Engineering Company. Formed in August 2005, the company is part of the Tianjin Construction Commission. It was authorized by the Construction Commission and the Environmental Sanitation Commission, both under the municipality of Tianjin, to implement the Shuangkou LFG recovery and utilization project as project developer and operator and as seller of the emission reductions. The operation of the Shuangkou landfill is managed by the Tianjin Solid Waste Treatment Center, a division of the Environmental Sanitation Commission.

World Bank. The World Bank is a trustee of 12 funds and facilities for which it negotiates long-term purchasing agreements and manages relations with the associated projects. In the case of the Tianjin LFG recovery program, the World Bank is the trustee for the Spanish Carbon Fund.

Project registration. The Tianjin project was registered with the CDM Executive Board on August 27, 2008, which means the project has been able to earn certified emission reductions since that date. (For specific information on the project, visit http://cdm.unfccc.int/Projects/DB/JQA1193375340.58/view.)

Capital investment

Investment in equipment is an important first step in this process. The identification of appropriate equipment begins before procurement after sector managers have determined the relevant specifications. It is important to attain a balance between capital expenditures for equipment, the operating and maintenance costs, and the useful lifespans, which are influenced by operating environments. Reforms to the waste management system may also have beneficial effects on capital and operating expenditures, as evident in Yokohama.

Following the procurement of capital assets, a sound program of preventive and scheduled maintenance should be established to maximize value. Preventive maintenance helps avoid unscheduled downtime, which has a monetary cost. It is important to keep a supply room well stocked with lubricants, supplies, and spare parts in line with recommendations from equipment manufacturers.

Operations services

Supporting private sector participation and re-engineered public operations may improve the cost efficiency of services in the waste management sector. In either case, it is important to establish goals and metrics to track performance. In addition, historical data and lessons from other cities and regions may be collected to compare options and devise strategies.

References

City of Yokohama. 2003. "Yokohama shi ippan haikibutsu shori kihon keikaku, Yokohama G30 plan" 横浜市一般廃棄物処理基本計画、横浜G30プラン [City of Yokohama, master plan for management of general waste: Yokohama G30 Plan]. City of Yokohama, Yokohama, Japan. http://www.city.yokohama.jp/me/pcpb/keikaku/kei1.html> (accessed February 2009).

———. 2006. "Yokohama G30 Plan: Kenshou to kongo no tenkai hi tsuite" 横浜G30プラン「検証と今後の展開」について [Yokohama G30 Plan: Verification and next steps]. Resources and Wastes Recycling Bureau, City of Yokohama, Yokohama, Japan. http://www.city.yokohama.jp/me/pcpb/keikaku/G30rolling/ (accessed February 2009).

———. 2008. "Kankyou model toshi teian sho" 環境モデル都市提案書 [Proposal for Eco-model cities]. Climate Change Policy Headquarters, City of Yokohama, Yokohama, Japan. http://www.city.yokohama.jp/me/kankyou/ondan/model/ (accessed February 2009).

Hattori, Keiro. 2004. "Ningen toshi Curitiba: kankyou, koutsuu, fukushi, tochiriyou wo tougou shita machizukuri" 人間都市クリチバ—環境・交通・福祉・土地利用を統合したまちづくり [Human city Curitiba: Urban planning integrating environment, transportation, social aspects, and land use]. Gakugei Shuppan Sha, Kyoto.

Pablo, Carlito. 2007. "Rats, Yes, but Bacteria Love Garbage Strikes Too." Health Features, July 26, Straight.com, Vancouver. http://www.straight.com/article-102902/rats-yes-but-bacteria-love-garbage-strikes-too.

Pagiola, Stefano, Roberto Martin-Hurtado, Priya Shyamsundar, Muthukumara Mani, and Patricia Silva. 2002. "Generating Public Sector Resources to Finance Sustainable Development: Revenue and Incentive Effects." World Bank Technical Paper 538, Environment Series, World Bank, Washington, DC.

Peterson, Charles. 1989. "What Does 'Waste Reduction' Mean?" Waste Age, January. http://www.p2pays.org/ref/10/09702.pdf.

Peterson, Charles, Zarina Azizova, Qi Wenjie, Liu Baorui, and Jane Huang. 2009. "Landfill Gas Capture and Electricity Generation and the Clean Development Mechanism (CDM): Shuangkou Landfill, Tianjin, China." Presentation at the 12th Annual Landfill Methane Outreach Program Conference, Baltimore, January 13.

Rabinovitch, Jonas, and Josef Leitmann. 1993. "Environmental Innovation and Management –°in Curitiba, Brazil." Working Paper 1, Urban Management Programme, United Nations Human Settlements Programme, Nairobi.

Vaz Del Bello, Giovanni, and Maria Terezinha Vaz. 2007. *A Convenient Truth: Urban Solutions from Curitiba, Brazil*, DVD. Directed by Giovanni Vaz Del Bello. Felton, CA: Maria Vaz Photography, in association with Del Bello Pictures.

Managing the Spatial Structure of Cities

Introduction

Cities enjoy high productivity because their large consumer and labor markets drive increasing returns to scale. The theoretical and empirical literature correlating the wealth of cities to urban spatial concentration is abundant and no longer controversial. National data show that the economic output of large cities is much greater than suggested by the population shares of these cities. The World Bank's *World Development Report 2009: Reshaping Economic Geography* and the 2009 report of the Commission on Growth and Development, *Urbanization and Growth,* summarize and document the theoretical and empirical arguments justifying the economic advantages of concentrating economic activities in large cities (see World Bank 2009; Spence, Annez, and Buckley 2009). Moreover, it is widely accepted that effective urban management and the spatial structure of cities are crucial to success.

Cities in the 21st century need to confront myriad development challenges because of rapid growth in income, population, and built-up areas. However, developing one-size-fits-all models or forms of urban development is unrealistic given the diversity in urban culture, history, economy, climate, and topography.

Nonetheless, cities face two fundamental spatial challenges as they absorb new populations and manage urban transitions: maintaining mobility and enabling the provision of affordable land and shelter for existing and new citizens. Mobility is important because the productivity of large labor markets is important for cities, and lack of mobility fragments labor markets and decreases productivity. *Affordability* is crucial because poor rural migrants often become middle-class citizens in cities. A city should have the capacity to shelter migrants during this transition. Ignoring the needs of migrants for adequate shelter impedes assimilation in the formal economy and has a high social cost.

Maintaining Mobility

Maximizing the economic advantages of spatial concentration hinges on the capacity of workers to find employment anywhere in a city and the ability of employers to select workers among a large and dispersed pool of labor. Maintaining the mobility of people and goods within a metropolitan area is one of the conditions for realizing economic benefits. Congestion acts as a tax on city productivity because it impairs the free movement of goods and people. In 2000, congestion in 75 metropolitan areas in the United States caused fuel and time losses valued at US$67.5 billion (Downs 2004). Those losses exceeded the value of Kenya's 2008 GDP of US$61.7 billion (CIA 2009). If cities are unable to maintain mobility, the tax of severe congestion may surpass the economic benefits of spatial concentration. In the long run, a city that is unable to sustain mobility is bound to decay economically.

Maintaining the mobility of people and goods should be a prime objective of land use planning and infrastructure investments. Mobility has two aspects: location mobility among firms and households and commuting mobility of workers and consumers.

Location mobility among firms and households

Location is important for residences and workplaces. People and firms should be able to buy or rent residences or business facilities anywhere in a city under as few restrictions as possible. The traditional principle of locating low-income housing close to industrial areas in distant suburbs is based on a 19th-century vision of labor. Relative to factories, the service sector in a modern city employs more unskilled labor. Poor people should have access to all areas of a city, and zoning plans should not segregate low-income housing into predesignated areas.

Real estate transaction costs should be as low as possible to ensure that households and firms are able to select the best locations they can afford and move quickly to better locations if economic circumstances or external conditions change.[1] Tying housing subsidies for poor households to low-income housing projects prevents mobility and tends to increase unemployment.

Zoning should not segregate land uses arbitrarily because this restricts firm and household mobility. Zoning in modern cities should segregate only those economic activities that create real hazards and nuisances. It should not apply inherited and arbitrary categories that needlessly curtail mixed use development.

The commuting mobility of workers and consumers

In all but the largest cities, workers and consumers should be able to reach any point in the metropolitan area within an hour. Ideally, multimodal transportation systems ensure sufficient mobility. However, urban planners often favor one transportation mode at the expense of others. For instance, in cities where underground metros are being built, the high capital costs of the metro often divert needed funds from other modes of transportation, such as buses. In other cities, cars are heavily subsidized by low gasoline prices, free parking, and street designs. In both cases, less mobility is the result.

Though planners in the past decade have tried to privilege one mode of transportation over others, they should acknowledge that every city needs a multimodal transportation system and that consumer safety, affordability, and convenience are the main aims of transportation modes.

The choice of a dominant transportation mode or other modes in a multimodal system should be linked to a city's spatial structure. City managers should not arbitrarily select a city's dominant transportation mode, whether transit, car, or bicycle. Depending on a city's spatial structure, among other factors, consumers decide for themselves which transportation mode is the most convenient in terms of speed, comfort, and cost. City managers, however, may influence a city's spatial structure through regulations and infrastructure investments. In high-density, mainly monocentric cities, mass transit may be an efficient choice for commuters. However, in low-density-polycentric cities, cars, motorcycles, and taxis are often the most convenient travel modes.

Mobility, spatial structures, and transportation networks

Land developments and transportation options determine the patterns of daily commuter trips to and from work. As household incomes rise, noncommuting trips to shop, pick up children, visit family, or undertake recreation become more important. The proportion of commuting trips over other trips thus decreases. Figure 3.63 illustrates the most usual trip patterns in metropolitan areas.

In monocentric cities (figure 3.63, part A) where most jobs and amenities are concentrated in central business districts (CBDs), transit is the most convenient transportation mode because most commuters travel from the suburbs to the CBD. Trip origins might be dispersed, but the CBD is the most common destination. Small collector buses may bring commuters to radials, where bus rapid transit or underground metro systems may usher them at high speed to the CBD. Monocentric cities are usually dense (that is, above 100 people per hectare).

In low-density, polycentric cities (figure 3.63, part B), few jobs and amenities are in the city center, and most trips are from suburb to suburb. There is a large number of possible travel routes, but few passengers per route. The trips have dispersed origins and dispersed destinations. In this type of city, individual transportation modes or collective taxis are more convenient. Mass transit is difficult and expensive to operate because of the multiplicity of destinations and the limited passengers per route. Polycentric cities usually have low densities be-cause car use does not favor concentration in any specific area.

Figure 3.63, part C illustrates the urban village model that is often included in urban master plans, but is difficult to produce in the real world. In this model, the city includes many centers, and commuters travel only to the center closest to their residences. Under this model, everyone may walk or bicycle to work, even in a large metropolis. For this model to function, urban planners must be able to match residences and workplaces perfectly. However, this notion often contradicts the economic justification for large cities. Employers in large cities do not select employees based on where they live, and spe-cialized workers do not select jobs based solely on proximity to residences. Moreover, the urban village model implies a systematic fragmenta-tion of labor markets that is not economically sustainable in the real world.

In certain suburbs of Stockholm, urban regulations permit developers to build new residential units only if they can show that a corresponding number of jobs exist in the

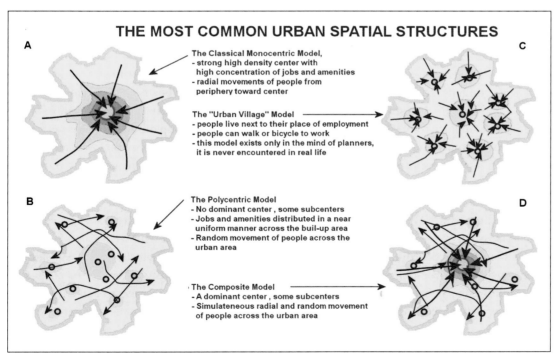

THE MOST COMMON URBAN SPATIAL STRUCTURES

A

The Classical Monocentric Model,
- strong high density center with high concentration of jobs and amenities
- radial movements of people from periphery toward center

The "Urban Village" Model
- people live next to their place of employment
- people can walk or bicycle to work
- this model exists only in the mind of planners, it is never encountered in real life

C

B

The Polycentric Model
- No dominant center , some subcenters
- Jobs and amenities distributed in a near uniform manner across the buil-up area
- Random movement of people across the urban area

The Composite Model
- A dominant center , some subcenters
- Simulateneous radial and random movement of people across the urban area

D

Figure 3.63 Spatial Structures and Trip Patterns

Source: Author compilation (Alain Bertaud).

neighborhood. Meanwhile, the five satellite towns around Seoul offer examples of the problems with urban village layouts. When the towns were built, jobs and inhabitants were being carefully balanced, and the satellite communities were expected to be self-contained in housing and employment. However, recent surveys show that most people living in the new satellite towns now commute to work in the main city, while most jobs in the satellite towns are taken by citizens from the main city.

The composite model in figure 3.63, part D is the most common urban spatial structure. This model includes a dominant center, but a large number of jobs in the suburbs. Under this model, most trips from the suburbs to the CBD use mass transit, while trips from suburb to suburb use individual cars, motorcycles, or collective taxis.

The composite model might be an intermediary stage in the progressive transformation of a monocentric city into a less dense and polycentric one. As population grows and the urban built-up area expands, the city center becomes more congested and loses its attractiveness, which was based on easy access and communication stemming from spatial concentration.

As a city grows, CBD decay from congestion is avoidable. Good traffic management, timely transit investments, strict parking regulations, urban environment investments (pedestrian streets), and land use reforms permitting vertical expansion contribute to reinforcing the city center and its attractiveness to new business and urban commuters. In New York, Shanghai, and Singapore, such measures have been reasonably successful. However, coordinating policies between investments and regulations is often difficult. Such coordination has to be executed consistently over long periods to enhance the viability of city centers.

One way to reduce noncommuting (nonwork-related) trips, while enhancing the vibrancy of a city is to develop mixed use neighborhoods that reflect intelligent urban design. Within these neighborhoods, planners may provide greater commercial street frontage, often with wide sidewalks and attractive window spaces, supporting an array of coffeehouses, cleaners, grocery stores, hardware stores, restaurants, and other local shops. Such a strategy encourages local commerce, supports walking and bicycling trips, and enhances urban safety and attractiveness. Moreover, shorter nonwork trips result because services are located close to households.

Failure to expand traditional city centers through infrastructure and amenities weakens transit systems in the long run, as the number of jobs in city centers stagnates or falls, while additional jobs are being created in suburban areas.

City structure is path dependent. Once a city has become mainly polycentric, the return of the city to a monocentric structure is nearly impossible. Monocentric cities, however, may become polycentric if the traditional centers decay. The inability to manage traffic and operate an efficient transit system is a main factor explaining the decay of traditional CBDs.

The space required by traffic and parking make transit indispensable for transportation to dense city centers

Transit is viable as a dominant mode only in servicing a dense business or commercial core. Not surprisingly, a car-dominant mode of transportation is incompatible with dense city centers. Cars occupy a large and incompressible amount of space if they are moving or if they are parked. In addition, cars require more street space as their speeds increase.

Unfortunately, many city managers consider public parking in downtown areas as a municipal responsibility that should be subsidized. The subsidies constituted by free or quasi-free parking in downtown areas are not trivial. A car parked on or off the street uses about 14 square meters of costly real estate, which might otherwise be rented or bought at market prices. The private parking spaces provided at the bottom of the Marina Towers in Chicago (figure 3.64) suggest that car parking is and should be considered commercial real estate and not a public good provided by the municipality.

Figure 3.65 shows the area of street space per person in selected neighborhoods in various cities. The street area as a percentage of total area varies from neighborhood to neighborhood (in the case studies shown in the figure, from 11 percent to 50 percent). The densities (or land area per person) vary across neighborhoods (in the case studies, from 18.00 people per hectare to 1.65 people per hectare). Residential density is used throughout because it is usually the only available density measure. However, for the two New York neighborhoods (Midtown and Wall Street), job density is also provided.

The figure shows the street area one car requires for on-street parking (dotted horizontal line) and for moving at 15 kilometers per hour and 30 kilometers per hour. It shows that, at certain densities, a parked car uses more space than

Figure 3.64 Parking Space as Real Estate at Marina Towers, Chicago

Source: Photo by Alain Bertaud.

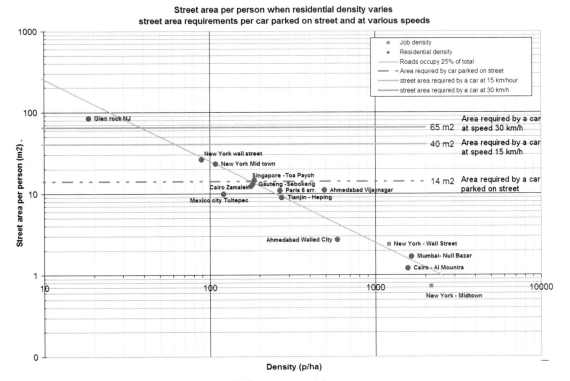

Figure 3.65 Car Space Requirements and Available Space Densities

Source: Author compilation (Alain Bertaud).

Note: km = kilometer; km/h = kilometers per hour; m² = square meter; p/ha = person per hectare.

the street space available per person. In Midtown, New York City, the available street space per worker is about 0.70 square meters. A parked car requires a much larger 14 square meters, and a car moving at 15 kilometers an hour requires about 40 square meters.[2] If every worker were driving a car to work, 3 hectares of parking would be needed for each hectare of land, which would require about six floors devoted solely to parking. In contrast, Glen Rock, New Jersey, a suburb of New York City, has sufficient street space to allow every resident to drive a car simultaneously at more than 30 kilometers an hour.

New York City is not unique. Many urban neighborhoods in developing countries have residential densities similar to Midtown's densities. Null Bazar in Mumbai, for instance, has a residential density of 1,600 people per hectare, which is similar to that in downtown Mumbai. The Al Mounira neighborhood in Cairo has a similar density. Limited street space in these neighborhoods restricts car ownership to around 40 cars per 1,000 people.

In these dense cities, transit is and will be the only way to provide adequate mobility. Some streets may be widened in strategic areas, but never sufficiently to allow cars to be the main mode of transportation. The failure to provide convenient transit services in dense cities will result in less mobility for a large share of the population. Without adequate mobility, labor markets in large cities will become fragmented, and these cities will be less productive.

Why does a city's spatial structure matter?

Low-density, polycentric cities may be viable, but are they efficient and socially desirable? These cities are viable only if household income is sufficiently high to allow people to buy and operate cars and if suburban densities stay low (below 50 persons per hectare). Moreover, those people who cannot afford or drive cars, such as the poor and older citizens, may be dislocated from job opportunities under this model, yet have little recourse to adequate transit alternatives.

Energy costs are more important for individual transportation options than for transit. If energy prices increase, low-density, polycentric cities reliant on car transportation see transportation costs increase in nearly the same proportion. Most low-density, polycentric cities in the United States were built when energy prices were generally below US$50 per barrel of oil in real terms. It is uncertain, absent a rapid technological breakthrough, that these cities could maintain a unified labor market if the price of a barrel of oil was, say, US$200 over a sustained period.

The spatial structure of cities is at the nexus of three fundamental urban objectives: mobility, affordability, and transportation-related energy efficiency. In low- and middle-income cities, maintaining a dominantly monocentric structure is a precondition for maintaining worker mobility, a larger share of transit trips, and adequate revenue in case energy prices spike.

The spatial distribution of populations in Gauteng, South Africa, and Jakarta, Indonesia (figure 3.66), illustrates different urban spatial structures and their consequences for transit operation. The classical profile of Jakarta, which has significant residential densities close to the city center, facilitates the operation of a transit system (trains and buses) that is convenient to users. In contrast, the dispersion of relatively high population densities in Gauteng explains the dominance of individual cars among middle- and high-income households and collective taxis among low-income households. The dispersed structure of Gauteng is partly attributable to the city's history of apartheid. Subsidies are offered to low-income households in distant settlements, but these subsidies cannot be readily used for transportation. Suppliers may have derived benefits from building these settlements on a large scale, but the settlements do not provide convenient access to work.

However, transit should not be considered the only means of transportation. Delivering freight necessarily involves cars and trucks, and cars will always comprise some share of com-

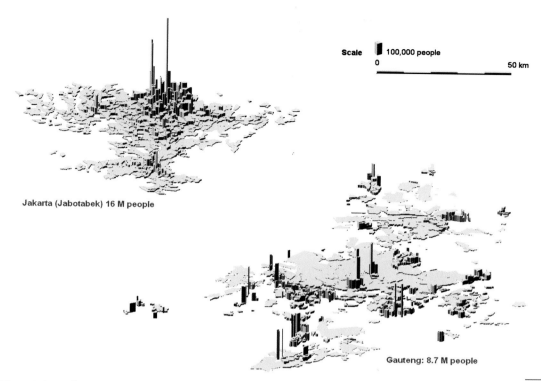

Scale ▮ 100,000 people

0   50 km

Jakarta (Jabotabek) 16 M people

Gauteng: 8.7 M people

Figure 3.66 A Three-Dimensional Representation of the Spatial Distribution of Population, Gauteng, South Africa and Metropolitan Jakarta, Indonesia, 2001

Source: Author compilation (Alain Bertaud).

Note: km = kilometer; M = million.

muting trips. The challenge is to maintain a balance across modes to minimize the vulnerability to a rise in energy prices and to reduce congestion and pollution by harnessing the available technology.

Cities have complex structures that constantly evolve. Technology also evolves, sometimes unpredictably. Introduction of cars like the Tata Nano (a very small, two-cylinder car) and research in China to develop a low-cost electric car may reverse some of these assumptions, but likely not all of them. Moreover, individual cars, because of their space requirements, will always be incompatible with high-density city centers, such as those in Cairo, Mexico City, Mumbai, New York, and Shanghai.

Spatial structures, regulations, and markets
Urban spatial structures matter. However, governments are not the sole actors that influence spatial structures. For example, the real estate market plays a fundamental role in determining urban form. Typically, a government's sphere of action is limited to infrastructure, land use regulations, and taxation. The sections below address how governments may use market forces to influence the shapes of cities.

The interaction of market forces with government taxation, transportation investments, and land and tenure regulations is complex. This interaction affects spatial layouts. Table 3.33 summarizes the impacts of government measures on spatially linked market factors (that is, land supply and prices in city centers and suburbs) and spatial development (that is, the dispersion and concentration of population and jobs).

It is neither good nor bad per se whether government actions favor concentration or dispersion. Assessing value depends on a city's long-range policy and the starting point. Job dispersion would be negative in a city that has invested heavily in a radial transit system be-

Table 3.33 The Impact of Government on Land Markets, the Informal Sector, and the Spatial Structure of Cities

Government action / Sector	Market reaction: Land supply		Land price		Impact on size of informal sector	Spatial Impact: Dispersion		Concentration	
	Center	Suburbs	Center	Suburbs		Population	Jobs	Population	Jobs
Transport infrastructure									
Improving or/ and building radial roads		(+)	(+)	(-)	(-)			(+)	(+)
Building ring roads		(+ +)	(-)	(-)	(-)	(+)	(+)		
Building transit in radial pattern		(+)	(+ +)	(-)	(-)			(+)	(+ +)
Building transit in grid pattern		(+)	(-)	(-)	(-)	(+)	(+)		
Land use regulations									
Low floor area ratio			(+ +)	(+ +)	(+ +)	(+)	(+)		
High minimum plot area		(-)		(+)	(+ +)	(+)			
High standard of land development		(-)		(+)	(+ +)	(+)			
Long approval process for building permits	(-)	(- -)	(+ +)	(+ +)	(+ +)			(+)	(+)
Restrictive zoning practices	(-)	(- -)		(+ +)	(+ + +)				
Setting up an urban growth boundary (UGB)		(- -)		(+ +)	(+ +)	(?)		(?)	
Land tenure									
Large government land holding	(- -)	(- -)	(+ +)	(+ +)	(+ +)	(+)	(+)		
Rent control	(-)		(+ +)		(+ + +)				
Restriction on land transactions in periphery		(- -)	(+ +)	(+ +)	(+ + +)			(+)	(+)
High stamp duty on land transactions		(-)			(+ + +)				

Source: Author compilation (Alain Bertaud).
Note: Increase = +; decrease = –.

cause many jobs would be out of reach of the transit system. However, dispersion might be a positive in a city that mostly relies on minibuses and cars because dispersion would likely relieve congestion and provide cheaper land for housing and businesses.

The government actions shown in table 3.33 are often implemented with insufficient consideration of long-range objectives and impacts on land and overall urban form. For instance, the goal of building ring roads is usually to alleviate congestion by allowing through traffic to bypass city centers. However, little thought is typically given to the impacts on land supplies and on prices along ring roads and in other affected areas.

Because local governments often support urban regulations and investments focused on short-term objectives, government actions and goals may contradict each other. For instance,

in Bangalore, India, the local government has financed a bus rapid transit system that tends to concentrate jobs in the city center. At the same time, the government limits the floor area in the CBD to a ratio lower than in the suburbs, thus preventing the concentration of jobs that justified the bus system.

This type of policy contradiction between two branches of local government—transport and land use planning—is typical. Transport engineers desire high job and population densities along transit routes to ensure a large number of transit passengers. Planners facing congestion in city centers find it easier to mandate decreased densities to alleviate congestion.

Though regulations significantly affect city shape, market forces have the most influence on urban spatial structures in the long run. Market forces particularly affect the spatial distribution of densities. In a monocentric city, land prices fall as one moves away from the city center. In polycentric cities, land prices tend to decrease from the centers of built-up areas, though usually at a slower pace than in monocentric cities. Where land prices are high, households and firms tend to consume less land. Population and job densities thus tend to be higher in CBDs or other urban centers and lower in suburbs.

In Bangalore, the regulatory floor area ratio (FAR) is lower in the city center than in the suburbs. However, population densities are higher in the center than in the suburbs because of the high cost of land. Households in Bangalore's city center consume much less floor space than they would consume if the city's FAR were higher. In Bangalore's case, FAR regulation has been unable to counteract market forces in shaping urban structures.

The density profiles of most large cities suggest that the traditional monocentric city model is still a good predictor of density patterns despite the fact that cities are becoming increasingly polycentric. These profiles demonstrate that markets remain the most important force for allocating land despite price distortions caused by direct and indirect subsidies and ill-conceived land use regulations. The profiles of population densities of 12 cities on four continents shown in figure 3.67 demonstrate that, despite the economic and cultural differences among these cities, markets have played an important role in shaping the distribution of populations around the centers. All the cities shown in figure 3.67 closely follow the negatively sloped gradient predicted by the classical monocentric urban model.

Moreover, the spatial structure of most cities tends to follow three trends: over the long run, average densities decrease; traditional CBDs tend to lose primacy (with notable exceptions); and the evolution of spatial structure is often adversely affected by household mobility.

Over the long run, average densities decline for the following reasons: improvements in transport networks (that is, length and speed)

increase land supplies more rapidly than required by the needs of growing populations; increases in household income and decreases in household size allow people to consume more land and floor space; and diversification and specialization in economic activities require more land and floor space.

In China, for instance, cities managers have become increasingly alarmed by growing per capita land consumption, and they have taken measures to curb sprawl by making land development more difficult or expensive or by imposing quotas on land development. Many cities in other countries have taken measures to constrain the supply of land by delineating boundaries for urban growth or establishing quotas for land development that boost the prices of land and housing, which often adversely affect urban dispersion.

It is not possible to establish optimal per capita rates of urban density or land consumption that are completely consistent over time. Land is an input in the production of floor space. Where land is expensive, developers (regulations permitting) will substitute capital for land by increasing FARs and densities. Where land is cheap, such as in suburbs, substituting capital for land is not justified, and FARs and densities are low. A financially optimum density may be temporarily achieved during an area's development provided the prices of inputs (that is, land and construction) are not too distorted by regulations or subsidies. Without price distortions and externalities, a financially optimum density would equal the economically optimum density. However, the prices of land, capital, and other inputs will eventually change, possibly in different directions, and the new optimum density will shift from the optimum achieved during development. If the new optimum density differs substantially from actual density, land will be redeveloped.

The periodic recycling of land into new densities is indispensable for the maintenance of land use efficiency and urban productivity. Unfortunately, many land use regulations, such as

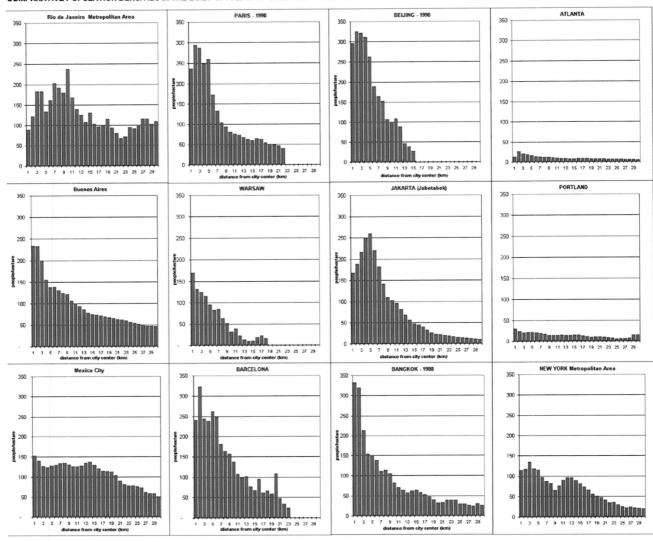

Figure 3.67 The Profile of Densities of Built-Up Areas in 12 Large Metropolises

Source: Author compilation (Alain Bertaud).

rent controls to maximize the FAR, tend to prevent the recycling of land.

Traditionally, CBDs become larger as cities develop. The inability of local governments to manage the traffic in CBDs and to maintain mobility often results in the deterioration of the CBDs (Mexico City) or in the progressive abandonment of the CBDs (San Salvador, El Salvador). As a result, jobs disperse into new locations, and transit networks are unable to link residential areas adequately to these new loca-

tions. Minibuses and collective taxis gradually become more efficient as ways to reach dispersed locations.

Coordination between land use and transit: The example of Singapore

Though most dense, monocentric cities have become progressively less dense and polycentric, a few cities have kept a large share of the jobs and amenities in CBDs despite population growth (for example, New York City, Shanghai, and Tianjin; Singapore has also pursued this strategy [its transit network is illustrated in map 3.9]). These cities have maintained consistent land use and transport policies that have extended the CBD area, increased FARs, and expanded radially oriented concentric transit networks serving the CBDs. In addition, these cities have included amenities such as theaters, pedestrian streets, and museums within or near city centers. These approaches are important in the development of dense cities, particularly in Asia, where urban built-up densities are often above 100 people per hectare. These strategies also facilitate efficient transit operations in the densest urban areas.

In these neo-monocentric cities, trips between suburbs might continue to grow. Suburb-to-suburb trips will be made by cars or collective taxis. In Singapore, car ownership is strictly controlled; an elaborate high-technology congestion pricing system is in operation; and land and transit policies are consistent. Nonetheless, private car trips as a share of all trips grew from 37.0 percent to 41.6 percent between 1990 and 2000,

while transit ridership fell from 55.0 percent to 52.4 percent.

Singapore's example shows that cities with significant transit investments and good coordination among land and transit policies may nonetheless experience growth in car trips, although more slowly than cities that do not pursue these strategies. Cities should thus continue efforts to reduce car traffic and congestion, along with running programs to increase transit share. In the long run, traffic and congestion may drive jobs and people to other areas, increasing trip length and making transit less viable.

Land and Housing Affordability

Ensuring that developed land and floor space are affordable for various income groups is a major challenge, particularly in large cities in rapidly urbanizing countries. Cities in countries with sizable rural populations must accommodate annual migrations of rural citizens with incomes below urban averages. Local governments

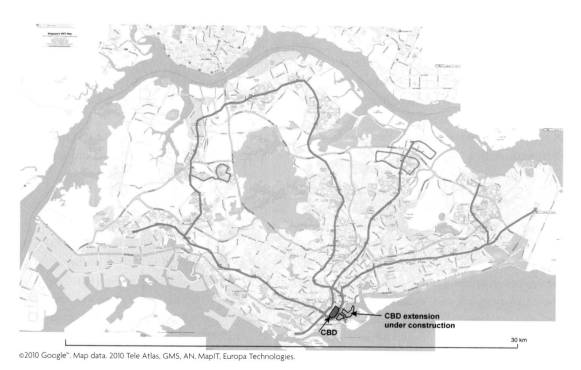

©2010 Google™. Map data. 2010 Tele Atlas, GMS, AN, MapIT, Europa Technologies.

Map 3.9 Singapore Metro Network: Centered on Expansion in the Central Business District

Source: Author compilation (Alain Bertaud), Google Maps.

should carefully monitor land supplies and audit land development regulations to ensure that the regulations do not establish thresholds below which low- or middle-income households are unable to buy or rent legal dwellings.

Infrastructure and land use regulations affect the supply and price of land and floor space

Variations in the supply of land and floor space typically drive variations in rents and land prices. The supply of land and floor space is highly dependent on primary infrastructure, transport networks, and land use regulations, which are usually under the jurisdiction of local governments. Unfortunately, local governments often ignore the links between land supply and land prices and rents. Unfamiliar with supply and demand, local officials often attribute high prices to speculation, without realizing that bottlenecks in land supply and FARs are fueling high prices. In confronting significant increases in land prices and rents, officials often fail to increase land supplies or FARs. Instead, they try to control prices by imposing higher transaction costs or tightening rent control legislation. These actions result in higher land and housing prices, which, in turn, generate more regulations in a vicious inflationary spiral.

Because government actions have such an important impact on real estate markets and urban spatial structures, it is worthwhile examining the direct and indirect impacts of infrastructure investments and land use regulations.

The aims of land use regulations are to avoid externalities linked to changes in land use. Preventing a negative externality is beneficial. However, regulations also have costs. Unintended side effects may increase the costs of many regulations beyond the supposed benefits. Many land use regulations are not tested for their impacts and may create social costs by artificially increasing the cost of developed land and floor space.

In reality, most regulations are not formulated to correct explicit externalities, but reflect the utopian vision of officials on the features of appropriate urban design. This explains why most land use regulations entail greater land consumption than would be the case if land use were driven by consumer demand. Normative regulations establish arbitrary and enduring consumption thresholds for land and floor space (for example, minimum plot sizes and maximum FARs) and systematically fail to adapt to changing incomes, technologies, and land and construction prices. In poor countries, regulations often impose minimum fixed values on variables such as plot size, apartment size, and FARs. In utility and affordability equations for housing and other land uses, these variables should not have fixed minimum values and should be dependent variables linked to independent variables, such as the price of land, interest rates, and the cost of construction.

Land use regulations for a given city may fill several volumes. However, only four types of regulations are really important. These regulations should be carefully audited owing to their impact on land demand and affordability. They are as follows:

- Regulations establishing minimum plot sizes and minimum apartment sizes (explicitly or implicitly)
- Regulations limiting FARs
- Zoning plans limiting the type and intensity of urban land use
- Land subdivision regulations establishing permissible ratios of developable and salable land in new greenfield areas

Regulations on minimum plot sizes and apartment sizes

The goal of mandating minimum plot sizes is to prevent excessive densities and ensure a high-quality urban environment. Households in all cultures tend to maximize the land and floor area they consume by taking into account the trade-offs between location and local real estate costs. For example, poor households may opt to live in dense, inner-city slums rather than in low-density, legal subdivisions in urban peripheries. These decisions are completely rational because the households are simply trying to maximize welfare. Minimum plot size regulations contradict those rational decisions and impose another trade-off: either live in a legal subdivision in a distant suburb or in an illegal slum closer to job opportunities. The trade-off is no longer between distance and density, but between legality and illegality. Not surprisingly, many households select the illegal solution.

Minimum plot size regulations establish a de facto cost threshold below which it is illegal to develop land. Though the minimum plot size may be fixed, the cost threshold often varies temporally and spatially. For instance, a minimum plot size of 200 square meters may be affordable up to 3 kilometers from the city center for 90 percent of households in a certain year. However, economic changes in a separate year may mean that only half of the households may be able to purchase such plots, pushing other households into the distant periphery or into illegal dwellings. The problems associated with minimum plot size regulations in a major African city, Addis Ababa, are illustrated in figure 3.68.

Addis Ababa's minimum plot size was 75 square meters in 2002. Most poor households close to the center (Kebbele housing) occupy, on average, about 35 square meters of land, which includes communal courtyards and passageways. Figure 3.68 illustrates the impact of the minimum plot size regulation on housing affordability, assuming a suburban location in which land is cheaper. Because of the regulation, 75 percent of households cannot afford to satisfy the minimum standards, as represented by the red bars on the left in the figure, with or without financing. With a 30-year financing package, only a quarter of the households can afford a

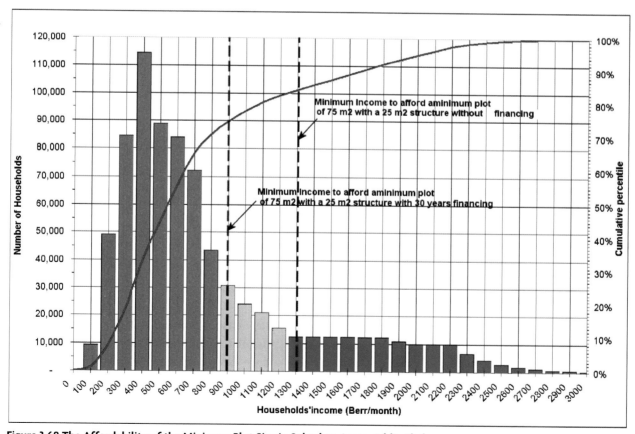

Figure 3.68 The Affordability of the Minimum Plot Size in Suburban Areas, Addis Ababa

Source: Author compilation (Alain Bertaud).

Note: m² = square meter.

house of 25 square meters on a minimum plot size of 75 square meters. Without financing, such a structure and plot are affordable for only 18 percent of the households, represented by the dark bars on the right in the figure.

Aware of this affordability challenge, the government developed a subsidized housing finance scheme for formal developers who develop land adhering to minimum regulatory standards. Even with a subsidized loan, 75 percent of the households cannot afford the minimum standards in Addis Ababa. Not surprisingly, about 75 percent of the households are in informal housing.

A similar study carried out in Kanpur in the State of Uttar Pradesh in India in 1988 showed that a minimum plot size of 100 square meters was unaffordable for 87 percent of the population.

The best solution is to repeal minimum household standards for land and floor space and to replace these standards with minimum standards for developed *parcels* of land, without specifying how many households may subdivide and live on the parcel. In this way, infrastructure issues related to small land parcels of unequal size would be solved. In other words, water and sewerage connections would be shared by households sharing a regulated parcel, but the number of households sharing a parcel would be left to the users and would be based on affordability.

This shared parcel design is common in many informal and sometimes formal settlements around the world. Informal houses in Kabul, Afghanistan, are usually built on a large lot of around 300 square meters and served by one street entrance. The number of houses built around a lot's central courtyard depends on household income and may vary from one in affluent areas to 20 in the poorest areas. If economic circumstances change, land and floor space per household will adjust as the number of households rises or falls, without any change in infrastructure or street layout.

In South Africa, the courtyards of many detached formal houses are subdivided into informal backyard shacks rented to lower-income members of the community. Informal subdivisions of this type are often discouraged by municipalities or by the housing institutions that have developed a settlement.

The argument against lot subdivision is that increased density risks overloading infrastructure. This may be a genuine risk, despite the fact that lower-income households consume only a fraction of the water and electricity consumed by higher-income households. In any case, it is normally easier and cheaper to upgrade infrastructure in such subdivisions than to develop new land on the urban fringe to accommodate new households. In addition, it is difficult to develop land for those households that rent backyard shacks.

Figure 3.69 shows the case of Sebokeng, a formally subsidized low-income settlement in a southern suburb of Gauteng. The beneficiaries of this subsidized housing project have built and rented one-room houses in their backyards. This approach is probably the most efficient way to provide land and housing to the low-income population in this area. This process should be encouraged rather than discouraged. It provides additional income for the plot owner, while allowing renters to share community facilities, including schools, with higher-income residents. As a result of these practices, the density of the neighborhood rose about 50 percent relative to the original design, but infrastructure should be sufficient. If not, it remains less expensive to run an additional water pipe along the road than to develop new land for the renters.

Regulations limiting FARs

Regulations limiting FARs are designed to reduce externalities by limiting densities and the bulk of buildings. Negative externalities are created by the shadows of taller buildings and the increased traffic and utility consumption implicit in higher densities. There is no optimum

Sebokeng sample density

Total area	17.51	ha		Back yard shacks	0.8	per stand
formal stands and houses	431	units		Total backyard shacks	345	units
people per formal house	5	people		People per backyard shack	3	people
Total population	2,155	people		people in backyard shacks	1,034	people
Design density	123	p/ha		**Total real density**	182	p/ha

©2007 Google™, ©2008 Europa Technologies. Image ©2008 Digital Globe.

Figure 3.69 Sharing Larger Plots among Lower-Income Households, in Sebokeng, Gauteng, South Africa

Source: Author compilation (Alain Bertaud), Google Earth.
Note: ha = hectare; p/ha = persons per hectare.

FAR that may be calculated for an entire city. However, it is possible to calculate FARs for given prices of land and construction that produce the lowest cost of construction per square meter. In many cities, such as Mumbai, FAR values are much too low, leading to urban decay in the city center and favoring sprawl.

Zoning plans

Zoning plans aim to separate incompatible land uses and prevent development in environmentally sensitive areas. In addition, zoning plans typically contain regulations for each zone that limit the intensity of land use, including regulations on plot coverage, maximum heights, setbacks, and FARs.

Zoning is an important tool for preserving sensitive areas. It may be used to mandate green areas that absorb excess water and runoff. However, zoning plans often do not clearly establish their objectives or the externalities they are designed to correct. Though zoning may be a useful tool, it must not obstruct the provision of affordable land and shelter in rapidly urbanizing cities.

Land subdivision regulations

Land subdivision regulations are mostly directed at new greenfield developments. They establish the standards for roads, block lengths, open spaces, the areas to be reserved for community facilities, and so on. The percentage of postdevelopment salable land is arguably the most important parameter set by the regulations. Unfortunately, this parameter is seldom explicit and must be derived from other regulatory parameters. In many countries, the share of such salable land, including residential and commercial lots, is below 50 percent. This figure should be higher, which means new developments need to be land intensive. Typically, the regulations are established for all new urban greenfield development and do not consider location or land prices.

Implicitly, the regulations also set maximum densities, but these are seldom explicitly calculated by regulators. Maximum densities are seldom compatible with the land prices and developments that are affordable for low-income households.

By imposing ill-conceived standards, the regulations are often responsible for an exceedingly high proportion of informal sector settlements, as in Cairo and in India and Mexico. The regulations should be tested using land and infrastructure price models to establish the minimum household income required to afford a minimum standard plot in a new greenfield development.

Land supply and high-intensity use are key factors in increasing affordable housing

Regulatory constraints affecting land supply are often ignored by governments that are otherwise eager to increase the supply of affordable housing. Subsidized interest rates are often the main tool used to create affordable housing. However, if regulatory constraints, such as the land subdivision regulations, prevent the elasticity of supply, more and cheaper housing is not the result; the result is housing inflation.

Unlocking land supply is a prerequisite to stimulating housing demand. An audit of land use regulations should be conducted at least every two years. The audit should calculate the cost of building a house or apartment under current minimum standards in various areas of a city and incorporate the current market prices for developed land and construction. The cost of this minimum house should be compared with the incomes of various socioeconomic groups. The results of the audit should be drawn on a map that clearly shows the parts of a city in which legally developed land and shelter are affordable for different income groups. Land use regulations should then be amended through an iterative process so that housing in sizable urban areas is affordable for unskilled migrants. The two key aims are to open the supply of land through regulatory reform and primary infrastructure investments and to allow poor households to consume as little land and floor space as needed in areas that are convenient for finding employment. The low-cost housing issue is at the nexus of mobility and affordability objectives.

A Plan for the Active Management of Constantly Evolving Urban Spatial Structures

A plan includes the following dimensions.

Methodology and training for urban planners and managers

Urban managers should change their urban planning methodologies. Rather than collect data once each decade to devise a development plan that often faces a long approval process, managers should constantly monitor urban demographic and physical changes and real estate prices. Infrastructure investments and regulations should also be adjusted on a regular basis.

Understanding the way markets work is essential for those people who make decisions on infrastructure investments and regulations. Because the acquisition of economic knowledge constitutes a departure from the traditional role of urban planners, academic requirements should be modified. All urban planners should have some formal training in urban economics regardless of their original fields.

The reorganization of urban planning departments

Urban planning units should become involved in actively managing cities. Planning units cannot replace line agencies that design and maintain infrastructure such as roads and water treatment plants, but they should have a say in the new infrastructure investments that affect mobility and land supplies.

Planning units should have monitoring and operational departments. The monitoring department should constantly collect and analyze data on the evolution of the city. Whenever possible, data should include spatial and demographic dimensions, such as coordinates and census identifiers. The data and indicators should cover real estate prices, household income, the issuance of building permits, car ownership, commuting times, the shares of transportation modes, and urban densities. A synthesis of data should be issued annually and describe trends in mobility and affordability.

The work of the operational planning department should be guided by clear goals set by the mayor or the city council. The operational department should build on the monitoring report to propose complementary actions with clear objectives and targets, including infrastructure investments and regulatory changes. After proposed actions are approved by the city government, program design and implementation should be passed on to traditional municipal line agencies (such as public works departments, water supply utilities, and transit companies).

Cities with decision processes fragmented across many departments that pursue their own agendas tend to produce undesired outcomes, including inconsistency and incoherence. Clear objectives and targets are key factors in successful urban planning.

Notes

1. Lowering real estate transaction costs includes decreasing stamp duties and excessive taxes on capital gains.
2. A Tata Nano, a rear-engine, four-passenger city car built by Tata Motors in India, would require only slightly less space.

References

CIA (Central Intelligence Agency). 2009. "Kenya." In The World Factbook. CIA, Washington, DC. https://www.cia.gov/library/publications/the-world-factbook/geos/ke.html (last modified September 3).

Downs, Anthony. 2004. *Still Stuck in Traffic: Coping with Peak-Hour Traffic Congestion,* rev. ed. Washington, DC: Brookings Institution Press.

Spence, Michael, Patricia Clarke Annez, and Robert M. Buckley, eds. 2009. *Urbanization and Growth.* Washington, DC: Commission on Growth and Development, World Bank.

World Bank. 2009. *World Development Report 2009: Reshaping Economic Geography.* Washington, DC: World Bank.

World Bank Group's Financial Instruments and Multidonor Funds

The World Bank Group offers subnational governments various products and services through its associated institutions: loans and concessional credits through the two arms of the World Bank, the International Bank for Reconstruction and Development (IBRD), and the International Development Association (IDA); commercial financing through the joint International Finance Corporation (IFC)–World Bank Subnational Finance Program; loan guarantees through the Multilateral Investment Guarantee Agency (MIGA); loans through multidonor funds in which the World Bank takes part (such as the Global Environment Facility); and "carbon finance" (such as the purchase of greenhouse gas emission reductions) through market-based instruments. This section discusses the major World Bank Group financial instruments relevant to financing Eco² initiatives at the municipal level.

World Bank: IBRD Loans and IDA Credits

IBRD and IDA credits include specific investment loans and subnational development policy lending. Specific investment loans are a major financial instrument of the IBRD and the IDA (table 3.34). They are disbursed to middle-income countries as loans in the case of the World Bank and to the world's 78 poorest countries on concessional terms in the case of IDA credits.

Development policy operations provide untied, direct budget support to governments for policy and institutional reforms aimed at achieving specific development results. Subnational development policy lending (DPL) typically consists of series of two or three single-tranche loans—referred to as DPL 1, DPI 2, etc.—that may be released against the delivery of policy and institutional reforms (table 3.35). The World Bank may provide this lending to a national government or to subnational divisions of a member country, which might include state and provincial governments with legislative and budget authority.

Table 3.34 World Bank IBRD Loans/ IDA Credits: Specific Investment Loans (SILs)

ELIGIBILITY	Eligible entities are subnationals of developing countries that are IBRD/IDA member countries.
	Application: Municipalities apply through national governments.
FUNDING OBJECTIVE	Financing a broad range of investment activities aimed at creating physical and social infrastructure: • Standalone projects with specific predetermined investment components and programmatic investment activities • Technical assistance activities related to investment projects and their sector reforms. *For a country's own Eco² program:* • Infrastructure services necessary for sustainable urban development, such as water supply, wastewater management, power generation and distribution, solid waste management, roads, public transport, etc. • National Eco² Fund programs (see chapter 3).
INDICATIVE AMOUNT/TERMS	IBRD Flexible Loan *Variable spread option:* LIBOR –0.5% –1.45% *Fixed spread option:* LIBOR +1.00% (–1.45%) based on average payment maturity (10 years or less/10-14 years/ greater than 14 years) and currency (USD, EUR, JPY). Local currency loan is available for currency conversion option. *IDA Credit:* No interest rate

Source: Author compilation.

Note: Information as of March 1, 2010. Rates are as of May 1, 2009 and are subject to change. For updated information, see http://treasury.worldbank.org/. EUR = European Union euro. JPY = Japanese yen. LIBOR = London interbank offered rate. USD = U.S. dollar.

Table 3.35 World Bank IBRD Loans/IDA Credits: Subnational Development Policy Lending (DPL)

ELIGIBILITY	Eligible entities are jurisdictions with legislative autonomy and independent budgetary authority immediately below the national government, including states, provinces, and other entities with similar status (e.g., republics and regions of the Russian Federation and the federal districts [capital cities] of federal countries in Latin America). Normally, municipalities and countries subject to state or provincial legislation and oversight are not eligible. (Countries that have experience with Subnational DPL include Argentina, Bolivia, Brazil, India, Mexico, Pakistan, Russia, and Ukraine.) Countries must be IBRD/IDA member countries.
FUNDING OBJECTIVE	Support of sector reforms through • development of specific policies and policy instruments; • enforcement of policy implementation with legal instruments; and • development of institutional capacities for effective implementation. *For a country's own Eco² program:* Subnational DPLs could address major policy and institutional reforms required for sustainable urban development, particularly in the areas of resource efficiency and energy saving.
INDICATIVE AMOUNT/TERMS	IBRD Flexible Loan: same as table 3.34. IDA Credit: same as table 3.34 *Disbursement.* Loans are provided directly to state or local government or agency, with a sovereign guarantee, or to the country with proceeds onlent to the subnational unit. IDA credits are provided to countries which lend on the proceeds.

Source: Author compilation.

Note: Information as of March 1, 2010. DPL = development policy lending. IBRD = International Bank for Reconstruction and Development. IDA = International Development Association.

Other World Bank Group Financing

Other financing by World Bank Group institutions includes joint subnational finance by the World Bank and the IFC, the financing and services of the IFC, and guarantees by the MIGA.

Joint World Bank–IFC subnational finance provides eligible states, provinces, municipalities, and their enterprises with financing and access to capital markets, without sovereign guarantees, for investment in essential public services (table 3.36). The IFC provides financing and services for investment in the private sector in developing countries (table 3.37).

MIGA promotes developmentally beneficial foreign direct investment in emerging economies by insuring investments against political risks, such as expropriation, breach of contract, war, and civil disturbance; by resolving investment disputes; and by helping developing countries attract private investment (table 3.38).

Table 3.36 World Bank Group Financing: Joint World Bank–IFC Subnational Finance

ELIGIBILITY	*Eligible applicants* • State, municipal, provincial, regional, or local governments and their enterprises (including water and sanitation utilities) • Financial intermediaries supporting local infrastructure • Nationally owned enterprises operating in natural monopoly, infrastructure sectors (selectively) • Public-private partnership entities (to cover commitments of the public partner) Eligible projects must • be located in a developing country that is a member of IFC; • be in the public sector; • be technically, environmentally, and socially sound; and • benefit the local economy. Eligible sectors are water, wastewater, solid waste, transportation, social infrastructure (e.g., health and education), power, gas distribution, district heating, and other essential public services.
FUNDING OBJECTIVE	Strengthening the borrowers' ability to deliver key infrastructure services (such as water, wastewater management, transportation, gas, and electricity) and improving their efficiency and accountability as service providers. Investment selection criteria include • financial (predictability of cash flows to service debt without sovereign guarantee), • socioeconomic (robust economic base), • institutional (operational efficiency), • regulatory (functional system), and • development impact (essentiality of investment and strong economic benefits).
INDICATIVE AMOUNT/TERMS	Products are commercially priced, tailored to client's needs, and can be delivered in 3 to 6 months. All products are provided without sovereign guarantee and may be available in local currency. Products: • Lending instruments (senior, subordinated, and convertible loans) • Credit enhancement (partial credit guarantees, risk sharing facilities, and securitizations) • Equity and quasi equity (other hybrid instruments)

Source: Author compilation.
Note: Information as of March 1, 2010. For details, see http://www.ifc.org/.

Table 3.37 World Bank Group Financing: IFC Financing and Services

ELIGIBILITY	Eligible projects must • be located in a developing country that is a member of IFC, • be in the private sector, • be technically sound, • have good prospects of being profitable, • benefit the local economy, and • be environmentally and socially sound, satisfying IFC standards and those of the host country.
FUNDING OBJECTIVE	*For Eco² catalyst projects:* Components such as private infrastructure, including industrial zones and development of energy efficiency industries, such as energy efficient buildings and production of light-emitting diodes (LEDs), are suitable. IFC also offers guarantees to local banks that invest in energy services companies (ESCOs).
INDICATIVE AMOUNT/TERMS	*Financial products and advisory services:* Financial products, the traditional and largest service of IFC, include loans, equity and quasi-equity finance, financial risk management products, and intermediary finance to finance private sector projects in developing countries. Products are commercially priced and tailored to client's needs. IFC typically does not invest in projects valued at less than 20 million dollars. IFC works with local banks and leasing companies to finance smaller projects. Advisory services are offered in such areas as privatization, business-related public policy, and industry specific issues for private businesses and governments in developing countries.

Source: Author compilation.
Note: Information as of March 1, 2010. For details, see http://www.ifc.org/.

Table 3.38 World Bank Group Financing: MIGA Guarantees

ELIGIBILITY	Eligible applicants: • Nationals of a MIGA member country other than the country in which the investment is to be made • Juridical persons if they are either incorporated in and have their principal place of business in a MIGA member country, or if they are majority owned by nationals of MIGA member countries • State-owned corporations if they operate on a commercial basis investing in MIGA member countries other than the country where they are incorporated • Nationals of the host country or juridical persons incorporated in said host country or whose capital is majority-owned by its nationals, provided that the invested assets are transferred from outside the host country
FUNDING OBJECTIVE	Offering political risk insurance against losses relating to currency transfer restrictions, expropriation, war and civil disturbance, and breach of contract for projects in a broad range of sectors (e.g., power, water, wastewater, transport and green infra- structure, energy, telecommunications, and finance) in developing countries that are MIGA member countries. MIGA can cover expropriation and breach of contract by a subnational entity. MIGA's contribution to reducing the adverse impact of climate change focuses on supporting green infrastructure investments in developing countries that build renewable energy capacity, encouraging resource conservation and distribution efficiency, improving sanitation, and off- setting GHG emissions.
INDICATIVE AMOUNT/TERMS	Per project insurance: Up to US$200 million (if necessary, more can be arranged through syndication of insurance). Duration up to 15 years (20 years if justified). Guarantee premiums based on country and project risk. Rates for the SIP guarantee (3 coverages): 0.45%–1.75% basis points per year. Types of foreign investments include • equity interests, • shareholder loans, • shareholder loan guarantees, and • other investments, such as technical assistance, management contracts, franchising, and licensing agreements.

Source: Author compilation.
Note: Information as of March 1, 2010. For details, see http://www.miga.org/. SIP = Small Investment Program.

Multidonor Funds

Multidonor Funds include climate investment funds and the Global Environment Facility. Climate investment funds consist of the Clean Technology Fund (CTF)(table 3.39) and the Strategic Climate Fund (SCF)(table 3.40). In keeping with multilateral development bank practice, investment projects and programs may include financing for policy and institu-

3.39 Multidonor Funds—Climate Investment: Clean Technology Fund (CTF)

ELIGIBILITY	Eligible entities are subnationals in developing countries with • eligibility for Official Development Assistance according to OECD/DAC guidelines), and • an active MDB country program.a *Application:* Municipalities apply through their countries, which express interest in and request a joint mission from the World Bank and a regional development bank for preparation of a country CTF investment plan. The investment plan is a country-owned document, prepared with the assistance of the MDBs, which outlines the country's priorities and strategy for the utilization of CTF resources. In 2009, the CTF Trust Fund Committee endorsed Investment Plans for Egypt, Mexico, Turkey, South Africa, Morocco, Philippines, Thailand and Vietnam, as well as a regional program for concentrated solar power in the Middle East and North Africa (MENA) region.
FUNDING OBJECTIVE	Promoting scaled-up financing for demonstration, deployment, and transfer of low-carbon technologies with a significant potential for long-term savings in GHG emissions, including programs in • the power sector (renewable energy and highly efficient technologies to reduce carbon intensity), • the transport sector (efficiency and modal shifts), and • the energy sector (energy efficiency for buildings, industry, and agriculture). Programs/projects selection criteria: • Potential for long-term GHG emission savings • Demonstration potential at scale • Development impact • Implementation potential • Additional costs/risk premium

INDICATIVE AMOUNT/TERMS	Indicative total amount: about $US 5 billion was pledged in September 2008.
	Indicative number of country/regional programs: 15–20 Investment Plans
	Through MDBs, CTF would seek to provide
	• concessional financing in the near-to-medium term to meet investment needs to support rapid deployment of low carbon technologies;
	• concessional financing at scale, blended with MDB financing, as well as bilateral and other sources of finance, to provide incentives for low carbon development;
	• a range of financial products to leverage greater private sector investments; and
	• financial instruments integrated into existing aid architecture for development finance and policy dialogue.
	
	Products and terms:
	• Concessional loans have 2 options:
	  1. Harder Concessional: Maturity, 20 years; grace period, 10 years; principal repayments (year 11–20), 10%; grant element: ~45%; service charge: 0.75%.
	  2. Softer Concessional: Maturity, 40 years; grace period, 10 years; principal repayments, 2% (year 11–20), 4 % (year 20-40; grant element, ~71% ; service charge, 0.25%.
	• Grant: up to US$1 million (for CTF project preparation)
	• Guarantees: partial credit guarantee and partial risk guarantee
	
	MDBs lend to national governments, national governments for on-lending to subnational entities, or subnational entities.

Source: Author compilation.

a. An "active" program is one in which an MDB has a lending program and/or ongoing policy dialogue with the country.

Note: Information as of March 1, 2010. For details on Climate Investment Funds, see http://www.climateinvestmentfunds.org. OECD = Organisation for Economic Co-operation and Development. DAC = Development Assistance Committee (OECD). MDB = Multilateral Development Bank. CTF = Clean Technology Fund. GHG = greenhouse gas. SCF = Strategic Climate Fund.

Table 3.40 Multidonor Funds—Climate Investment: Strategic Climate Fund (SCF)

Pilot Program for Climate Resilience (PPCR): The PPCR is a program under the Strategic Climate Fund (SCF) designed to pilot and demonstrate ways to integrate climate risk and resilience into developing countries' core development planning. The pilot programs implemented under the PPCR are country led, build on National Adaptation Programs of Action, and are strategically aligned with other sources of adaptation finance, such as Adaptation Fund, UNDP, and other donor-funded activities.

ELIGIBILITY	MDB eligibility (Regional Development Banks, International Development Association (IDA)
FUNDING OBJECTIVE	Supporting scaled-up action and transformational change in integrating climate resilience in national development planning of a few highly vulnerable countries.
INDICATIVE AMOUNT/TERMS	Indicative total amount: about US$900 million Grants and concessional lending for technical assistance and programs of public and private sector investments

Program for Scaling-Up Renewable Energy in Low Income Countries (SREP): The SREP program aims to demonstrate in a small number of low income countries how to initiate energy sector transformation by helping them take renewable energy solutions to a national programmatic level. SREP offers a unique two-pronged approach. It is designed to support developing countries in their efforts to expand energy access and stimulate economic growth through the scaled-up deployment of renewable energy solutions, and it provides a trigger for transformation of the renewables market in each target country through a programmatic approach that involves government support for market creation, private sector implementation, and productive energy use.

ELIGIBILITY	To be eligible, a country • must be low-income and eligible for MDB concessional financing (i.e., through IDA or a regional development bank's equivalent); and, • be engaged in an active country program with an MDB.[a]
FUNDING OBJECTIVE	Pilot and demonstrate, as a response to the challenges of climate change, the economic, social, and environmental viability of low-carbon development pathways in the energy sector by creating new economic opportunities and increasing energy access through the use of renewable energy.
INDICATIVE AMOUNT/TERMS	Indicative total amount: about US$292 million (as of February 2010).

Source: CIF (2009a,b; 2010) and personal communication with CIF in March 2010.

a. An "active" program is one in which an MDB has a lending program and/or ongoing policy dialogue with the country.

Note: Information as of March 1, 2010. For details on Climate Investment Funds, see http://www.worldbank.org/cif.

tional reforms and regulatory frameworks. This is the role of the Clean Technology Fund. The Strategic Climate Fund is broader and more flexible in scope. It serves as an overarching fund that supports various programs to test innovative approaches to climate change. It consists of three programs: the Pilot Program for Climate Resilience, the Forest Investment Program, and the Scaling-Up Renewable Energy Program for Low-Income Countries. It provides financing to test new development approaches or to scale up activities aimed at specific climate change challenges through targeted programs.

The Global Environment Facility (table 3.41) is a global partnership among 178 countries, international institutions, nongovernmental organizations, and the private sector to address global environmental issues, while supporting national sustainable development initiatives. It provides grants for projects related to six focal areas: biodiversity, climate change, international waters, land degradation, the ozone layer, and persistent organic pollutants. It works with seven executing agencies and three implementing agencies, including the World Bank. At the end of 2007, the active portfolio of Global Environment Facility projects implemented by the World Bank included 219 projects with total net Global Environment Facility grant amount commitments of US$1.6 billion. In terms of approval, the grant amount approved by the World Bank Board in fiscal year 2007 was US$220 million (22 projects).

Table 3.41 Multidonor Funds: Global Environment Facility (GEF)

ELIGIBILITY	A country shall be an eligible recipient of GEF grants if it is eligible to borrow from the World Bank (IBRD and/or IDA) or if it is an eligible recipient of UNDP technical assistance through its country Indicative Planning Figure (IPF).
	An eligible project must
	• be undertaken in an eligible country;
	• be consistent with national priorities and the GEF operational strategy;
	• address one or more of the GEF focal areas, improving the global environment or advancing the prospect of reducing risks to it;
	• seek GEF financing only for the agreed incremental costs on measures to achieve global environmental benefits;
	• involve the public in project design and implementation; and
	• be endorsed by the government(s) of the country(ies) in which it will be implemented.
	Application: Municipalities in member countries apply in consultation with Country Operational Focal Point.
FUNDING OBJECTIVE	Providing grants for projects related to six focal areas: biodiversity, climate change, international waters, land degradation, the ozone layer, and persistent organic pollutants.
INDICATIVE AMOUNT/TERMS	Per project size: *Full-sized project:* Grant over US$1 million. *Medium-sized project:* Grant up to US$1 million. *Enabling activities:* Grant up to $0.5 million in GEF financing, but varies across focal areas.

Source: Author compilation.

Note: Information as of March 1, 2010. For details, see http://www.gefweb.org/. UNDP = United Nations Development Programme. GEF = Global Environment Facility.

Market-Based Instruments

Market-based instruments that are relevant for Eco^2 initiatives include carbon finance, which consists of 11 funds and one facility, the Carbon Partnership Facility (table 3.42). The World Bank Carbon Finance Unit uses funds contributed by governments and private companies in the countries of the Organisation for Economic Co-operation and Development to purchase project-based greenhouse gas (GHG) emission reductions in developing countries and countries with economies in transition. The GHG emission reductions are purchased through one of the unit's carbon funds on behalf of the contributor and within the framework of the Kyoto Protocol's Clean Development Mechanism or joint implementation. Unlike other World Bank financial instruments, the Carbon Finance Unit does not lend or grant resources for projects. Rather, it contracts to purchase emission reductions on a basis similar to a commercial transaction, paying for them annually or periodically once the emission reductions have been verified by a third-party auditor. The selling of emission reductions (carbon finance) has been shown to increase the bankability of projects by adding an additional revenue stream in hard currency that reduces the risks of commercial lending or grant finance. Thus, carbon finance provides a means of leveraging new private and public investment in projects that reduce GHG emissions, thereby mitigating climate change, while contributing to sustainable development. The Carbon Finance Unit has several carbon funds aimed primarily at fulfilling commitments under the Kyoto Protocol by 2012.

Carbon finance and the Carbon Partnership Facility represent a new generation of carbon finance that is being developed to scale up emission reductions and their purchase over the longer term beyond the regulatory period

Table 3.42 Market-Based Instruments: Carbon Finance, Carbon Partnership Facility (CPF)

ELIGIBILITY	Eligible entities: Seller Participants should be public or private entities committed to develop one or more emission reduction (ER) programs and sell a portion of the ERs to the Carbon Fund, one of the trust funds under CPF; they should also be acceptable to the World Bank in accordance with established criteria.
	Buyer Participants should be public or private entities committed to contribute to the Carbon Fund. For the first tranche of the Carbon Fund, €35 million is the minimum required contribution from a public or private entity (a group of entities can form a pool/consortium to participate as a group).
	Eligibility of an ER program includes • reduction of 6 GHG covered under the Kyoto Protocol or under any future climate change regime; • demonstration of value added to the programs by World Bank's involvement (e.g. power sector development, energy efficiency, gas flaring, transport sector, and urban development programs); and • suitability for scaling up, i.e., can be replicated as part of a larger program or in another country.
	Prioritized programs • are aligned with Country Assistance Strategy/Country Partnership Strategy and UN Framework Convention on Climate Change/Kyoto Protocol, • are built on World Bank lending pipeline and other operations, • use commercially available technology, and • are expected to have significant ERs (preferably several million tons over 10-15 years)
FUNDING OBJECTIVE	Facilitating the development of low-carbon investments with a long-term impact on mitigating GHG emissions under the UNFCCC Framework or Kyoto Protocol and any future agreement under the UNFCCC or other regime deemed appropriate by the Trustee in consultation with the Participants.
INDICATIVE AMOUNT/TERMS	*Indicative total amount:* First tranche of the CPF Carbon Fund to become operational with a target capitalization of €200 million, and could grow to about €400 million. Expected to become operational in first half of CY2010. *Indicative per project size:* several million tons of ER/program over 10-12 years. *Price of ERs:* Transparent CPF pricing approach, based on market prices; may allow for upside and downside sharing between buyers and sellers (to be confirmed). While the first tranche of the CPF Carbon Fund is denominated in €, subsequent tranches could be denominated in other currencies as well.

Source: Author compilation.

Note: Information as of March 1, 2010. For details, see http://go.worldbank.org/9IGUMTMED0. ER = emission reduction. CPF = Carbon Partnership Facility. UNFCCC = United Nations Framework Convention on Climate Change. CY = current year.

of the Kyoto Protocol, which ends in 2012. The objectives and business model are based on the need to prepare large-scale, potentially higher-risk investments with long lead times that require durable partnerships between buyers and sellers and significant capacity building for program development. They are also based on the need to support long-term investments in an uncertain market environment, possibly spanning several market cycles. Learning-by-doing approaches are an essen-

tial aspect of the Carbon Partnership Facility as the program moves from individual projects to programmatic approaches, including methodologies for such approaches. It is expected that the size of the facility will be €5 billion for the period 2012–16.

The various climate change funds may be used simultaneously or sequentially (box 3.24). Carbon finance, in particular, offers attractive opportunities for cities to focus on reducing GHG emissions (box 3.25).

Using Various Climate Change Funds Simultaneously or Sequentially

The major financial instruments available at the World Bank Group to help mitigate climate change are the Global Environment Facility, the Clean Technology Fund, and the Carbon Partnership Facility. These tools share a similar objective: reducing the growth of GHG emissions by creating favorable market conditions for GHG reduction. They are also compatible. They may therefore be pieced together to serve the same project, as long as the coverage does not overlap. The Global Environment Facility focuses on the removal of barriers by providing grant funding for innovative projects in energy efficiency, renewable energies, and sustainable transportation to establish the proper conditions for market transformation. The Clean Technology Fund focuses on support through investment to fill financing gaps by providing grants, concessional finance, and guarantees to scale up markets. By supporting efforts to reduce the cost of investment or the provision of guarantees, the fund aims to reduce the risks. The Carbon Partnership Facility, a new type of carbon fund, provides performance rewards or output-based revenue sources to create incentives for carbon-reducing investments.

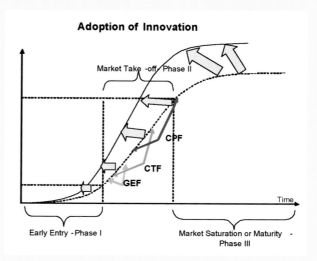

CPF = Carbon Partnership Facility. CTF = Clean Technology Fund. GEF = Global Environment Facility.

Note: Projects need to be planned to avoid the double or triple counting of the same quantities of GHG among the Global Environment Facility, the Clean Technology Fund, and the Carbon Partnership Facility.

Citywide Greenhouse Gas Emission Reduction and Carbon Finance

Emissions in urban areas arise from a wide range of sources, including transportation, electrical and thermal energy consumption in buildings and industries, water and wastewater management, municipal waste, and various public services. Under the Clean Development Mechanism, there are around 20 methodologies relevant to the needs of urban authorities. The waste sector is addressed the most frequently. The methodologies enable the projects to measure GHG reductions relative to baselines or business-as-usual trends for specific sources of emissions, and they help monitor the emission reductions. However, because the GHG impact of individual emission sources, such as a single waste site or streetlighting, is small, many of these projects are unable to access carbon finance because of the high transaction costs. Furthermore, many sectors, including building emissions, are not effectively addressed through current methodologies.

The Carbon Finance Unit of the World Bank is developing a framework methodology that attempts to aggregate the GHG impact of all individual sources into a single administrative area, thereby allowing for the simplification and streamlining of measurement and monitoring and enabling the development of a citywide program for GHG mitigation. For a typical existing city, the proposed baseline is the current and projected future levels of service provision. For a new city, the baseline might be the average emission level in the region. Emission sources are categorized according to waste,

transport, and energy use. The emissions may be reduced through a range of activities. In the waste sector, methane avoidance, biogas generation, and recycling facilities are key sources. Increasing the share of public transportation may have significant GHG mitigation benefits. Energy efficiency opportunities include buildings, lighting in public areas such as streetlights, water pumping, district heating, and integrated planning for heating and cooling supplies. Significant emission reductions may also be achieved by using energy from renewable sources through wind, solar, and geothermal technologies.

GHG emission reductions in cities may be realized in each sector through projects or through regulatory and incentive-based initiatives that facilitate the participation of the private sector and the general public. A typical city program would be managed by city authorities. Projects may be implemented by contractors or by city authorities through public-private partnerships. GHG mitigation projects are typically implemented in the three sectors (waste, transport, and energy use) over a period of time and generate emission reduction credits based on performance. Depending on the acceptance of a citywide aggregated methodology, emission reduction credits may be traded or sold for use by industrialized countries to meet a part of their emission reduction targets under the Kyoto Protocol or in the voluntary market for use by industries, governments, or cities.

References

Climate Investment Funds (CIF). 2009a. PPCR Fact Sheet. http://www.climateinvestmentfunds.org/ cif/sites/climateinvestmentfunds.org/files/PPCR_ fact_sheet_nov09.pdf

———. 2009b. SREP Fact Sheet. http://www. climateinvestmentfunds.org/cif/sites/ climateinvestmentfunds.org/files/SREP_fact_ sheet_nov09.pdf

Climate Investment Funds (CIF). 2010. "Criteria for Selecting Country and Regional Pilots under the Program for Scaling up Renewable Energy in Low Income Countries." Washington, DC: Climate Investment Fund. http://www. climateinvestmentfunds.org/cif/sites/ climateinvestmentfunds.org/files/SREP%20 Criteria%20country%20and%20region%20 program%20selection_SCmeeting_Feb3_ 012010.pdf

Index

risks on the landscape, 136–137

Stern Review on the Economics of Climate Change
loss of global GDP in business-as-usual scenarios, 17–18

Stockholm, Sweden
brownfields development, 185
case study, 183–193
city-owned land, 185
cityscape (figure), 183
congestion and road pricing schemes, 271
demographics of, 184
development strategies (box), 185
ELP-related achievements in Hammarby Sjöstad (figure), 159
environmental load profile and, 103, 157–160, 183, 189–190, 192
future plans, 191–192
Hammarby Model, 20, 187–192
Hammarby Sjöstad project to improve sustainability, 20–21, 67, 92, 157, 158–159, 183, 185–192
initial first-phase results of Hammarby Sjöstad according to the Environmental Load Profile Life-Cycle Analysis Tool (figure), 21
inner city of Stockholm and adjacent development areas (figure), 186
lessons learned, 192
local investment subsidy program funding across types of projects in Sweden (figure), 191
location of (map), 184
profile of, 184
public transportation, 275
Royal Institute of Technology, 103, 157, 189
Stockholm Royal Seaport project, 191–192
target for all new construction to be carbon neutral by 2030, 116
urban village model and, 313–314
Vision 2030, 184

Storm water management
flood management, 258
integrated storm water management (figure), 72
rainwater harvesting, 257–258

Strategic Climate Fund
Forest Investment Program, 333
Pilot Program for Climate Resilience, 333
Scaling-Up Renewable Energy Program for Low-Income Countries, 333–334

Sustainable Cities Program
European Awareness Scenario, 161

Sweden. *See* specific cities

T

Tamil Nadu, India
streetlamp retrofits, 245

Thailand
urban area contribution to GDP, 15

Tianjin, China
landfill gas capture project, 306, 308

Tokyo, Japan

Chuo Ward reserve fund for school facilities, 91–92

Tokyo Waterworks, 91
value-based property taxes, 276

Towards Sustainable Urban Infrastructure: Assessment, Tools and Good Practice, 94

Transportation. *See also* Mobility of people and goods
amount of roadway used by the same passengers traveling by car, bicycle, or bus (figure), 291
average new vehicle fuel economy standards (figure), 271
balancing investments in, 283
Bangalore, India, bus rapid transit system, 82–83
basic and advanced stakeholder interests (table), 286
basic and advanced transport interventions (table), 274
Beijing: travel demand management and the legacy of the Olympic Games (box), 280–281
benefits of, 267
the benefits under speed conditions of select highway applications of intelligent transportation systems (figure), 284
bicycle- and pedestrian-friendly pathways, 24, 169, 172, 179, 277, 279
Brisbane, Australia, case study, 216–217
building block approach, 268
bus rapid transit (box), 288
carbon taxes, 270, 290
charging users for costs of travel and parking, 271
cheap or free parking subsidy effects on car use, 75
city structure and development patterns and, 275–276
classification of intelligent transportation system market packages (figure), 284
CO_2 emissions from a range of vehicle types (table), 286
coordination of land use with transit, 320–321
cost-benefit analyses, 287
Curitiba, Brazil, public transit system, 22, 169, 170–174, 179
Curitiba: terminal Carmo, adjacent shops, and Citizenship Street (figure), 289
demand forecasts and, 288–289
design and orientation of buildings and, 277
desired outputs, 268
economic and financial aspects, 287–290
elements of a public transportation network (table), 283
elements of utility in models for choosing a transportation mode (figure), 289
emission-based road pricing in Milan, Italy (box), 280
emissions inventories and, 290
energy consumption and, 290–291
environmental protection laws and, 270–271

an example of microdesign and walking iso-
    chrones (figure), 279
financing mechanisms, 276–277
financing options, 289–290
flat-rate "social" bus fares, 174, 179
the four pillars of sustainable urban transportation
    institutions (box), 273
four-step transportation model, 290
the framework of transportation intervention
    (table), 274
fuel taxes and, 270
high-density activity and job sites and, 275
high-quality public transportation corridors, 275
the impact of government on land markets, the
    informal sector, and the spatial structure of
    cities (table), 318
independent inputs, 267
infrastructure and services, 267, 281–285
the input-output framework of transportation
    interventions (figure), 268
institutional context, 272–273
institutional functions and jurisdictions in
    transportation (table), 272
integration opportunities, 290–291
intelligent transportation systems, 283–285
intersections and crossings and, 277
land development and value capture and, 289
land management and, 276
land use, densities, connectivity, and access and,
    75–76
land use and travel demand, 267, 275–280
locality in the state of Colorado (figure), 277
macroplanning, 277, 290
managing speed, 284–285
metropolitan and local government responsibili-
    ties, 271–272
microplanning and, 277, 279
mobility infrastructure hierarchy (table), 282
mobility management, 281, 283
new technology and, 24–25
overview, 267–268
parking and access management and, 279
peak load management and, 65
pedestrian-friendly street in Curitiba, Brazil
    (figure), 279
physical systems, technology, and spatial planning,
    273–285
policies, legislation, and regulations affecting the
    transport sector (table), 270
policy, legislative, and regulatory framework,
    269–272
predict-and-provide transportation planning
    models, 53
prioritization of public transportation, 269–270,
    274–275
public-private partnerships and, 289, 291
ring road model and, 275, 276
road, vehicle, and fuel standards, 270

road patterns and design and, 277
road space allocation, 272, 290
sector notes, 267–292
sensitivity analyses, 287–288
Singapore metro network: centered on expansion
    in the central business district (figure), 321
stakeholder dynamics, 285–286
the structure of the integrated public transporta-
    tion network in Curitiba, Brazil (figure), 276
summary of cross-sector integration opportunities
    (table), 292
summary of select vehicle and fuel interventions
    (table), 285
tax increment financing, 290
time dimension and, 281
transit-oriented development (box), 278
travel demand management, 279
type of development and the implications for
    transportation (table), 282
typical objectives or desired outputs of transporta-
    tion interventions (table), 269
urban densities and, 277
urban density and transport-related energy
    consumption (figure), 75
urban expansion and land management policies,
    271
urban transportation outcomes in selected cities
    (table), 269
vehicle fleet and fuel supply, 267, 285
World-Bank financed project requirements, 291 n1

U

UN-Habitat. *See* United Nations Human Settlements
    Programme
United Kingdom. *See also specific cities*
    water regulations, 255
United Nations Human Settlements Programme
    estimate of the number of slum dwellers in
        developing countries, 16
United Nations Population Fund
    study of the link between urbanization and
        poverty, 15
United States. *See also specific states and cities*
    fuel efficiency standards for vehicles, 270
    fuel taxes, 270
    integrated transport, air quality, and land use
        plans, 271
    LEED-certified office and school buildings, 244
    low-density, polycentric cities and, 316
Urban areas. *See also* Urban planning; *specific cities*
    boundaries for, 58–59
    challenges cities face, 30–32
    collaborative design and decision making platform
        and, 53–55, 58–59
    collaborative working group for, 54–55
    description, 53–54
    emergence of the regional city as a crucial scale
        for long-term planning, 54

ECO-AUDIT

Environmental Benefits Statement

The World Bank is committed to preserving endangered forests and natural resources. *Eco² Cities: Ecological Cities as Economic Cities* is printed on 60# Chorus Art, a recycled paper made with 25-percent post-consumer waste. The Office of the Publisher follows the recommended standards for paper usage set by the Green Press Initiative, a nonprofit program supporting publishers in using fiber that is not sourced from endangered forests. For more information, visit www.greenpressinitiative.org.

Saved:

- 16 trees
- 5 million BTUs of total energy
- 1,489 pounds of net greenhouse gases
- 7,170 gallons of waste water
- 435 pounds of solid waste